The Neurobiology of Memory

The Neurobiology of Memory

Concepts, Findings, Trends

YADIN DUDAI

Professor of Neurobiology
The Weizmann Institute of Science
Rehovot, Israel

OXFORD UNIVERSITY PRESS

Oxford University Press, Walton Street, Oxford OX2 6DP
Oxford New York
Athens Auckland Bangkok Bombay
Calcutta Cape Town Dar es Salaam Delhi
Florence Hong Kong Istanbul Karachi
Kuala Lumpur Madras Madrid Melbourne
Mexico City Nairobi Paris Singapore
Taipei Tokyo Toronto
and associated companies in
Berlin Ibadan

Oxford is a trade mark of Oxford University Press

Published in the United States by
Oxford University Press Inc., New York

First published 1989
Reprinted 1990 (with corrections), 1991, 1994

British Library Cataloguing in Publication Data
Dudai, Yadin
The neurobiology of memory
1. Man. Memory. Neurophysiological aspects
I. Title
612'.82

Library of Congress Cataloging-in-Publication Data
Dudai, Yadin.
The neurobiology of memory: concepts, findings, trends/Yadin Dudai.
1. Memory—Physiological aspects. 2. Learning—Physiological
aspects. 3. Neurobiology. I. Title.
QP406.D844 1989 153.1'2—dc19 89–2915
ISBN 0 19 854229 1 (Pbk)

Printed in Great Britain by
Bookcraft (Bath) Ltd
Midsomer Norton, Avon

Preface

This book was written for students taking advanced undergraduate and graduate courses on the neurobiology of learning and memory or other courses in neuroscience that cover this topic. I hope that the book may also be of interest to additional audiences.

I am very grateful to many colleagues who devoted their precious time to review various versions of selected chapters of the manuscript. They include Tom Abrams, Nachum Alon, Amos Arieli, Tim Bliss, Tom Brown, Yossi Buxbaum, Tom Carew, Terry Crow, Nelson Donegan, Eitan Domany, Rina Dudai, Tamar Flash, Joaquin Fuster, Al Galaburda, Howard Gardner, Amiram Grinvald, Haim Kahana, Al Kaye, Daphne Kidron, Mark Konishi, Rafi Malach, Heike Mitschulat, Haim Sompolinski, Menachem Segal, Yossi Shaul, Benny Shilo, Bolek Srebro, Avi Susswein, Vivian Teichberg, Shimon Ullman, Klaudius Weiss, Bob Werman, and Yossi Yarom. I alone am to blame for not always accepting their suggestions, for the resulting mistakes, and for the idiosyncrasy of some of the views. I am also grateful to Libette Chernobrov, from the Wix Library of the Weizmann Institute, for assisting me in obtaining books and journals, and to Oxford University Press for an enjoyable collaboration.

January 1989   Y. D.

Credits

Permission granted by authors and publishers to adapt material for the preparation of figures is gratefully acknowledged; sources are accredited in the appropriate figure legends. Material published by the following publishers was used with permission in preparation of the following figures: **Fig. 1.3.**, Edward Arnold, London; **Fig. 2.1.**, Harcourt Brace Jovanovich, London, and American Association for the Advancement of Science (AAAS), Washington, DC; **Fig. 2.4.**, American Psychological Association, Washington, DC; **Fig. 2.5.**, Carol Donner, Scientific American, Inc. New York; **Fig. 3.2.**, Elsevier Publications, Cambridge; **Fig. 4.1.**, W. H. Freeman and Co., New York; **Figs 4.3.**, **4.4.**, AAAS, Washington, DC; **Fig. 5.1.**, Chapman and Hall, Ltd., and Oxford University Press, New York; **Fig. 5.2.**, Jerome Kuhl, Scientific American Inc., New York; **Fig. 5.3.**, J. Wiley and Sons Inc., New York; **Fig. 6.1.**, Springer-Verlag, Heidelberg; **Figs 6.3.**, **6.4.**, Macmillan Magazines Ltd., London; **Fig. 6.4.**, Elsevier Publications, Cambridge; **Fig. 7.1.**, Springer-Verlag, New York; **Fig. 8.1.**, Macmillan Magazines Ltd, London; **Fig. 8.3.**, AAAS, Washington, DC; **Fig. 8.4.**, Alan Liss Inc., New York and Springer-Verlag, Heidelberg; **Fig. 9.2.**, Elsevier Science Publications, Amsterdam; **Fig. 9.3.**, Elsevier Publications, Cambridge; **Fig. 10.1.**, The University of Chicago Press, Chicago, Illinois; **Fig. 10.3.**, Modern Medicine Publications, Cleveland, Ohio; **Fig. 10.4.**, The American Physiological Society, Bethesda, Maryland; **Fig. 11.1.**, Cambridge University Press, Cambridge; **Fig. 11.2.**, Alan Liss Inc., New York; **Fig. 11.3.**, AAAS; **Fig. 11.4.**, Elsevier Publishing Co., Inc., New York; **Fig. 11.5.**, Cambridge University Press, Cambridge; **Fig. 11.6.**, Springer-Verlag, Heidelberg, and Psychonomic Society Publications, Austin, Texas; **Figs 12.1.**, **12.2.**, **12.3.**, Oxford University Press, Oxford; **Fig. 13.1.**, Annual Reviews Inc., Palo Alto, California; **Figs 13.2.**, **13.3.**, Oxford University Press, New York; **Fig. 14.1.**, Scientific American Inc., New York; **Fig. 14.2.**, American Medical Association, Chicago, Illinois; **Fig. 14.3.**, Carol Donner, Scientific American Inc., New York; **Fig. 14.4.**, Oxford University Press, New York, and The American Physiological Society, Bethesda, Maryland; **Fig. 14.5.**, Churchill Livingstone, Edinburgh; **Fig. 14.6.**, The American Psychological Association, Washington, DC; **Fig. 14.7.**, Scientific American Inc., New York; **Fig. 15.1.**, British Medical Journal, London; **Fig. 15.2.**, Pergamon Press, Elmsford, New York; **Fig. 15.3.**, AAAS, Washington, DC. Also, for the first and last sentences of this book I am indebted to *A la recherche du temps perdu* by Marcel Proust. 'What the frog's eye tells the frog's brain' (in the opening paragraph of Chapter 1) is the title of the classic paper by Lettvin, J. Y., Maturana, H. R., McCulloch, W. S., and Pitts, W. H. (1959). *Proceedings of the IRE* **47**, 1940–51.

Contents

PART II
WINDOWS TO MOLECULAR AND CELLULAR MECHANISMS

4 Cellular mnemonic devices detected in relatively simple nervous systems: the case of *Aplysia*   **49**

5 Cellular mnemonic devices detected in relatively simple nervous systems: the case of *Hermissenda*   **77**

6 A cellular mnemonic device in the mammalian brain: long-term potentiation   **88**

PART III

WINDOWS TO THE ARCHITECTURE OF MEMORY SYSTEMS

PART I

The Conceptual and Experimental Framework

1 Some basic notions and their ontogenesis

Innate and acquired behaviours

Life, in many respects, is a remembrance of things past. Genetic memory, neuronal memory, and cultural memory shape species, individuals, and cultures. The most elementary memory, and the prerequisite for them all, is the genetic. This collective past of the species provides organisms with prefixed, innate responses, which are an indirect translation of a linear chemical code into neuronal- and muscular-based algorithms. What the frog's eye tells the frog's brain is based on a memory established during millions of generations; so is the escape of the fly from the frog's tongue. What is in a fly that makes the frog extend its tongue, and what is in a tongue that makes the fly escape, is the domain of studies of biological semiotics, dealt with by ethologists.*

Worlds may exist in which the behaviour of organisms is based solely on the slowly accumulated memory of their species. However, the behavioural repertoire of most, if not all, organisms around us can be modified by individual experience. This depends on the ability to learn.

In all probability, learning has developed in evolution because it endows organisms with an immense adaptational potential. Innate responses do have their adaptive value, too. They ensure a basic 'survival kit', independent of prior exposure of the individual to the world (Table 1.1). They also provide an infrastructure for further development of skills, eliminating the need to learn everything from scratch in each generation. Therefore, even the most sophisticated species do not give up the innate components in their behavioural repertoire.

Note that an organism capable of learning must possess at least one type of a biological 'learning apparatus', coded by genes. Our memory capabilities are thus but one kind of information transmitted by genetic memory.

What is learning, and what is not?

Although learning is intuitively understood by everyone, a formal definition of it is not easy to formulate and defend. Most definitions of learning are based on inference of the process

* References for this and other research fields mentioned *en passant* in this introductory chapter are listed under Further reading.

Table 1.1. Comparison of idealized, pure instances of innate and acquired behaviours

Innate behaviour	*Acquired behaviour*
1. Correct responses to appropriate stimuli are ensured a priori. This might be critical for survival.	1. Trial and error (albeit sometimes very quick) are necessary to obtain the correct response.
2. Response rigid, may lead to energy waste and even to deleterious consequences if the environmental signal is misleading, or if the situation changes after the response was initiated.	2. Response can be altered in reaction to quick environmental changes; thus a better fit is often ensured between the behaviour and its goals.
3. Every step of the response is dictated by information encoded in the genes.	3. The ability to acquire, store, and retrieve information is encoded in the genes, but not the information itself.
4. The behavioural repertoire is limited, since there is a limit to the amount of genetic information that can be allocated for coding specific, detailed behavioural responses.	4. The behavioural repertoire is much more flexible.
5. Modifications of the behavioural repertoire evolve slowly (time course of generations) and are shaped by the accumulated experience of the species, and shared by all its members.	5. Behavioural modifications occur during the lifetime of the individual, but perish afterwards, unless transmitted by culture.

Major properties of idealized cases of innate and acquired behaviours are compared. Such ideal, pure instances of either fully innate or solely acquired behaviours may not exist in multicellular organisms; innate responses are modified by developmental and/or environmental experience, and acquired responses have innate components.

from behaviour. The following are three representative examples: 'a process which manifests itself by adaptive changes in individual behaviour as a result of experience' (Thorpe 1956); 'development of responses to limited aspects of the environment' (Maier and Schneirla 1964); 'relatively permanent change in behaviour which occurs as a result of experience' (Tarpy 1975).

It is of interest to note what the above definitions, and similar ones, contain, and what they do not.

1. Explicit in some definitions, and implicit in others, is the basic notion that learning is a modification of behaviour based on *individual experience* (as opposed to species experience). Some authors state that the experience must be repetitive (Bower and Hilgard 1981), but this is an unnecessary restriction, since learning can clearly occur following an isolated, brief experience.

2. Some definitions emphasize that learning is a *process*, not an instantaneous change.

3. Some definitions emphasize that learning is a *lasting* change, distinguishing it from transient reactions to sensory stimuli.

4. Some definitions assume that learning must have an ecological *purpose*. Here is a

classical ethologist's stand on this point: 'Learning selects from many possibilities contained in an open program the one that seems to fit current circumstances best. . . . However one wants to define learning, the definition must include teleonomic function' (Lorenz 1981). Other scholars regard this argument as philosophically wrong, and/or unnecessarily restrictive.

5. Some definitions of learning do not *differentiate* between learning and development, injury, fatigue, and poisoning, all of which also lead to experience-dependent, lasting behavioural alterations. Explicit exclusion of purely developmental changes, injury, fatigue, or drug reactions must, therefore, be included (Bower and Hilgard 1981).

6. Since not every instance of learning directly materializes into behaviour, it is appropriate to consider also alterations in the *potential* to react, or in *knowledge* (James 1890; Bower and Hilgard 1981). Radical behaviourists oppose the use of terms that dissociate learning from performance, whereas students of the cognitive sciences regard such use as compulsory. The following is a 'cognitive' definition of learning: 'organized knowledge, which grows and becomes better organized' (Charniak and McDermott 1985). Note here the requirement for a base of a priori knowledge. Research in several branches of cognitive sciences has revealed that 'for an organism to learn anything, it must already know a lot' (ibid.); in other words, without a minimal set of rules and cues, learning is not feasible. This a priori knowledge is provided, in the first place, by genetic memory. Here, thus, is another facet of the intimate link between neuronal learning and genetic memory.

Internal representations

What do we mean by 'knowledge'? The term is, as expected, loaded with philosophical meanings, which far exceed the scope of our discussion (e.g. see Dretske 1983). Suffice it to say that in the present context, the term 'knowledge' is used in a most elementary sense and is devoid of anthropomorphic connotations of consciousness. We define it as structured bodies of information possessed by the organism about the world, and capable of setting the organism's reactions to the world. The 'world' means both the external milieu and internal states of the organism, and the 'organism' means, specifically, the nervous system. We term these neuronally encoded structured versions of the world which could potentially guide behaviour, as 'internal representations'. Treatments which are more formal refer to internal representations in terms of maps of event spaces in neuronal spaces (Cooper 1973), or sets of symbols with rules for putting them together (Marr 1982). The above definition of internal representations, i.e. *neuronally encoded structured versions of the world which could potentially guide behaviour*, suffices for our purposes, and captures the meaning of the notion as used throughout this book.

It is important to note that the term 'representation' is used here in a most general sense and is not restricted, as is the common practice in cognitive psychology and mammalian neuropsychology, to complex mental phenomena. According to the reductionist view presented here, representation is an inherent and fundamental function of all nervous systems. Therefore, internal representations are expected to vary tremendously in their complexity. Some are very simple. For example, a neuronal circuit subserving limb withdrawal in response to pain, encodes a representation of 'no pain' or various intensities of pain, and the appropriate motor response programme. (In the intact organism, even in this case, the representation encoded in the reflex circuit is in fact but one element in a structured body of information encoded by the nervous system *in toto*.) Other internal representations are far more complex, and many, for example the representations of cognitive states, are expected to be extremely complex. We will briefly readdress the complexity of internal representations later in this chapter, and then again, after reviewing experimental data on the simplest representations, in Chapter 9. However, regardless of their complexity, all biological internal representations: (1) are encoded in some way or another in neuronal systems; (2) may determine the behavioural output to a sensory input; and (3) when altered, may modify the potential to react rather than the immediate reaction to an input.

A redefinition of learning

Following the above arguments, *learning* is here defined as an experience-dependent generation of enduring internal representations, and/or experience-dependent lasting modification in such representations. 'Enduring' and 'lasting' mean at least a few seconds, but in most cases much longer, and in some cases up to a lifetime. The effects of rigid developmental programmes, injury, disease, and drugs are excluded. *Memory* is the retention of experience-dependent internal representations over time. *Retrieval* is the use of memory in neuronal and behavioural operations.

Recapitulation of fundamental assumptions

A tenet of neurobiology, implicit in the above discussion, should be clearly spelled out at this point: *behavioural states correspond to brain states*. This may look self-evident, but it is not; philosophers of the mind debate ardently the issue of correspondence of brain states to mental and behavioural states (reviewed in Churchland 1986).

Moreover, neurobiologists assume that *the hardware properties of the system are important and useful for understanding its function and output*. Again, there are opposing views. For example, radical behaviourists (Skinner 1938) as well as functionalistic philosophers of the mind (Pylyshyn 1980) deem the structure and properties of the neuronal substrate practically irrelevant to studies of the operations of the mind. Internal representations, being structured bodies of information, can, of course, be treated as formal entities, independent of the hardware on which they are implemented. This is the domain of studies of branches of mathematics, computer sciences, and artificial intelligence (Lewis and Papadimitriou 1981; Marr 1982; Charniak and McDermott 1985; Rumelhart and McClelland 1986a; Anderson and Rosenfeld 1988). However, neurobiologists think that since nervous systems and the capacity to encode internal representations in these systems co-evolved in phylogeny, the hardware places critical constraints on the relevant codes and computations, and provides clues for their understanding. Therefore, *the neurobiology of learning investigates the neuronal substrates that are expected to subserve internal representations, and, specifically, experience-dependent modifications in these substrates.*

Learning as a multilevel phenomenon

From what has been said so far, it is evident that learning and memory can be analysed at different levels of biological organization. The highest level is that of the *behaving organism*. In its purest form, analysis at this level treats the organism as a black box, and the properties of learning are inferred from the input–output relationship. This level is the interface between neurobiology and behavioural psychology, and the properties of the neuronal substrates are not yet explicitly taken into consideration. Next comes the *organ*, or the integrated *neuronal system* level. Here the question is how does the brain (or its equivalent in lower organisms) function as an integrated organ during learning. Next comes analysis at the suborgan or *circuit* level, which determines the role of specific brain parts, neuronal circuits, and neuronal pathways in learning. Still next is the *cellular* level: here the function of individual neurons and their connections is studied. Finally, there is the *molecular* level: the analysis of the biochemical and biophysical events associated with learning and memory. We should expect to elucidate mechanisms of learning and memory at each of these levels.

The meaning of reductionism

Our reductionist stand deserves clarification. The term 'reductionism' is often used with different meanings (Mayr 1982). In its most common use, reductionism means the mere process of analysing a complex phenomenon by dissecting it into elementary components. This is *constitutive* reductionism. In the context of our discussion, it means that one attempts to find brain, neuronal, and molecular *correlates* of learning. Such correlates might also be shown to be *necessary* for learning. This type of reductionist approach is accepted, and practised, probably by all neurobiologists.

In the process of practising constitutive reductionism, neurobiologists take *reductive steps*. These are shifts in the level of analysis from the level of a system as a whole to the level of its components. For example, a shift in analysis from molar electrical activity of the cortex to single cortical neurons, or from single neurons to individual molecules in the neuronal membrane, is a reductive step. In addition, constitutive reductionism often involves *simplifying steps.* These are procedures taken to facilitate experimental analysis, without altering intentionally the level of analysis. For example, proceeding in the analysis of single-neuron activity from *in vivo* to brain slices, or removing part of the tissue and hence decreasing the number of cells in a ganglion, while still maintaining the cellular level of analysis, is a simplifying step. Detailed examples of reductive and simplifying steps, and their integration into a comprehensive research programme, will be provided in Chapter 4.

More rigorous than constitutive reductionism is *explanatory reductionism*. It assumes that, ultimately, the knowledge of the components will explain properties of the system as a whole. This means, in our case, that having once understood the properties of neurons and molecules, one should be able to show how these properties are necessary and *sufficient* to explain learning. Many neurobiologists practise constitutive reductionism in the hope of ultimately achieving explanatory reductionism. This is the stand followed in this book, and examples of partial, heuristic explanatory reductionism are provided in the following chapters. However, some neurobiologists doubt whether complete explanatory reductionism is feasible.

Most demanding is *theory* reductionism, i.e. reducing a theory, including all its concepts and laws, into another, more inclusive or basic theory (Nagel 1961). This would mean that, having a biological theory (such as a theory of brain function, which is not yet available), one would be able to reduce it with no residue into a physical theory. This has not yet been done, and severe doubts are often expressed as to whether it is at all possible. For further discussions of reductionism in the neurosciences, see Fodor (1975, 1981), Block (1980), Popper and Eccles (1981), and Churchland (1986).

Fundamental questions in the study of learning

Several fundamental questions guide, sometimes explicitly and more often implicitly, neurobiological research on learning at all levels of analysis. The following is a selected set of questions.

How are internal representations encoded in the nervous system?

Ultimate comprehension of the neurobiological changes which occur in learning requires understanding the codes of internal representations, the computations made over these representations, and their neuronal realization. These are the fundamental concerns of the neurosciences in general, and are not unique to learning research. As we shall see later, in some simple reflexive systems, partial answers to the above questions are already available. In contrast, in complex systems, the identity of circuits, neurons, and connections, and, most importantly, the codes, algorithms, and computations, are not yet clear. For example, is the firing pattern of individual neurons in such systems decisive, or only the co-ordinated molar activity of many or all neurons? Are there emergent properties that cannot be detected at the single-cell level? Is the system hierarchically organized, so that single neurons store highly complex information, e.g. whole percepts and concepts? (Such hypothetical neurons are dubbed 'grandmother', 'pontifical', or 'gnostic' cells.) Or, taking the opposite extreme view, is the information stored in a distributed manner, so that no single cell or connection in the network is crucial? And are the connections in complex neuronal systems deterministic or probabilistic? These questions will haunt us throughout our discussion. For a selected spectrum of views, ranging from a single–percept–single–neuron to ensemble encoding in distributed networks, see Hebb (1949), John (1967, 1972), Konorski (1967), Barlow (1972), Hinton and Anderson (1981), Abeles (1982), Palm and Aertsen (1986), Rumelhart and McClelland (1986*a*), and Churchland (1986). Some of the views reflected in the above references will be further elaborated in the following chapters.

What differentiates between a 'learning' and a 'no learning' situation?

Not every experience 'teaches', i.e. results in a new enduring internal representation or in a lasting modification in an existing internal representation. We should like to know what conditions determine whether or not the interaction with the input results in learning. These conditions might include, for example, a certain 'preparatory' state of the neuronal system, input repetition, or coincidence of different inputs. We would also wish to understand how these conditions are realized at various levels of neuronal organization, and how events at one level relate to those at another level.

Is 'teaching' instructive or selective?

Since learning involves generation of, or changes in, representations, the question could be asked: does the 'teaching' input impose the new order in the neuronal system by somehow directly instructing the system to be modified in a certain way? Alternatively, does the input induce the new structure indirectly by selecting a representation among several endogenous variations, or spontaneous 'pre-representations'? In the latter case, the internal activity of the neuronal system has a decisive role in determining the repertoire of acquired representations. Both instructive and selective mechanisms can, of course, coexist in nervous systems. If learning by selection does exist, one might expect it to be more prominent in neuronal systems that encode complex representations, because these systems should have a larger number of potential functional permutations. The instructive mechanisms carry a Lamarckian flavour; the selective mechanisms are Darwinian, and have gained the favour of many neuroscientists (Young 1979; Changeux 1985; Edelman 1987; see Chapter 9 for further discussion).

Is memory retained in the system that learns?

One possibility is that memory is stored in the same loci that change in learning. Another possibility is that the primary sites of change remain altered, but changes spread to other parts of the system as well. And finally, the possibility exists that following learning, the primary sites of change relax into their naïve state, and the new information is deposited in special memory store(s).

What structural and functional modifications take place in neuronal systems during learning and memory?

This question refers to the actual structure and operation of the neuronal hardware, the nuts and bolts of biological 'learning machines'. Again, it can be applied to different levels of biological organization, from individual neurons and connections to the entire ganglion or brain. What are the actual molecular, cellular, and multicellular mechanisms involved in acquisition, retention, retrieval, extinction? And how many types of learning and memory mechanisms are there? Assuming that there is a multiplicity of learning and memory mechanisms and systems, is there also universality? In other words, are there components and operational rules common to different memory systems?

On the evolution of notions and experiments

The modern search for memory machines in brains was shaped by a scientific evolution interspersed by conceptual and technological revolutions. Paraphrasing Hermann Ebbinghaus's remark on psychology (cited in Boring 1950), one may say that the neurobiology of memory has a long past, but only a short history. As a scientific discipline, it is about 100 years old. Its roots, though, are much older. A comprehensive historical account of the study of learning far exceeds the scope of our discussion (see Boring 1950; Bower and Hilgard 1981; Boakes 1984). However, while risking simplifications and omissions, a few brief notes concerning some conceptual and experimental developments are appropriate. These developments have contributed to shaping the fundamentals of modern learning research.

• The following remarks have a threefold purpose: (1) to serve as leads for further reading; (2) to provide a background sketch for some leitmotifs recurrent in the coming chapters; (3) to serve as a reminder that the concepts and experimental infrastructure underlying the research described in this book were themselves gradually, and mostly rather recently, acquired and added to our cultural memory.

A conceptual framework: the brain as a machine

Whatever his or her philosophical stand on the mind–brain issue, the neurobiologist treats brains as electrical and chemical machines. The search for the operation of the memory parts of the mind can be traced back to ancient times, but the idea that one should look for them in the brain was at first an exception, not a rule. For example, Aristotle (fourth century BC), who reflected deeply on memory, internal representations, and imagery (Sorabji 1972), held that the heart is the seat of mental function. Similarly, the Prophets considered the kidney and the heart, not the brain, as the most intimate niches of the human soul; not surprisingly the brain is not mentioned in the Bible even once. Alexandrian anatomists in the third century BC did suggest localization of mental function to the brain; but it was left for Galen (second century AD) to leave a more enduring impression on the history of physiology, by challenging Aristotle, and by demonstrating that nerves originate in the brain and that motor and sensory functions are abolished by brain injuries.

In the centuries that followed, and in spite of the dominance of the Aristotelian dogma, assignment of mental faculties to the brain became common. This was followed by attempts to localize function in the brain. Medieval scholars such as Costa ben Luca (ninth century) and Albertus Magnus (thirteenth century) thought that memory was stored in brain ventricles. Variations on this theme were entertained by other contemporary scholars in Europe and the Middle East (Thorndike 1923).

Models of biological memory always reflect the technological frontier of their time; what the computer is for twentieth-century cognitive sciences, hydraulic gadgets were for medieval philosophy. Indeed, medieval descriptions of memory systems included complex pipes and fancy valves. Costa ben Luca, for example, suggested that a 'spiritus', which perishes when separated from the body, causes mental images and memory by flowing in the brain. The opening between the anterior and the posterior parts of the brain is closed by a valve, and when a memory is recalled, the valve opens and the spiritus passes from the anterior to the posterior ventricle. The speed with which the valve operates determines the speed of memory; inclining the head stimulates the opening of the valve, and hence memory (Thorndike 1923).

It was, however, René Descartes (1596–1650) who, by sharply defining animals as reflexive automatons, became the founding father of objective animal psychology. Man was portrayed as an automaton uniquely equipped with a rational soul, with the pineal gland being the locus of soul–body interaction. Descartes contemplated memory in several of his writings (Descartes 1649, 1664). His accounts of human memory included naïve versions of notions which were later to become essentials of modern memory models: representations subserved by physical traces in the brain, and activity-dependent facilitation of connections in these traces.

Thus, when the soul wants to remember something . . . volition makes the gland lean first to one side and then to another, thus driving the spirits towards different regions of the brain until they come upon the one containing traces left by the object we want to remember. These traces consist simply of the fact that the pores of the brain through which the spirits previously made their way, owing to the presence of this object, have thereby become more apt than the others to be opened in the same way when the spirits again flow towards them. And so the spirits enter into these pores more easily when they come upon them, thereby producing in the gland that special movement

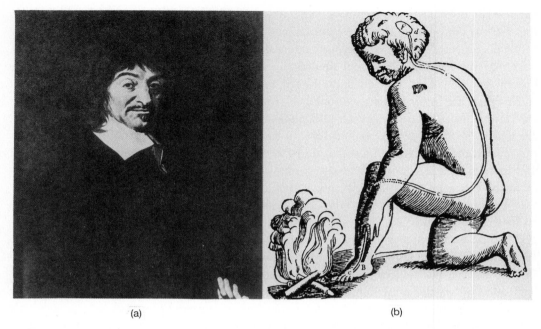

(a)   (b)

Fig. 1.1. Rene Descartes's notion of animals as reflexive automatons had a crucial, seminal impact on experimental psychology, including learning research. (a): Descartes. (b): A reflex. Sensory input (A) pulls a fibre in the nerve *cc*, opening the pore *de*, allowing 'animal spirits' from the brain cavity F to flow back through the nerve to the foot. This causes muscular contraction and the withdrawal of the foot from the fire. Descartes's memory models included the modern concept of experience-dependent modification of connectivity (Descartes 1664).

which represents the same object to the soul, and makes it recognize the object as the one it wishes to remember. (Descartes 1649)

A methodological breakthrough: quantification of learning

Two hundred years had to elapse between Descartes's accounts of reflexes and a memory system, and the emergence of modern psychology. Later philosophical inquiries, for example British associationism, did contribute tremendously to the maturation of psychology (Murphy 1949; Boring 1950; Russell 1961; Boakes 1984). However, in order to be transformed into an experimental science, psychology in general, and learning research in particular, had to acquire experimental research methods.

Experimental psychology began to shape up in the mid-nineteenth century, at the same time as a major revolution took place in the life sciences. The years 1830–60 saw the development of organic chemistry, embryology, cell biology, zoology, physiology, and the theory of evolution (Mayr 1982). At that time, biologists already had at their disposal experimental techniques, ample observations, and heuristic conceptual frameworks. Alas, in contrast, pioneers of the new science of the mind were still searching for objective methods and data.

Most of this search took place, initially, in Germany, where the second half of the nineteenth century saw the emergence of psychophysics, sensory physiology, and psychology, and the first formal psychological laboratory in the world (established by Wilhelm Wundt in Leipzig, 1875). Memory became a systematic subject of research somewhat later than sensory psychology. One could measure reactions to temperature or light, using rather simple physical instruments, but how could one quantify reminiscences? Two major developments changed this picture: the development of objective, quantitative measures of human learning and memory, and the emergence of experimental animal psychology.

Initiation of the first trend is associated with Hermann Ebbinghaus (1850–1909). Influenced by British associationists, atomism, and the physiological psychology of the senses, Ebbinghaus decided to look for measures of the acquisition and recall of atoms of information by the human mind. For that purpose, he invented the use of nonsense syllables (two consonants interspersed by a vowel) as atoms of objective data, and measured that rate at which they were learned and forgotten. As a genuine scientist, he experimented on himself. There is much one can learn from Ebbinghaus-type experiments: for example, the kinetics of acquisition, retention, and extinction; the existence of memory phases; constraints on the capacity of short-term memory; and the kinetics of data transfer from short- to long-term memory. 'What number of syllables can be correctly recited after only one reading? For me, the number is usually seven. Indeed I have often succeeded in reproducing eight syllables, but this happened only at the beginning of the tests and in a decided minority of the cases' (Ebbinghaus 1885). Well, he had no reason to be ashamed of his performance; it was entirely within the normal human range (Miller 1956).

Had the breakthrough in experimental memory research been confined to Ebbinghaus's methods, most, if not all, of the material covered in the following pages would not have come into existence. Research treating the brain as a 'black box' can hint at the properties of biological systems, and reveal constraints on their performance, but cannot disclose how the physical apparatus is constructed and how it functions. To elucidate that, one must dissect the brain. For such a purpose, the human being is not a very convenient experimental subject. Even the most devoted and enthusiastic scientists would be reluctant to sacrifice their one and only brain for the sake of their profession. The only relevant information on human brain and memory could be obtained from rare pathological cases. Animal models were, therefore, required.

A conceptual revolution: Darwin and the phylogeny of the mind

The fact that animals learn was, of course, obvious long before psychologists turned their attention to it. The development of systematic research on animal learning drew from two

(a)   (b)   (c)

Fig. 1.2. The concept of biological evolution has paved the way to studies of higher brain functions, learning included, in animals. (b): Charles Darwin. (a) and (c): Lynx and *Homo*, respectively, expressing emotions that hint at the phylogenesis of the mind. (From *The expression of the emotions in man and animals*, Darwin 1872.)

major conceptual sources. The first was Descartes's philosophy, advocating dissection of animal behaviour into elementary mechanistic processes (see above). The second was the meta-theory of evolution. Species share biological structures and mechanisms with their phylogenetic ancestors. Man is no exception. We should, therefore, expect to find learning in organisms other than man. 'There is no fundamental difference between man and the higher mammals in their mental faculties . . . With respect to animals very low on the scale . . . their mental powers are much higher than might have been expected' (Darwin 1871).

When nineteenth-century scientists became aware of the possibility that evolution applies to the mind, some of them became so carried away that they started assigning to beasts almost all the mental faculties of university professors. The data were at that time mostly anecdotal, and anthropomorphism, not only metaphorical but also explanatory, was common. Whereas research on human memory was moulded primarily in Germany, that on animal memory flourished at first in England. Works by Charles Darwin and George John Romanes are examples of the early tradition in animal psychology (Darwin 1872; Romanes 1882).

Although Romanes was a capable physiologist who experimented on reflexes, his book, *Animal intelligence*, was the pinnacle of the anecdotal tradition in psychology. He relied on a variety of accounts of animals' psychic life, generously volunteered by respected English gentlemen and gentlewomen. Some of these accounts provided 'evidence' even of the ability of animals to associate words with deeds; for example, a parrot which, having been trained to offer its claw in response to a human request, was said to have acquired the habit of saying 'Give me your claw' to itself, then complying with the request by placing one of its claws in its own beak. Many other animals were credited with similar intelligence.

Clearly, knowledge on animal learning could not have been advanced by adhering to anecdotes. The data were at best descriptive and qualitative, and the interpretations were at the mercy of the eyes of the beholders. The problem was clearly alluded to by Darwin's countryman, Conway Lloyd Morgan, in his famous 'Lloyd Morgan's canon' (Morgan 1894): 'In no case may we interpret an action as the outcome of the exercise of a higher physical faculty, if it can be interpreted as the exercise of one which stands lower on the psychological scale.' Morgan's canon was a derivation of a general maxim of logic, Occam's razor: 'Entities are not to be multiplied without necessity.' Research on animal learning at the close of the nineteenth century was, thus, in much need of Occam's razor–Morgan's canon, and of reproducible, *quantitative*, and *objective* methods, similar to those invented by Ebbinghaus for analysing human memory. Ebbinghaus's methods were not, however, applicable to animals, because they depended on human intelligence and language. Other paradigms were badly required.

A methodological breakthrough: experimental paradigms for quantifying animal learning

The first reproducible paradigms for quantifying animal learning were soon developed, but not in England or Germany. The few attempts to establish such paradigms in Britain (by Lloyd Morgan, Romanes, Donald Spalding) did not materialize into schools (Boakes 1984). In Germany, the main cradle of physiological psychology, physiologists were not yet attentive enough to the relevance of animal behaviour to memory research. Salvation came from two other scientific communities, on both sides of the Atlantic: in Russia and in the United States. The Russian solution was based on the analysis of reflexes; the Americans relied on the analysis of behavioural initiatives taken by animals in new situations.

By the latter part of the nineteenth century, Russia already had a strong tradition of research on reflexes, and strong academic ties with physiologists in Germany and France. The leading figure was Ivan Mikhailovich Sechenov (1829–1905), who considered the physiology of reflexes to hold the solution to the problems posed by psychology (Sechenov 1862). Among the younger Russian physiologists who followed his trail was Ivan Petrovich Pavlov (1849–1936). He started to work on learning *per se* only in his fifties, using methods employed in his earlier research on digestive reflexes in dogs, for which he was awarded the Nobel Prize (1904). Pavlov's work was a systematic and successful attempt to reduce learning to the study of readily quantifiable, restricted reflexive components. This approach, and the conclusions, have gained much attention in the West, especially following Pavlov's celebrated report on conditioned reflexes to the Medical Congress in Madrid in 1903, and the Huxley lecture delivered by him in London in 1906 (Pavlov 1906).

Pavlov presented clearly his view on how higher brain functions should be investigated:

In astounding contrast with the unbounded activity of the cerebral hemispheres stands the meagre content of present-day physiological knowledge concerning them. . . . These nervous activities have never been regarded from the same point of view as those of other organs. . . . This is an anomaly

Fig. 1.3. The development of experimental paradigms for analysing animal learning was a turning point in the neurobiological study of learning. In a rudimentary attempt to monitor trial-and-error learning, Morgan (1900) trained his dog, Tony, to operate the latch of a gate. However, only later systematic experimentation by Pavlov, Thorndike, and their colleagues resulted in robust, widely used paradigms.

which placed the physiologist in an extremely difficult position. On one hand it would seem that the study of the activities of the cerebral hemispheres, as of the activities of any other part of the organism, should be within the compass of physiology, but on the other hand it happens to have been annexed to the special field of another science—psychology. What attitude then should the physiologist adopt? It is logical that in its analysis . . . physiology should base itself on the more advanced and more exact sciences—physics and chemistry. But if we attempt an approach from this science of psychology . . . we shall be building our superstructure on a science which has no claim to exactness as compared to physiology. It would be more natural that experimental investigation of the physiological activities of the hemispheres should lay a solid foundation for a future true science of psychology. (Pavlov 1927)

Pavlov regarded himself, appropriately, as continuing the tradition of Descartes, and sharply contrasted his attitude with that of his leading American contemporaries who were not educated as physiologists. The American stand was a functional and a pragmatic one, shaped to a large extent by William James (1842–1910), author of the 'bible' of American psychology, *Principles of psychology* (James 1890). It is to one of James's students, Edward Lee Thorndike (1874–1949), that we owe the existence of another major prototype of experimental paradigm which altered the course of learning research. Thorndike was interested in the objective study of behaving, moving animals, and, particularly, in animal 'intellect' as measured by problem-solving (Thorndike 1911). The methodologies and terminology introduced and coined by Thorndike and Pavlov, and by their students and followers, are fundamental assets of learning research, and will be described in Chapter 2.

It is of course presumptuous to claim that Pavlov, Thorndike, and their colleagues were the first to succeed in training animals. The art of animal training was well known to man since the dawn of history. Pets ranging from birds to elephants were trained for fun or profit long before Pavlov's dogs salivated in Petrograd and Thorndike's cats escaped problem-boxes in New York. So-called 'gifted' animals, such as horses performing 'mathematical' tricks, evoked the amazement of many of the first professional psychologists (Watson 1914); the conclusion could not have been avoided that lay people and circus trainers often have a more successful approach than scientists to the minds of animals. In addition, as noted above, rudimentary experiments on animal trainability had already been performed during the 1860s by Spalding, and then by Romanes and Lloyd Morgan (Boakes 1984). What Thorndike and Pavlov did accomplish so successfully for the first time, was to incorporate animal training securely into the domain of science.

The neurobiology of learning: an interdisciplinary synthesis

More or less in parallel with the emergence of research on animal learning, exciting developments were taking place in clinical neuropsychology. Extending and synthesizing previous neurological findings, Theodule Armand Ribot in France (1882), Sergei Korsakoff in Russia (1887), John Hughlings Jackson in Britain (1888), and others, published seminal monographs on behavioural syndromes of brain lesions. These included disturbances in memory (see Chapter 15). During the same period, neuroanatomical, electrophysiological, chemical, and pharmacological methods had begun to revolutionize our understanding of the structure and function of the nervous system (see Chapter 3). The foundations for the neurobiology of memory have thus been laid from different, yet converging directions: philosophy, psychophysics, animal behaviour, clinical neurology, and biology.

If one has to pinpoint a turning point for the modern synthesis of the neurobiology of memory, the late 1940s seem an appropriate choice. The life sciences were on the verge of the great molecular revolution, electrophysiology and neuroanatomy had already yielded exciting results, and noted neuropsychologists sensed that it was time to integrate all levels of analysis of learning into a meta-approach. One manifestation of this attitude was the publication in England of *Conditioned reflexes and neuron organization* by the Polish neuro-physiologist Jerzy Konorski, one of Pavlov's students (Konorski 1948). Besides serving as a critique of some of Pavlov's theories, this book was an attempt to recruit neuropsychological and cellular mechanisms for the exploration of behavioural phenomena, and to extrapolate from the function of a lower level of neuronal organization to higher levels of brain organization and function. It was also a pioneering, declared attempt to break the barriers between psychology and neurophysiology and to interest neurophysiologists in neuronal and behavioural plasticity.

A year later, in 1949, the Canadian neuropsychologist Donald Hebb published *The organization of behaviour: A neuropsychological theory.* This book was bound to become one of the most, if not the most, seminal publications on the neurobiology of memory. 'I have tried', wrote Hebb, 'to bring together a number of different lines of research, in a general theory of behaviour that attempts to bridge the gap between neurophysiology and psychology, as well as between laboratory psychology and the problems of the clinic.' Hebb's monograph is an early example of what modern learning research is all about: a synthesis which draws from the neural sciences, clinical sciences, and psychology. This inter-disciplinarity endows the field with beauty and power, but also places demands on all parties involved. Psychologists must remember that beyond a behaviour there is a brain. Molecular biologists must remember that the animal behaved prior to homogenization.

The neurobiology of learning borders with, and draws from, classical psychology (Baddeley 1976; Bower and Hilgard 1981; Mackintosh 1983), cognitive psychology (Glass and Holyoak 1986), including formal analysis and modelling (Rumelhart and McClelland 1986*a*; Anderson and Rosenfeld 1988), and artificial intelligence (Charniak and McDermott 1985). But, of course, the neurobiology of learning is primarily based on hard-core neurobiology (Kuffler *et al.* 1984; Kandel and Schwartz 1985; Shepherd 1988). In the course of our discussion, we shall concentrate on selected biological aspects of learning research, but will also, when appropriate, briefly refer to notions from other disciplines. Relevant specific references will be provided in each case. Naturally, the scope of discussion of tangential disciplines will be very limited, and the reader may find each of the afore-mentioned books, as well as additional references mentioned under Further reading, excellent sources for extensive background in the appropriate discipline.

2   *Paradigms and research tools*

The neurobiology of learning requires paradigms for quantifying behavioural output, as well as methods for analysing the biological apparatus at various levels of organization from molecules to brain. The neurobiological tools are common to other branches of neurobiology. The behavioural methods, however, are unique to learning research, and link the neurobiology of learning to experimental psychology.

Categories of learning and behavioural paradigms

Behavioural observations clearly show that learning occurs in very many species under very different circumstances. Even prokaryotes such as chemotactic bacteria (Koshland 1980), and unicellular eukaryotes (Jennings 1906; Harris 1943) display some kinds of experience-dependent behavioural modifications; but since we explicitly confined our definition of learning to changes in nervous systems, these cases will not be discussed. To introduce some order into a very rich and heterogeneous body of observations, it is useful to classify learning into a few categories. This classification should be regarded, at this stage, merely as a way to start organizing our thinking about learning and the methodologies employed to investigate it. The following is a parsimonious and convenient classification:

1. *Non-associative learning*, in which the organism habituates or is sensitized to a stimulus.

2. *Associative learning*, in which the organism learns relations among events in the world—either among stimuli, or among stimuli and actions.

3. *Incidental learning and related processes*. Often, organisms acquire information about the world while attending 'incidentally' to sensory inputs (Hebb 1949; Glass and Holyoak 1986). For example, in everyday life we all acquire information which does not seem associated with notable events. 'Incidental' simply means the absence of an identifiable instructive situation. The term reflects the point of view of the observer; in all probability, and similarly to associative learning, here, too, learning occurs because the situation signifies potential importance. This point will be further elaborated in later chapters.

The behavioural consequences of incidental learning are usually latent. This type of learning can be manifested in a number of ways. For example, it may alter the perceptual distinctiveness of sensory attributes (perceptual learning) (Hebb 1949; Walk 1978). Or, it may contribute to the reorganization of internal representations, resulting in sudden solutions to previously unresolved problems (insight) (Kohler 1925).

Animals thus learn the structure of the world by different processes which result in the generation and modification of internal representations of different complexities. Processes involving complex internal representations are naturally more interesting because they are more relevant to the operation of the human mind. On the other hand, the more complex the memory, the more difficult is its neurobiological analysis. Many efforts in research on the neurobiology of learning are devoted to relatively simple learning, partly because some early experimenters were influenced by the radical behaviouristic stand that all learning can be reduced to primitives of stimuli and responses, but mostly because relatively simple learning is easier to study in laboratory animals.

Non-associative learning is the most simple and elementary form of learning, occurring in every eukaryote studied to date. It is, therefore, appropriate to consider it first.

Habituation: learning to withhold a response

Habituation is a gradual diminution of response to a stimulus, following repeated presentation of the same, or a very similar, stimulus. Response decrement due to sensory receptor adaptation, fatigue, injury, or drugs is excluded. Habituation, hence, involves decline of existing responses rather than acquisition of novel responses.

Behavioural examples of habituation

The marine polychete worm *Nereis pelagica* lives in mud burrows. It contracts in response to a sudden vibration or shadow. Clark (1960) maintained *Nereis* in glass tubes immersed in an aquarium, and quantified the effect of mechanical or visual stimuli on the contraction reflex. At first, the worms responded vigorously, but the response to each stimulus type diminished specifically on repeated, monotonous presentation of that stimulus (Fig. 2.1(a)). The original response was instantaneously restored following presentation of one of the other stimuli, or gradually restored following a rest period. The behaviour of *Nereis* illustrates: (a) habituation; (b) dishabituation; (c) spontaneous recovery.

Although shaped millions of years after that of *Nereis*, the human nervous system has retained the potential for some behavioural responses that characterize the worm. We all habituate to monotonous stimuli. Joseph Haydn was, probably, aware of that while incorporating the dishabituating full orchestra *fortissimo* chord in the Andante of Symphony No. 94, *The Surprise*, thus unexpectedly disrupting the very quiet, slow, and habituating opening theme. Twentieth-century psychologists make use of habituation and dishabituation to study the ontogenesis of sensory and cognitive faculties in infants (Fagan 1985). It is impossible to get a verbal or intentional gestural reply from a newborn but easy to observe

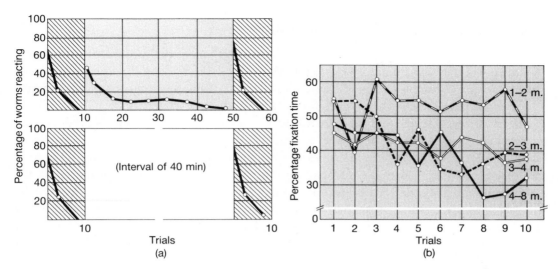

Fig. 2.1. Habituation in two extremes of the phylogenetic scale. (a): Habituation and dishabituation of *Nereis* contraction reflex. Top: The response to a repetitive moving shadow (hatched background) declined rapidly, but the response to a mechanical shock remained intact (curve plotted on a grey background). Following habituation to the shock, the worms were again dishabituated by the moving shadow. Habituation to the shadow recovered spontaneously after rest (bottom). (From Clark 1960.) (b): Habituation of infants to familiar photographs, measured by the duration of visual fixation. A decline in the relative amount of time spent in fixation on the familiar pictures was evident in infants older than 2 months, but not in younger infants. (From Fantz 1964.)

his or her sensory and motor reactions to stimuli immediately after birth. As every parent discovers anew, infants devote more attention to novel stimuli than to familiar ones. This behaviour makes it possible to determine whether the infant distinguishes between two dissimilar stimuli, for example visual patterns.

A set of experiments by Fantz (1964) illustrates this point. Infants were placed face up in a small crib inside a test chamber which provided a uniform background for two visual stimuli. The latter were cards with different photographs or advertisements from magazines, placed over holes in the ceiling. One of the photographs served as a constant pattern, and was exposed ten times for 1 minute each. During each exposure it was paired with another card, shown only once. Visual fixation was measured by observing corneal reflections of the patterns in the infant's eyes.

Infants with ages ranging from 6 to 25 weeks were tested. As depicted in Fig. 2.1(b), a complex visual pattern, presented successively, was fixated progressively less than a novel stimulus, by infants older than 2 months. The results could not have been attributed to fatigue, since the decreased fixation on the familiar pattern was accompanied by an increased fixation on the novel stimuli. The infants have, thus, habituated to the familiar stimulus and been dishabituated by the novel one.

Parametric characteristics of habituation

Habituation can be demonstrated in all the species that bridge the gap between *Nereis* and man. Clearly, it is not a unitary process (Harris 1943; Glaser 1966; Williams *et al.* 1975), but both simple and complex cases of habituation display a basic phenomenological similarity. Thompson and Spencer (1966) have formulated parametric characteristics that apply to many instances of simple and complex habituation alike. These are as follows. (1) Following habituation by repeated application of a stimulus, if this stimulus is withheld, the response recovers over time. (2) If repeated series of habituation training and spontaneous recovery are given, habituation becomes successively more rapid. (3) Other things being equal, the more rapid the frequency of stimulation, the more rapid and/or pronounced is habituation. (4) The weaker the stimulus, the more rapid/pronounced is habituation. Strong stimuli may fail to produce habituation. (5) The effects of habituation training may proceed below the naïve response level. (6) Habituation of response to a given stimulus may exhibit generalization to other stimuli. (7) Presentation of another (usually noxious) stimulus results in recovery of the habituated response (i.e. dishabituation).

Not all the above criteria are satisfied in every experimental system. Of all the criteria, that of dishabituation is the most critical, and is commonly employed as the major criterion in demonstrating that habituation has indeed occurred.

Is dishabituation a disruption of habituation, or is it an independent, facilitatory process, superimposed on habituation? If dishabituation is a disruption of habituation, then a dishabituated response should not rehabituate unless additional habituating stimuli are provided. If, on the other hand, dishabituation is a facilitatory process superimposed on habituation and having its intrinsic decay time, then the dishabituated response might gradually decrease towards the habituated level, even in the absence of additional habituation. Results from several behavioural experiments showed that habituation and dishabituation both have an intrinsic time course, thus supporting a 'dual-trace hypothesis': habituation and dishabituation develop independently but interact to yield the final behavioural outcome (Groves and Thompson 1970). Habituation and dishabituation of complex behaviours require more complex interpretations than the interaction of only two elementary independent processes (Coombs 1938; Sharpless and Jasper 1956; Sokolov 1963; Glaser 1966; see also the discussion of attention in Glass and Holyoak 1986).

Habituation is so common in the animal kingdom because it contributes to fitness. It is a mechanism for the selective analysis of input by limited-capacity, information-processing systems of organisms, enabling adaptation to the milieu, and preventing superfluous defensive responses. Kandel (1976) cites a fable by Aesop, which beautifully illustrates this point: 'A fox, who had never yet seen a turtle, when he fell in with him for the first time in the forest was so frightened that he was near dying with fear. On his meeting with him for the second time, he was still much alarmed but not to the same extent as at first. On seeing him the third time, he was so increased in boldness that he went up to him and commenced a familiar conversation with him.' Lorenz (1981) contrasts habituation with fatigue, by pointing out that the former is a mechanism for avoiding the latter.

Sensitization: learning to intensify a response

Sensitization is an augmentation of a response to a stimulus, following presentation of the same or another stimulus. The sensitizing stimulus is usually strong or noxious. Sensitization is termed non-associative because it does not result from specific associations between particular stimuli (as opposed to associative learning, see below); rather, a sensitizing stimulus increases the responsiveness to a wide variety of stimuli. From a behavioural perspective, dishabituation could be regarded as a case of sensitization. Whether an organism is said to undergo sensitization or dishabituation may depend on what we know about its history. However, as we shall see below, the biological mechanisms underlying sensitization and dishabituation are not necessarily the same.

Sensitization, like habituation, is also very widespread throughout the animal kingdom. It is a priming mechanism, increasing arousal and attention, and lowering the threshold of defensive responses. 'An earthworm', comments Lorenz, 'that has just avoided being eaten by a blackbird that has taken a peck at it, is indeed well advised to respond with a considerably lowered threshold to similar stimuli, because it is almost certain that the bird will still be nearby for the next few seconds' (Lorenz 1981). Whether the threshold stays low for seconds only depends, as in the case of habituation, on the instructing stimulus: repetitive exposure to noxious stimuli may lead to long-term sensitization, lasting weeks or even longer.

Associative learning: learning relations among events

Associative learning is a more discriminatory and complex form of learning than non-associative learning. The major classification and paradigms of associative learning are based on the seminal work of Pavlov (Fig. 2.2) and Thorndike (Fig. 2.3).

The two main categories of associative learning are *classical conditioning* and *instrumental conditioning*. Each of these categories is frequently denoted by additional terms. Moreover, authors often make fine distinctions and assume special connotations anchored in specific theoretical assumptions, when referring to terms that other authors use in a different way. On entering the domain of psychological terminology, one cannot refrain from citing Mackintosh (1983): 'As generations of students have complained, one of the most repellent features of the study of conditioning in animals is the barbarous terminology it has developed.' We refer here briefly to only a limited selection of basic characteristics of each type of learning, and to major features of the relevant experimental paradigms. (For comprehensive treatments of paradigms and theories, see Bower and Hilgard 1981; Mackintosh 1983; Flaherty 1985.)

Classical conditioning: learning relations between stimuli

Classical conditioning is also termed Pavlovian conditioning, conditioned reflex type I, respondent conditioning, and type S conditioning. In this type of learning, the organism is

Fig. 2.2. Pavlov, assistants, visitors, and dog. Pavlovian (classical) conditioning is the simplest associative learning paradigm.

(a)   (b)

Fig. 2.3. (a): Thorndike. (b): A puzzle box used by him in his seminal experiments on instrumental learning.

presented with a stimulus that evokes a certain behavioural response. This stimulus, chosen and controlled by the experimenter, is named the *unconditioned stimulus* (US). The response to the US is the *unconditioned response* (UR). Another stimulus, neutral or almost neutral with respect to the UR, is then presented in association with the US. This stimulus is called the *conditioned stimulus* (CS). Again, the CS is chosen and controlled by the experimenter. The association between the CS and US alters the response to the CS, resulting in a *conditioned response* (CR), which is similar to the UR.

Classical conditioning can be illustrated by Pavlov's classical experiments (Pavlov 1927). Pavlov concentrated on two reflexes of the dog. The first he termed the food, or the alimentary, reflex: this is the innate reaction of the dog to an edible substance that is placed either in front of it or in its mouth. The second reflex, which Pavlov termed a mild defence reflex, is the innate reaction of the dog to a repellent substance placed in its mouth. Both these reflexes were described as consisting of two components, motor and secretory. First, the dog attempts to get the food (in the alimentary reflex) or to get rid of it (in the defence reflex). This is followed by an enhanced secretion of saliva to digest the food (alimentary reflex) or wash the noxious substance out (defence reflex).

Pavlov studied the secretory part of the reflexes. He freed the terminal portion of the salivary duct from its surrounding tissue and carried it out, via a small incision, to the outside of the cheek or under the chin. As a result the saliva could be collected and quantified. Pavlov then linked either the alimentary reflex or the defence reflex with a 'neutral' stimulus: the click of a metronome, or the sound of a buzzer, or a shadow, or mechanical, thermal, or electrical stimulation of the skin. It is evident that none of these stimuli was indeed 'neutral'. Each elicited a complex set of responses. However, attention was a priori restricted to salivation, which these stimuli affected only slightly, if at all.

For example, in part of the experiments, the alimentary reflex was used as the UR, and the metronome as the CS. By using the appropriate pairing schedule, e.g. preceding a meat-and-bread mix by 5 s with the auditory click, the dog had learned to salivate in response to the metronome. Food was thus the US, salivation the UR and later the CR, and the auditory click the CS.

The possible combinations of various types of US–UR–CS–CR are practically limitless. Note that the behaviour of the experimenter, who chooses the US and CS, is almost as important as the animal's behaviour in determining the outcome of the experiment.

The specificity of the CR

Very rarely, if at all, do animals emit solely the CR monitored by the experimenter. The latter may intentionally disregard part of the CRs. In addition, specific CRs are frequently superimposed on 'non-specific' ones. Especially common are cases in which specific motor CRs, obtained after pairing of a CS with an aversive US, are superimposed on visceral reactions characteristic of arousal and fear. These 'diffuse', visceral reactions comprise a 'conditioned emotional response' or 'conditioned fear' (Mowrer 1939, 1947; Estes and Skinner 1941; Konorski 1967; Rescorla and Solomon 1967). Elements of fear are also typical of a normal reaction to novelty, even to a seemingly neutral CS ('orienting response'; Sokolov 1963). Interestingly, conditioned fear has been blamed for contributing to the ontogenesis of very common and devastating human afflictions—anxiety neuroses (Mowrer 1939; Klein 1987; Marks 1987).

Conditioned fear develops rapidly and has multiple manifestations, such as altered heart rate, blood pressure, and respiration. Sometimes these reactions are described as 'non-adaptive', because they do not seem to adapt the organism to deal with the US. This term is misleading; fear has its own adaptive value in the ethological context of 'fight or flight'. It is non-adaptive only in the artificial experimental situation; for a restrained dog exposed to a shock, flight and fight are both impractical.

Subtypes of classical conditioning

In 'true' classical conditioning, the CR is supposed to be a novel response to the CS. Often the CR is clearly only an intensification of a pre-existing response to the CS. This is termed *alpha conditioning*. For example, a weak tactile stimulus to the skin, that usually evokes only a mild withdrawal response, may come to evoke an intense withdrawal response following alpha conditioning. However, the validity of the traditional distinction between classical and alpha conditioning is questionable, because careful examination of cases considered as 'true' conditioning may show that a weak CR is evoked by the CS in naïve individuals.

Pseudoconditioning

A CS–US temporal contiguity is a critical factor in classical conditioning (Rescorla 1967; but see a clear exception below, and also discussion of the importance of contingency and predictability vs. mere contiguity in Mackintosh 1983; Rescorla 1988). If random pairing of the CS and US, or presentation of the CS alone, elicit a response similar to the CR obtained by the appropriate contiguity schedule (see below), then conditioning is, by definition, non-associative. Non-associative elicitation of a CR to a previously 'neutral' CS is termed pseudoconditioning. Non-associative alpha conditioning is sensitization.

The role of timing and order of stimuli in successful conditioning

As noted above, successful conditioning is critically dependent on the training schedule. Such a schedule can be defined in terms of the *duration* of the US and CS, their *rate*, the *interval* between them, and the *order* of their presentation. Commonly encountered training schedules are given below.

Simultaneous conditioning

In this protocol, CS onset and US onset coincide in time. Usually this is not an effective associative conditioning procedure. However, if the CS offset precedes the US offset, the CS may come to evoke the CR.

Delay conditioning

Here, the CS is presented first, and US onset precedes CS offset or coincides with it. The CS in this case usually comes to evoke the CR.

Trace conditioning

Here, the CS starts and ends before the US, i.e., there is a time interval between the two. Trace conditioning is often less efficient than delay conditioning in evoking the CR, especially when the interval between the CS and the US is longer than a few seconds. This is not too surprising, because why should an animal associate two remote, seemingly independent events? However, cases are known in which even an extensive delay does not prevent effective conditioning. These cases involve gustatory reflexes, in which food is followed hours later by vomiting and cramps (Garcia *et al.* 1966). Which stimulus is really associated with the adverse response in such trace conditioning is not clear. Nevertheless, these experiments suggest that nature has evolved a mechanism to ensure that when survival is very much at stake, conditioning tolerates a long delay between cause and effect. This is not very novel to farmers, who know well that rats and foxes become bait shy after surviving poisoning.

Backward conditioning

Here US onset precedes CS onset, and the US terminates before the CS terminates. Backward conditioning is usually not successful.

• Note that neuronal and molecular models of classical conditioning must explain why some parameters of temporal proximity and order dependency are crucial for successful conditioning, e.g. why backward conditioning fails.

Instrumental conditioning: learning the impact of actions on the world

Instrumental conditioning is also termed type II conditioning, type R conditioning, trial-and-error conditioning, or operant conditioning. In this type of learning, the probability or intensity of a spontaneous behavioural response is altered by a reinforcing stimulus. In contrast with classical conditioning, where the experimenter directly manipulates the response to a distinct CS by pairing it with a distinct US that yields a selected UR, here the experimenter manipulates the relationship between a spontaneous response and a reinforcer, so that delivery of the latter depends on the occurrence of the former.

Instrumental conditioning can be illustrated by the seminal experiments of Thorndike. He built puzzle boxes, small crates hammered together from wooden slats (see Fig. 2.3). In a typical experiment, he placed a hungry cat inside such a box, and a piece of fish visibly outside. The boxes had different systems of pulleys, strings, and catches, so arranged that pulling a loop or pressing a lever would allow the door to fall open. The mechanism was not simple even for a naïve human observer. Nevertheless, the cats did manage to open the door. Their performance improved with time; at first it would take a cat several minutes, but after several successive trials only a few seconds would elapse before it would jump out and devour the fish. The cat, restlessly exploring the box, would operate the releasing mechanism by chance, and since the door immediately fell open, the animal would apparently learn to associate its deeds with the outcome, leading to an increase in the frequency of the behaviour that had resulted in obtaining the reward. The behaviour, it should be emphasized, need not be mechanistically related to the solution of the problem; Thorndike even manipulated his animals (though with great pain to both parties involved) to scratch or lick themselves to escape from the box.

Thorndike's paradigm served as a prototype for many other trial-and-error learning paradigms. One famous descendant was the Skinner box (Skinner 1938), a problem box in which an animal learns to get a reward by pressing a key. Skinner termed the learning observed in such paradigms 'operant conditioning', since the spontaneous behavioural acts operate on the environment and are in turn affected by the environmental effects. He contrasted the operant responses, emitted rather than elicited, with the 'respondent' responses of classical conditioning, which originate with the stimulus that elicits them. Skinner became famous for his intentional as well as unintentional contributions to a social and educational philosophy based on operant behaviourism (Skinner 1984). Interestingly, his training methods found their way not only into experimental educational programmes, but even into top-secret defence projects: during World War II he was engaged in conditioning pigeons to guide missiles (Skinner 1960).

The term 'conditioned reflex type II' was coined by Konorski, who modified Pavlov's paradigm and trained a dog to flex a leg in response to a buzzer in order to get food (Miller and Konorski 1928). 'Instrumental conditioning' is a term denoting the fact that the organism's behaviour is instrumental in the materialization of the reinforcer.

Reward, punishment, and avoidance

Three major categories of instrumental conditioning are commonly discerned: (1) reward conditioning, in which an appetitive reinforcer (e.g. food) increases the probability or intensity of behaviours on which it is contingent; (2) aversive conditioning, in which an aversive reinforcer (e.g. electric shock) decreases the probability or intensity of behaviours on which it is contingent; (3) avoidance learning, in which an aversive reinforcer is delivered in the absence of a particular response, but is omitted, or delayed, in the presence of that response. For example, a rat would learn to press a lever if this prevents a shock.

Classical vs. instrumental conditioning

The relationship between classical and instrumental conditioning is a key issue in the psychology of memory. The voluminous literature relevant to this issue is not discussed here (see Konorski 1967; Rescorla and Solomon 1967; Tarpy 1975; Bower and Hilgard 1981; Mackintosh 1983; Flaherty 1985). We should be alerted, though, to one point; namely, that in many learning situations responses that could be denoted both 'classical' and 'instrumental' are intermixed. For example, classical fear conditioning may be a preparatory phase in the acquisition of instrumental responses mediated by aversive reinforcers (Mowrer 1947; Konorski 1967; Rescorla and Solomon 1967). Moreover, a closer look at situations considered to be either classical or instrumental sometimes evokes doubts about the real nature of the situation. Does a Pavlovian dog salivate because salivation is indeed an UR to food? An alternative explanation is that food is more tasty when moistened by saliva, and there is an instrumental contingency between salivation just prior to food delivery and the taste of the food (Perkins 1968).

Or, take an opposite situation: over the years many thousands of pigeons have been trained in hundreds of laboratories meticulously to peck small illuminated disks in Skinner boxes to obtain food pellets. The interpretation was that food delivery was contingent on pecking, and the probability of pecking changed appropriately. It turned out, however, that pigeons peck the disk in the absence of instrumental contingency. If the experimenter simply illuminated the disk before each delivery of food, but delivered the food automatically irrespective of the pigeon's behaviour, the pigeon would end up pecking as if there were an instrumental contingency (Brown and Jenkins 1968; Jenkins and Moore 1973; see also the stereotyped behaviour described as 'superstitious' by Skinner 1948). Pecking is a major component of the pigeon's consummatory response, and the contingency was possibly between a stimulus (the illuminated disk) controlled by the experimenter and a positive reinforcer (food). This is a classical conditioning situation. Thus, in a way Skinner's pigeons might have actually been disguised Pavlovian dogs.

In general, it could be said that both classical and instrumental conditioning involve manipulation of stimuli, responses, and reinforcers. However, whereas in classical conditioning the organism learns that external events predict the occurrence of a reinforcer, in instrumental conditioning the organism learns which of its actions are responsible for the occurrence of a reinforcer. Classical conditioning may, therefore, be regarded as a more

primitive, automatic consequence of certain associations, whereas instrumental conditioning is rather more complex because the organism must reach conclusions on the basis of
associations (Mackintosh 1983). However, the more complex instrumental learning
processes might be based on mechanisms and steps shared with the more 'primitive' classical
conditioning process. Using reduction *ad absurdum*, one may conclude that acquisition of
even the most complex cognitive abilities, e.g. language, can be described in terms of
elementary stimuli and reinforcements (Skinner 1957); this radical behaviourist view has
encountered extremely strong criticism (Chomsky 1959; see discussion in Skinner 1984).

Paradigms for investigating other forms of learning subserved by complex internal representations

Irrespective of whether the acquisition of some complex behaviours can be described in
terms of elementary associations of stimulus–reinforcement pairs, the problem-solving
potential of many species is clearly not exhausted by conditioning procedures conforming
to the classical Pavlov or Thorndike prototypes. Cognitive processes in higher-order
mammals are clearly heavily dependent on processes of incidental, perceptual, and insight
learning (see above), all of which involve the formation and retention of multimodal
complex internal representations, and their use for *ad hoc* planning and acting. Over the
years, several types of robust learning paradigms have been developed for assessing complex
learning in various species. A few examples will be described in the text. Only two
prototypes are mentioned here, because they recur in later chapters.

Maze learning

Willard Small (1901) introduced the systematic use of mazes into the psychological
laboratory, by testing the performance of rats in retrieving food in a reduced model of the
Hampton Court garden labyrinth. Since then, mazes, in many variations, have become a
powerful and standard tool for studying learning in rodents. Mazes are adapted to the rats'
innate sensory-motor dispositions, or, as Small put it: 'The experiments were couched in a
familiar language' (to the rat at least). The 'rat in a maze' monopolized experimental
psychology for decades. It led Tolman (1938) to state that: 'Everything important in
psychology (except such matters as building of a super-ego . . .) can be investigated in
essence through . . . determinants of rat behaviour at a choice-point in the maze.' Following
the same rationale, the classic *Handbook of psychological research on the rat* (Munn 1950)
became, for a while, a standard introduction to human psychology.

The enthusiasm concerning rats in mazes as a universal solution to the mechanisms of
learning has, appropriately, declined. Nevertheless, sophisticated maze experiments remain
very powerful tools for analysing learning, especially spatial learning and the formation of
'cognitive maps' of the world, in rodents. Examples are provided in later chapters.

Delayed responses

Higher-order mammals can learn to respond to a sensory stimulus even if prevented from making the response until the stimulus has been removed. Furthermore, they can do that even if the stimulus eliciting a correct response varies from trial to trial, so that information from one trial has little or no use in the next trial. This requires a system for temporary storage and retrieval of *ad hoc* information. Note that such a 'working memory' system (Baddeley 1986) does not have to be invoked in conventional classical conditioning or instrumental conditioning, in which all the information is in the test stimulus. The first delayed-response learning paradigms were introduced by W. S. Hunter (1913), and refined by Robert and David Yerkes in their studies of memory in the chimpanzee (Yerkes and Yerkes 1928).

Delayed-response tasks are very useful in analysing a gamut of sensory-cognitive faculties in monkeys and humans. Monkeys are routinely tested in a Wisconsin General Testing Apparatus (Harlow and Bromer 1938; Fig. 2.4). In this apparatus, the animal is confined to a cage with a stimulus tray in front of it. The tray can be temporarily hidden from the subject's view by an opaque screen. In a widely used version of a visuospatial delayed-response task, the screen is raised, and the monkey is shown the position of a food reward, e.g. a raisin, in one of two food wells on the stimulus tray. Following that, the wells are covered with identical pieces of cardboard. The screen is then lowered, usually for 10

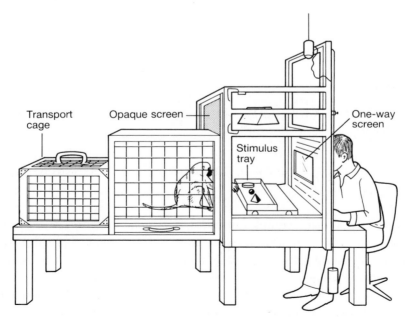

Fig. 2.4. The Wisconsin General Testing Apparatus. (Adapted from Harlow 1949.)

seconds, and then raised again. To obtain the reward, the monkey must choose the position at which the food had been placed prior to the delay (Fig. 2.5). The development of sophisticated delayed-response procedures was instrumental in analysing working memory and visual recognition in primates. The use of these procedures in investigating brain systems that subserve the acquisition and retention of complex internal representations is described in Chapter 14.

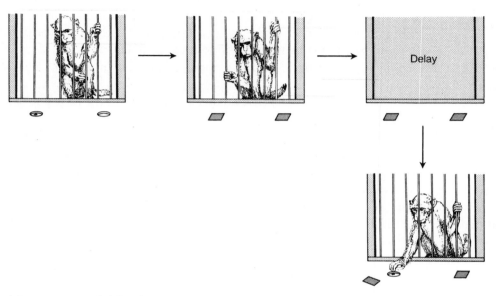

Fig. 2.5. A visuospatial delayed response task. The monkey is placed in the Wisconsin General Testing Apparatus, and observes the position of a raisin in one of two food wells on the stimulus tray. The wells are covered with identical pieces of cardboard. The screen is then lowered for 10 s ('delay'), and raised again. The monkey must retain the information about the location of the food during the delay, in order to choose the position at which the food has been placed. Variants of delayed response tasks are used for quantification of memory of sensory impressions and its *ad hoc* use for the planning and execution of responses in primates. (Modified from Mishkin and Appenzeller 1987.)

Special learning processes: innately predisposed learning in sensitive periods

Some experience-dependent complex internal representations are clearly innately predisposed, and established by exposure to sensory impressions only during restricted ontogenetic periods. These periods are aptly called 'sensitive periods'. The establishment of internal representations in such 'programmed learning' situations casts light on the intimate relationship between development and learning. Two examples are the restriction of social preferences in precocial birds, called imprinting, and the acquisition of bird song. Both processes are discussed in detail in Chapters 12 and 13.

On the universality of learning situations

The ease with which rats solve mazes, and the advantage that monkeys have over more primitive mammals in delayed-response tasks, illustrate how important it is to devise the right experimental problem for each species. In many cases, animals do not learn, not because they cannot do so, but because the experimenter does not ask the right questions. Usually, animals excel in tasks that have an ecological significance and are related to innate-response programmes and tendencies (Garcia and Koelling 1966; Bolles 1970; Lorenz 1981). Selected examples, in addition to rats in mazes, include the learning of landmarks, colours, and odours by bees (Menzel and Erber 1978; Gould 1984; Menzel and Mercer 1987), spatial memory in food-hoarding birds (Shettleworth 1983), social learning in primates (Cheney *et al.* 1986), and face encoding in humans (Carey and Diamond 1980). Caution is always advisable: some innate species-specific reaction patterns, displayed in training, might be erroneously interpreted as learned responses; for example, a greeting body-rubbing behaviour of a feline, evoked by the sight of the experimenter, was mistaken for instrumental conditioning (Moore and Stuttard 1979). The existence of innate constraints, or 'biological boundaries' of learning in different species, and the problematics of inferring learning universals from studies of a limited number of 'favourite' species, are discussed by Bitterman (1975); Johnston (1981); Domjan (1983); Terrace (1984); and Macphail (1987).

The neurobiological research tools: inference of function from dysfunction, correlation, and mimicry

Although, as noted above, methods and experimental tools used in the neurobiology of learning, other than the behavioural paradigms, are shared by other branches of neurobiology (Kuffler *et al.* 1984; Kandel and Schwartz 1985; Shepherd 1988), a concise taxonomy is appropriate here. The following are the main types of experimental approaches employed in this field; many concrete examples recur in later chapters.

Inference of function from dysfunction

In this commonly employed approach, the *necessity* of biological structures and processes for normal function is inferred from the observation that obliterating these structures and processes results in dysfunction. The advantage of the method stems from the availability of a plethora of invasive agents and methods that cause dysfunction. The drawback stems from the lack of appropriate specificity of these agents and methods, and from their secondary effects. The experimental tools used are *anatomical, electrophysiological, pharmacological,* and *genetic.* Thus, anatomical ablations may indicate whether a certain part of the nervous system is necessary for normal function; circumscribed electrical lesions may provide

similar information; inhibitors of identified enzymes, receptors, and other cellular constituents implicate the appropriate molecular entities and processes in normal function, so
also do specific mutations.

Correlation

This approach attempts to correlate structural and functional changes in the nervous system
with learning. The methods used include essentially non-invasive monitoring of cellular
activity, or *post-factum* analysis of the tissue. Again, the level of analysis may be *anatomical*,
either macroscopic or microscopic, implicating identified alterations in brain regions,
neurons, and synapses in acquisition or retention; *electrophysiological*, investigating the
electrical activity of multiple or single neurons; or *biochemical*, measuring changes in the use
of metabolites, and/or the abundance or activity of enzymes, receptors, neurotransmitters,
second messengers, and other molecular entities. Correlative studies, similarly to ablation
studies, commonly start from gross mapping of the system suspected to be involved in
learning, and then proceed to fine-grain analysis. Correlative measurements are often rather
easily performed, but their interpretation is problematic, as it commonly involves *post-hoc*
arguments.

Mimicry

The argument here is that if experimental activation of a distinct neuronal system, pathway,
cell, or molecular process facilitates memory formation or, even better, encodes or retrieves
a memory, then these structures and processes function in normal learning and memory *in
vivo*. The approach is usually difficult to practise, because of technological limitations, and
because it may require detailed mechanistic models of learning and memory systems. It is,
nevertheless, the only approach that may suggest processes *sufficient* for learning and
memory. Mimicry experiments may be supplemented by simulation, or modelling, either
on paper or by using *in vitro* reconstitution of cellular networks and molecular complexes.
Note, however, that even if a certain biological process is shown to be sufficient for learning,
the *necessity* and *exclusiveness* of this process *in vivo* must still be established.

The information available on the structure and function of biological learning and
memory systems usually stems from combined interventional and correlative methods, and
is rarely supplemented and tested by mimicry experiments. For further discussion of the
problems inherent in the identification of processes necessary and sufficient for learning, see
Rose (1981) and Yovell and Dudai (1987).

3   *Biological universals*

Criteria for choosing a starting point for neurobiological analysis of learning

Since learning is a multilevel, multifaceted phenomenon, different starting points for its neurobiological analysis can be conceived. Two major criteria are relevant here: the complexity of the learning situation, and the level of analysis.

The complexity of the learning situation

The habituation of the contraction response of *Nereis*, and visual recognition in a rhesus monkey, are both examples of learning, yet, of course, they differ markedly in their complexity. As noted in the previous chapter, one could choose to analyse learning in very complex systems, because they are more 'interesting' and relevant to learning in humans. On the other hand, one might prefer to analyse very simple learning in a very simple nervous system, because it is simpler to dissect and hopefully also to understand. Bridging the gap between these two choices are almost limitless compromises between the complex and the simple, for example analysis of classical conditioning of reflexes in vertebrates.

The level of analysis

Be it simple or complex learning, one could still start the analysis at different levels of organization of the biological apparatus: anatomical and functional topography of relevant neuronal systems; electrical activity of neurons in these systems; or, for example, molecular mechanisms in neurons and synapses.

● The choice of the experimental system and the level of analysis depend on a personal bias, which is anchored in philosophical attitudes and professional expertise. Very different approaches are, of course, legitimate, valuable, and complementary. Almost always, analysis of learning transcends a given level of biological organization. In some cases, information obtained from simple systems is applicable to more complex systems, and vice versa.

35

In this book, the exposition of the subject corresponds to the following step-by-step strategy: at first, we concentrate on the microlevels, i.e. on the cellular and molecular mechanisms of learning. This will be done mainly by addressing simple learning in relatively simple organisms. Operational rules used in these simple learning systems, and the biological hardware harnessed by evolution for realization of these rules, will be delineated. Subsequently, we address the macrolevel of biological learning systems, i.e. the topography and functional architecture of brain systems that subserve learning. This will be done by discussing more complex forms of learning in more complex organisms, culminating in the most complex learning system of all, *Homo*. In certain respects, our *overall treatment* of the subject could, therefore, be described as a *bottom–up* strategy.

There are, however, three points to be noted.

First, the term 'bottom–up' is used in the cognitive sciences to denote the successive recoding of information in a hierarchical processing system, from low-level to high-level analysis (Marr 1982; Glass and Holyoak 1986). The output of the low-level analysis is the input for the higher-level analysis, and low-level analysis is not affected by higher-level analysis. (In contrast, a procedure in which the output of high-level analysis in the system influences lower-level analysis is termed 'top–down'.) Here, the term 'bottom–up' is borrowed as a convenient metaphor for describing a procedure in which the micro is discussed in detail before the macro, and/or 'simple' systems are addressed prior to complex ones. This procedure epitomizes a preference of certain schools of neurobiology. It also characterizes the global scheme of exposition of the subject in this book. Most importantly, no a priori assumptions are being made here about the use of bottom–up or top–down information-processing procedures by biological learning systems *in vivo*.

Second, as noted above, in most experimental systems employed for studying learning, including very simple systems, the analysis transcends levels. Frequently, even when the overall intention is to concentrate on the molecular level, in practice analysis initially shifts from the molar to the molecular, so that behaviour is actually studied first, followed by an outline of the relevant topography of the nervous system, and only then by the cellular and molecular studies. In other words, many research projects in the neurobiology of learning, which are considered here as 'bottom–up', involve a brief 'top–down' prelude. Simple nervous systems are usually intimately associated with cellular and molecular approaches to learning, because this is the level of analysis mostly emphasized in these systems. Analysis of simple systems, however, also necessitates behavioural and anatomical studies, and analysis of complex systems also involves cellular, sometimes molecular, studies.

And *third*, last but not least: often, simple systems are referred to as 'model systems'. This is wrong. Simple systems are not model systems for anything—they are biological systems that have evolved to fulfil a certain function. Their investigation may, however, generate concepts, working hypotheses, and experimental tools that are useful in the investigation of more complex systems.

Rationales for a 'bottom–up' strategy

Why start from simple systems and the cellular and molecular correlates of learning? The following rationales can be offered:

Simple systems are expected to be more amenable to experimental analysis

Brains are extremely complex. At least 10^{11} neurons and 10^{14} connections exist in the human brain. They form many regions, nuclei, and tracts. Even a minute, densely packed insect brain may contain 10^5 neurons and 10^8 connections. Learning may involve many centres and neurons in the brain. It is, therefore, advantageous to analyse learning in systems that contain relatively few neurons and connections. These could be either nervous systems of primitive organisms, or neuronal subsystems of complex organisms. In systems with relatively few neurons and a simple architecture, subserving a simple behaviour, there might be a better chance of elucidating biological mechanisms of learning.

Ample information is available on cellular and molecular processes in general, and experimental techniques for their analysis are highly developed

It is in the analysis of molecular and cellular activity that the life sciences have made their most remarkable breakthroughs in recent years. Sophisticated methodologies, experimental tools, and data are available on cellular and molecular processes. We can, therefore, expect to elucidate successfully mechanisms and concepts on this level of analysis. These may be useful later in the analysis of higher levels of biological organization and function.

Simple systems and cellular studies may reveal universal components of learning apparatuses

Different behaviours are clearly expected to be based on the execution of different neuronal and muscular algorithms. The neuronal systems that execute these algorithms, and that are somehow modified during learning, should also be expected to vary tremendously between tasks and species. However, due to evolution, it is also likely that different biological learning systems share elementary building blocks. These could be, for example, somatic and synaptic elements that serve as basic structural and functional components of the neuronal network, and local neuronal circuits that operate as logic gates. Cellular and molecular analysis is likely to reveal these elementary, 'universal' building blocks. Moreover, the most elementary constituents and operations of learning and memory systems should be expected to be found in the simplest nervous systems. In searching for learning universals one is thus justified in taking advantage of nervous systems and behavioural repertoires far simpler than those of Pavlov's dog or Ebbinghaus's Ebbinghaus.

Caveats concerning the transition from simple to complex systems

Two caveats are pertinent here. We do not assume a priori that exactly the same mechanisms revealed in simple systems, and described in Chapters 4, 5, and 7 of this book, will also be revealed in the more complex systems described in later chapters. Frankly, given the expected richness of biological mechanisms, and the limited selection of experimental systems as well as the limited information we currently have on these systems, it would be rather surprising if exactly the same micro mechanisms revealed in the simple systems would fit smoothly in the description of the complex system.

Moreover, even if simple and complex learning systems are found to share molecular and cellular mechanisms, this should not be taken to imply that components of primitive systems suffice to explain all the properties of more advanced and complex systems. One would clearly expect evolution to develop new molecular and cellular tools for subserving learning as the phylogenetic demands become more and more complex. For example, it is plausible to assume that working memory, in which complex representations are modified transiently, requires a more sophisticated machinery than a simple reflexive memory, in which a simple representation is altered permanently. All we assume is simply that the types of building blocks revealed in simple systems may still serve in more complex cases, and that the solutions that phylogeny found for simple learning may shed light on operations of neuronal systems that subserve complex learning.

Keeping these caveats in mind, we may now proceed to enquire what are the universal building blocks of nervous systems, and what is their potential relevance to learning?

The universal building blocks of nervous systems

The search for the most elementary structural and operational elements of memory systems is well anchored in the conceptual and methodological evolution of both psychology and biology. Certain relevant trends were mentioned in Chapter 1. Among these, the acquaintance with phylogeny and the resulting emphasis placed on animals as experimental subjects, turned attention to the intelligence of primitive organisms. The observations made by Elisabeth and George Peckham on the mental powers of spiders (Peckham and Peckham 1887), by Jacque Loeb on insect tropism (Loeb 1900), and by Herbert Jennings on the psychic life of paramecia and other simple organisms (Jennings 1906) are selected examples. These studies reinforced the notion that learning and memory systems in primitive organisms share common features with certain types of memory in complex organisms.

In the evolving neurosciences, microscopic histochemical techniques (described in Barka and Anderson 1963; Bloom and Fawcett 1968) had paved the way, at the close of the nineteenth century, to the realization that nervous systems are made up of a limited number of fundamental building blocks. Following the discovery by the German anatomist Theodor Schwann, about 50 years earlier, that animals, like plants, are made of cells (reviewed in Mayr 1982), the Spanish neuroanatomist Santiago Ramon y Cajal introduced the *neuron theory*. This theory states that the nervous system is made of neurons, which are individual, cellular structural and functional units (Cajal 1911). Later studies have clearly demonstrated that neurons in primitive ganglia and complex brains are essentially similar in their morphology. Around the turn of the century, the British physiologist Charles Sherrington had coined the term *synapse* to denote the structural and functional loci of interaction between neurons in all nervous systems (Sherrington 1906).

It also became evident that the similarity between nervous systems of animals remote on the phylogenetic scale is not confined to morphology, but also applies to function. In 1843, the German physiologist Emil Du Bois-Reymond established that electric currents are involved in nerve conduction (and not only, as previously demonstrated, in muscle contraction; for an account of the work of Du Bois-Reymond, his contemporary, Hermann von Helmholtz, and other founding fathers of electrophysiology, see Brazier 1988). Several generations later, following the seminal studies of Edgar Adrian on the electrical properties of individual neurons (Adrian 1932), electrophysiologists succeeded in deciphering elements of the language of electrical signalling used by nerve cells for communicating and integrating information (Katz 1966; Hodgkin 1977; Aidley 1978). The elementary features of these signals are indentical in different nerve cells and organisms. As Kuffler *et al.* (1984) put it: 'Signals are so similar in different animals that even a sophisticated investigator is unable to tell with certainty whether a photographic record of a nerve impulse is derived from the nerve fibre of a whale, mouse, monkey, worm, tarantula, or professor.'

Furthermore, following the pivotal findings of John Langley, T. R. Eliot, and later Henry Dale, Otto Loewi, David Nachmansohn, and many other pharmacologists and biochemists (Dale 1953; Nachmansohn 1959; Feldberg 1977), it became clear that similar chemical substances are used by neurons for communication and regulation in different parts of the nervous system and in different organisms. The above quotation concerning electrical recordings from neurons, applies to chemical extracts of neurons also.

● In the past two decades, neurochemical and molecular-biological studies have confirmed that underlying the structural and functional universality of neurons is a molecular universality. Nerve cells, be they in the ganglion of a leech or the cerebellum of a cat, use similar, often identical, types of molecular mechanisms in their development, maintenance, and function. The *structural, functional, cellular,* and *molecular* universality of nervous systems supports the 'bottom–up' strategy in learning research. If different neurons in different contexts use common signalling systems, why should there not also be some common molecular and cellular mechanisms of learning? Moreover, by extending the argument, one could postulate that studies of neurons in general, not necessarily in the context of learning, might suggest candidates for the components and operations of elementary learning machines.

Introductions to neurophysiology, neuroanatomy, neurochemistry, and neuropharmacology far exceed the scope of this book. The comprehensive textbooks mentioned in Chapter 1 should, therefore, be consulted: Kuffler *et al.* (1984), Kandel and Schwartz (1985), and Shepherd (1988). See also Bradford (1986), Hucho (1986), and Siegel *et al.* (1989) for neurochemistry, and Cooper *et al.* (1986) for neuropharmacology. Some basic notions relevant to neuronal mechanisms are, however, appropriate here.

The common language of neurons

When searching for potential constituents of biological learning apparatuses, it makes sense to consider first those properties and macromolecules that make neurons what they are. Nerve cells are unique in their highly developed ability to process and transmit information. A generalized nerve cell is actually six devices in one (Fig. 3.1). It is an *input* device, receiving chemical and electrical messages from other neurons and/or sensory receptors, and chemical signals from other systems in the body. It is an *integrative* device, integrating and manipulating the input. It is a *conductive* device, conducting the integrated information over distances. It is an *output* device, sending information to other neurons or to effector cells and organs. In the process of manipulating information, the neuron functions as a *computational* device, mapping one type of information into another, either by itself or in concert with other neurons in a network. And it is a *representational* device, subserving the formation of internal representations (see Chapter 1).

For the purpose of manipulating information, neurons employ a language composed of electrical and chemical signals. These two types of signals are interactive and interdependent. The electrical signals are transient changes in membrane potential produced by currents flowing across the membrane. These currents are carried by Na^+, K^+, Ca^{2+}, or Cl^- ions. Every living cell maintains a potential difference across its membrane, but what distinguishes excitable cells, i.e. nerve, muscle, or some glandular cells, from non–excitable cells is the ability of the former to alter their membrane potential transiently and reversibly. This property is made possible by the presence, in membranes of excitable cells, of several types of gated ion channels. These are specific pores that, by opening and closing, selectively regulate the traffic of the appropriate ion(s) across the cell membrane. It is customary to classify gated ion channels on the basis of their gating mechanism. Such classification distinguishes voltage-gated channels from chemically gated channels (Hille 1984). However, in

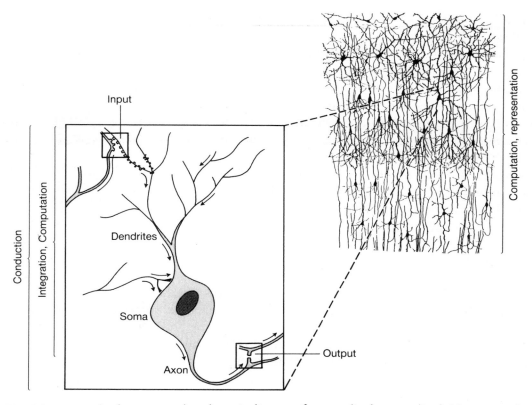

Fig. 3.1. A network of neurons, and a schematic diagram of a generalized neuron (*inset*). Neurons receive, integrate, compute, conduct, and emit information, and subserve internal representations. Their properties are subject to experience-dependent modifications, which serve as bases for neuronal plasticity. Neuronal plasticity occurs during development, maintenance, and response to injury, and is expected to subserve learning. The network is a Golgi-stained preparation of pyramidal cells from the mouse cortex (Cajal 1911).

recent years it has become evident that voltage-gated channels can also be gated chemically (Siegelbaum and Tsien 1983; Levitan 1985; Dudai 1987).

Neurons use two basic types of electrical signal, generated and regulated by the afore-mentioned channels. One type of signal is termed the 'local potential'. It is used to transmit information along short distances (typically less than 1 mm). Local potentials are small (0.1–10.0 mV), can be graded continuously in size, propagate passively, and are rapidly attenuated, due to the initial small amount of current which encounters resistance and is dissipated by leakage. Local potentials are generated by the neuronal input and used for integrative purposes. Local potentials generated in synapses are termed 'synaptic potentials'. They can be either excitatory (depolarizing) or inhibitory (usually hyperpolarizing). Local potentials generated in sensory endings are termed 'receptor potentials'. Both synaptic and receptor potentials are chemically gated, by neurotransmitters or their second messengers in chemical synapses, and by sensory signals or their resulting biochemical cascades in sensory receptors.

The second type of electrical signal used by neurons is the action potential (spike). This is a large (60–110 mV), brief (1–10 ms), all-or-none, regenerating signal, travelling essentially unattenuated for long distances. Action potentials are generated and sustained by voltage-gated channels. The neuron integrates local potentials, and if the summation of these potentials depolarizes the neuronal membrane beyond a certain threshold at the appropriate cellular trigger zone, an action potential results. Action potentials are used for conduction of information.

When an action potential invades the presynaptic terminal of a chemical synapse, it releases molecules of chemical transmitter(s), which cross the synaptic gap and bind to specific receptors on the postsynaptic membrane. This binding leads, either directly or indirectly, to the generation of synaptic potentials. The latter may initiate an action potential in the postsynaptic terminal. A chemical synapse is, therefore, a miniature, sophisticated transducer, converting electrical signals into chemical signals, and vice versa. Receptors for neurotransmitters, neuromodulators, or sensory signals, together with their signal-transducing molecular machinery, are thus input devices, and neurotransmitters (together with the molecular machinery that synthesizes and releases them) are output devices (Fig. 3.2). Transmitter molecules released from the presynaptic terminal often also bind to autoreceptors on the same terminal, regulating the release of the same or other transmitters. Furthermore, substances diffusing from the postsynaptic terminal regulate presynaptic activity. Local microchemical circuits are, thus, nested in cellular, neuronal networks.

Potential elements of representational codes

Since our ultimate concerns are the representational properties of neurons, it is pertinent to ask already at this stage, what is the nature of the potential representational codes? Here we mention briefly only elementary encoding principles used by single neurons. (The issue of neuronal encoding of representations, including possible collective encoding by neuronal populations, is further discussed in Chapter 9.)

We are already familiar with the two basic properties of neurons: they are functionally connected in networks, and they use spikes for fast, long-range communication. These two cellular properties, i.e. connectivity and spiking, are potential codes. They are aptly termed *place code* and *spike code*, respectively.

In spike code, the information is in spike timing, and/or frequency, and/or pattern of firing, and/or in the rate of change of these parameters. In place code (also termed labelled line code), the information is in the position of the neuron in the network. Spike code and place code are not mutually exclusive, and the same neuron may use both, each for encoding different information (see examples of codes in Konishi 1989, and in Chapter 9).

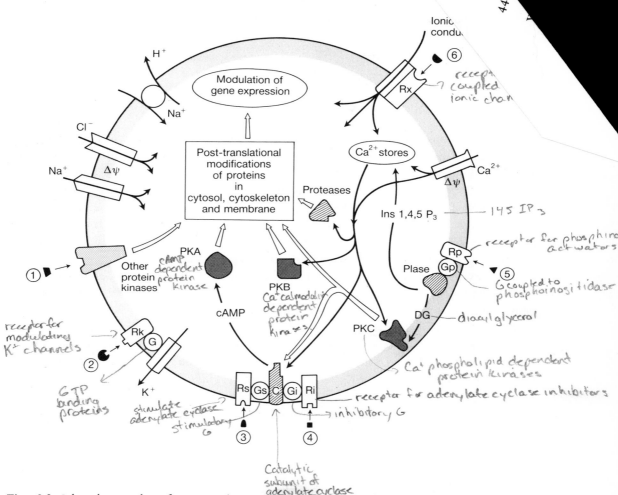

Fig. 3.2. Selected examples of transmembrane signalling in a hypothetical neuron. Activation and modulation of each of these systems may cause enduring alterations in neuronal properties. Concrete examples of the recruitment of transmembrane signalling systems in learning and memory are provided in Chapters 4–8. C, catalytic subunit of adenylate cyclase; DG, diacylglycerol; G, GTP-binding proteins; Gi, inhibitory G; Gp, G coupled to phosphoinositidase; Gs, stimulatory G; Ins 1,4,5 P_3, 1,4,5-triphosphoinositol; PIase, phosphoinositidase; PKA, cAMP-dependent protein kinase; PKB, Ca^{2+} calmodulin-dependent protein kinases; PKC, Ca^{2+} phospholipid–dependent protein kinases; Ri, receptor for ligands inhibiting adenylate cyclase; Rk, receptors for ligands modulating K^+ channels; Rp, receptors for ligands activating phospho-inositidase; Rs, receptors for ligands stimulating adenylate cyclase; Rx, receptors coupled directly to ionic channels. Examples for transmembrane signals are: 1, growth factors; 2, serotonin, GABA, acetylcholine (muscarinic); 3, adenosine, serotonin; 4, noradrenaline, serotonin; 5, glutamate, histamine, acetylcholine (muscarinic); 6, glutamate, GABA, acetylcholine (nicotinic). Also shown are voltage-dependent ion channels, and a Na^+–H^+ exchanger. The overall scheme is modified from Michell (1988). Other sources are: Nestler and Greengard (1984), Aronson (1985), Andrade *et al.* (1986), Berridge (1987), Dudai (1987), Dunlap *et al.* (1987), Hunter (1987), Yoshimasa *et al.* (1987), and Nishizuka (1988).

When searching for neuronal mechanisms of learning, we should pay attention to experience-dependent modifications in both the spiking properties and the connectivity of the neurons in question.

The universality of plasticity

Studies of neurons clearly establish that they may change their integrative, communicative, and hence, probably, computational and representational properties with time. Morphological and functional changes occur in neurons during development, maintenance, and response to injury (Changeux and Danchin 1976; Hubel *et al.* 1977; Lund 1978; Cotman 1985; Burke 1987; Merzenich 1987). The characteristics of a neuron at any given time depend on its genetic programme and on past and present interactions with its milieu, i.e. other cells, the extracellular matrix, ions, growth factors, hormones, and transmitters. Neurons are thus endowed with structural and functional *plasticity.*

Plasticity, in its broadest sense, is not unique to neurons. Other cell types, too, may alter their morphology and activity in response to external stimuli (Darnell *et al.* 1986). But cellular plasticity in neurons has unique physiological consequences, because it alters the representation of information in neuronal systems.

The notion of plasticity has been with the neurosciences since the late nineteenth century (James 1890), and was reinforced in the interdisciplinary synthesis that occurred in the late 1940s (see Chapter 1). Konorski, for example, explicitly promoted it as a principal, universal property of the nervous system (Konorski 1948). Since neuronal plasticity is an experience-dependent alteration of cellular properties, it follows that individual neurons can 'learn'. Whether this cellular learning contributes to a modification in internal representations, i.e. whether it contributes to learning as defined by us earlier, depends on the context in which the neuron functions.

• The notion of neuronal plasticity raises again the question of the relationship between learning and development. It is not merely a coincidence that the distinction between learning and development created a difficulty in the definition of learning (see Chapter 1). The two processes are intimately related; learning is often regarded as an ongoing ontogenesis of the nervous system (Hebb 1949; Changeux *et al.* 1984; Bear *et al.* 1987). As we shall see in Chapters 12 and 13, it is indeed sometimes very difficult to dissociate the contribution of development from that of learning to the ontogenesis of a behavioural change.

Potential molecular and cellular sites of neuronal plasticity

Neurons are equipped with multiple molecular systems, distributed at multiple subcellular loci, that might be recruited in processes of neuronal plasticity. These systems operate at the level of input, integration, and output, which are all graded processes. Conduction *per se*, being an all–or–none process, is a less probable site for plasticity.

The most thoroughly studied mechanism for altering neuronal excitability is modulation of the gating of ion channels. Such modulation can regulate excitability in synapses, soma, and neurites. Neuronal plasticity may also be mediated by additional kinds of modification. These include changes in the availability or activity of enzymes and regulatory proteins, such as protein kinases and guanyl nucleotide-binding proteins, known to affect channel gating; alterations in the metabolic processes that generate transmitters and regulate their release; alterations in the availability and sensitivity of receptors for neurotransmitters and neuromodulators, and of the molecular machinery transducing the receptor-mediated signal downstream into the cell; and modification of other cellular components, e.g. cytoskeleton and structural membrane elements. Selected examples of transmembrane signalling systems are depicted in Fig. 3.2.

Indeed, many studies performed in recent years clearly demonstrated that all the aforementioned molecular changes may contribute to the plasticity of single neurons and synapses (reviewed in Nestler and Greengard 1984; Schramm and Selinger 1984; Levitan 1985; Miller 1986; Burke 1987; Dudai 1987). The question addressed in some of the following chapters is: *how are such molecular and cellular changes recruited in learning?* One should also be tempted to expand the constitutive reductionism into explanatory reductionism, enquiring *how might events on the molecular and cellular level explain some phenomena detected on the behavioural level*, such as temporal specificity and differential efficacy of associative conditioning procedures, or the time course of memory?

A methodological compromise

An additional word of caution is appropriate here. Constitutive reductionism is very powerful and sometimes yields heuristic explanatory reductionism. One should, however, beware of being carried away enthusiastically, *ad absurdum*, so far as to imply that the analysis of single, isolated cells suffices to explain learning. We emphasized that learning involves receiving, processing, and emitting information in the context of a multicellular representational system. Therefore, even though studies of isolated neurons, and even non-neuronal cells (Pfenninger 1986), may cast light on some mechanisms involved in learning, reference must always be made to higher levels of organization, namely to the integrative properties of the neuronal system and its coherent output.

The reductionist attitude on the one hand, and the boundary conditions imposed by the complexity of the phenomena of learning on the other hand, guide our bottom–up treatment, in the first phase of our exposition, into a methodological compromise: a search for experimental systems in which individual, identified neurons can be studied in small networks subserving learning. Such systems are encountered in the next chapter.

PART II

Windows to Molecular and Cellular Mechanisms

4 *Cellular mnemonic devices detected in relatively simple nervous systems: the case of Aplysia*

Why *Aplysia?*

An ideal subject for studying the cellular and molecular bases of learning should, probably, have ten large nerve cells, ten genes, a generation time of 1 week, and the ability to play a cello and recite Shakespeare (W. G. Quinn, personal communication). Since such organisms have not yet been identified in nature, or generated by genetic engineering, one must settle for radical compromises. The marine mollusc, *Aplysia*, is such a compromise.

Aplysia spends its days creeping along the sea floor by the beach, eating seaweed. In season, the hermaphrodite animal engages in a slow ritual of group sex and, later, in egg laying. Often it just sits in quiescence adhered to a rock between the seaweed. Not surprisingly, a noted electrophysiologist once described *Aplysia* as an analogue of the mammalian hippocampus: they both sit there quietly and think about sex. The biology and ethology of *Aplysia* are interesting research topics *per se* (Kandel 1979); but it is the combination of a simple behavioural repertoire and a relatively simple nervous system easily accessible to neurophysiological analysis that has brought *Aplysia* its fame in the field of learning research.

The central nervous system of *Aplysia* consists of a few ganglia, containing all together only *c.* 20 000 neurons. The main ganglia are four paired head ganglia (cerebral, buccal, pleural, and pedal), forming a ring around the oesophagus, and an abdominal ganglion, made up of a number of fused ganglia. The members of each pair of ganglia are connected to each other by nerve fibre tracts termed 'commissures', and to other ganglia by fibre tracts termed 'connectives'. The head ganglia innervates the head and foot and control the major somatic functions: locomotion, feeding, and mating. The abdominal ganglion innervates the mantle and the visceral hump and controls the major visceral functions: circulation, respiration, excretion, and egg laying. *Aplysia* also has a peripheral nervous system, which specializes in the mediation of local responses rather than in the transfer of information over a distance (Kandel 1979).

Similarly to ganglia of other invertebrates, *Aplysia* ganglia are divided into two major zones: a peripheral cortex containing cell bodies and a central neuropil containing synaptic fields and fibres of passage. As opposed to vertebrate neurons, invertebrate neurons do not usually synapse onto cell bodies; the synapses are mostly on neurites and confined to the dense neuropil.

Some cells in *Aplysia* ganglia are extremely large, reaching a diameter of 0.5–1.0 mm. These and other, somewhat smaller cells are essentially invariant in their shape and position, and can be identified on the translucent surfaces of the ganglia. Angelique Arvanitaki was probably the first to realize the experimental potential that these neuroanatomical properties offer for studying relatively simple neuronal networks, and identified several cells on the basis of their position and spontaneous firing pattern (Arvanitaki and Cardot 1941; Arvanitaki and Chalazonitis 1958). Following additional studies by Ladislav Tauc on the biophysical and pharmacological properties of *Aplysia* neurons and their connections (Tauc 1966), Frazier *et al.* (1967) applied additional criteria for the identification of individual nerve cells. These criteria included the pattern of spontaneous postsynaptic potentials, the pathway of the efferent axon, synaptic responses to orthodromic stimulation, the depolarizing or hyperpolarizing response to iontophoretic application of a transmitter, and connections with other identified cells. Since then, many individual *Aplysia* neurons have been catalogued (Kandel 1976, 1979; Hawkins *et al.* 1981).

The gill and siphon withdrawal reflex

Eric Kandel and his colleagues have combined the neuroanatomical and electrophysiological advantages of *Aplysia* with behavioural, pharmacological, and molecular methodologies in an extensive interdisciplinary investigation of the cellular and molecular bases of simple learning (Kandel and Schwartz 1982). They have concentrated on a simple reflex, the gill and siphon withdrawal reflex, and its modification by experience. John Byrne and his colleagues have followed a similar strategy in the analysis of another *Aplysia* reflex, the tail withdrawal reflex (Byrne 1985). In this section, we focus on the gill and siphon withdrawal reflex; but, as we shall see later, much of the discussion actually applies also to the tail withdrawal reflex.

Aplysia has an external respiratory organ, the gill, housed in the mantle cavity on the dorsal side of the animal (Fig. 4.1(a)). The cavity is a respiratory chamber covered by the mantle shelf. At its posterior end, the shelf forms a fleshy spout, called the siphon. The siphon normally protrudes out of the mantle cavity between winglike extensions of the body wall, called parapodia. If a tactile stimulus is applied to the siphon or mantle shelf, a two-component reflex is elicited. One component is contraction of the siphon and its withdrawal behind the parapodia. This is the siphon withdrawal reflex. The other component is contraction of the gill and its withdrawal into the mantle cavity—the gill withdrawal reflex.

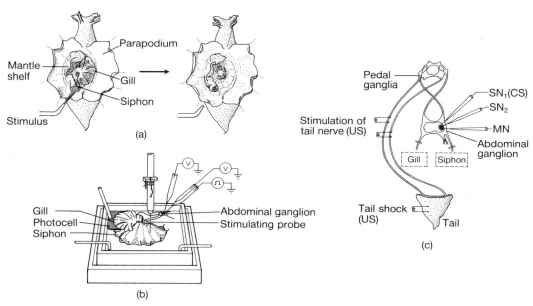

Fig. 4.1. The gill and siphon withdrawal (GSW) reflex in *Aplysia*, and preparations used for its analysis at various levels of organization and simplification. (a): A dorsal view of an immobilized *Aplysia*, with the respiratory chamber exposed by pinning the edges of the parapodia which normally cover it. In the intact animal the siphon protrudes out between the parapodia. A tactile stimulus to the siphon (delivered here by a gentle stream of water) causes the gill and the siphon to contract. Gill contraction can be monitored by a photocell placed under the gill (omitted for simplicity, but see (b) below). (After Kandel 1976.) (b): A simplified isolated-reflex system, used for studying cellular correlates of habituation. The siphon and gill were removed from the organism together with the abdominal ganglion and the connective nerves, and placed in a perfusion chamber. The reflex is elicited by a tactile stimulus to the siphon and the retraction of the gill quantified with a photocell. Recordings were made from sensory and motor cells in the ganglion. (After Kandel 1976.) (c): A simplified preparation used for studying differential classical conditioning of the GSW reflex. The central ganglia and the abdominal ganglion, together with the appropriate nerves and a piece of the tail, but without the gill and siphon, were removed from the animal. The conditioned stimulus (CS) was electrical stimulation of a siphon sensory neuron (SN_1) in the abdominal ganglion. Stimulation of this neuron was paired with the unconditioned stimulus (US). Stimulation of a second siphon sensory neuron (SN_2) was unpaired with the US. The US was an electric shock to the tail, or electrical stimulation of the tail nerves. The measured responses were postsynaptic potentials from a motoneuron (MN). (After Abrams 1985)

The gill also contracts in response to direct stimuli. Both the gill and the siphon withdrawal are innate elementary defensive reflexes.

The gill and siphon withdrawal (GSW) reflexes are controlled in intact *Aplysia* by both the central and the peripheral nervous systems (Kupfermann *et al.* 1971; Lukowiak and Jacklet 1972; Jacklet *et al.* 1975; Lukowiak 1979). The peripheral nervous system plays a major role in certain localized responses, for example the retraction of individual gill pinnuli in response to direct gill stimuli. It is also recruited in response to strong stimuli applied to the siphon or mantle shelf. The central nervous system plays a major role in controlling the overall reflex, including its peripheral components. Although both peripheral and central components of the reflex are often recruited concurrently *in vivo*, they can be treated, for the sake of simplicity, as independent processes.

The portion of the central nervous system directly controlling the GSW reflex in response to weak-to-moderate stimuli, is located in the abdominal ganglion. The connectivity of this system has been partially elucidated by electrical recording of (1) the responses of identified ganglion cells to tactile stimulation of the siphon or the mantle; (2) the responses of neurons to electrical stimulation of other neurons within the ganglion; and (3) the responses of the siphon and gill muscles to electrical stimulation of ganglion cells (Kandel 1979; Hawkins *et al.* 1981). In the abdominal ganglion there are at least 20 identified motor cells of different types. Some innervate the gill, others the siphon, and a few innervate both the gill and the siphon. The motor cells are activated by two populations of sensory neurons, each containing about 24 cells. One population innervates the siphon skin, and the other the mantle shelf. The system also contains at least six types of excitatory and inhibitory interneurons, and is connected to other ganglia. The sensory neurons synapse onto interneurons and also directly onto motoneurons. The motoneurons synapse onto the muscles that perform the behaviour (Fig. 4.2(a)).

Even a reflex as simple as the GSW reflex is thus executed and controlled by many nerve cells. This means that neurobiological analysis of the reflex could greatly benefit from concentrating on only part of the circuits involved (see below). The siphon withdrawal component of the reflex is more easily monitored in the intact animal. On the other hand, the gill withdrawal is more easily quantified in some semi-intact preparations (see below). Moreover, historically, motoneurons controlling the gill were discovered before moto-neurons innervating the siphon. All together, this is why in the following account much of the data on GSW reflex actually refer to the gill withdrawal component.

Analysis of learning in the reflex: a classic example of a research programme employing reductive and simplifying steps

The studies of Kandel and his colleagues on the GSW reflex are, to date, the most extensive pursuit of the cellular and molecular mechanisms of both non-associative and associative learning in a relatively simple system. These studies involve a series of *reductive steps*, associated with operational *simplifying steps* to lessen the complexity of the experimental system at each level of organization.

At this point in our discussion, we should recapitulate our definitions of the terms 'reductive steps' and 'simplifying steps' (see Chapter 1). A 'reductive step' is a shift in the level of analysis from the level of a system as a whole to the level of its components. In bio-logical systems, the delineation of some levels is not unequivocal, because all levels are intimately integrated. In the following discussion, we delineate three levels: the behaving

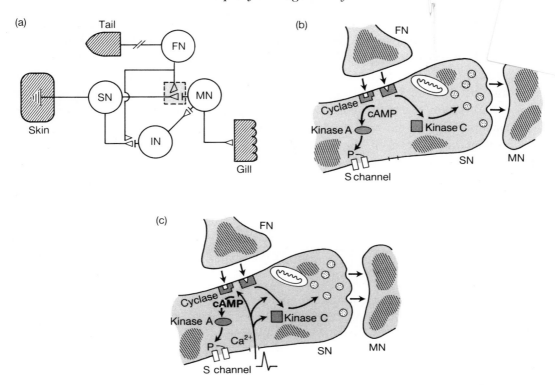

Fig. 4.2. (a): A highly simplified diagram of a neuronal module in the abdominal ganglion of *Aplysia*, subserving the GSW reflex. SN, sensory neuron; FN, facilitatory neuron; IN, interneuron; MN, motoneuron. There are several classes of neurons of each type, for example, excitatory and inhibitory interneurons (not shown). (Adapted from Kandel and Schwartz 1982.) (b): A molecular model for some of the processes that occur during sensitization in the sensory-to-motor synapse of the monosynaptic component of the module depicted in (a) (this synapse is the grey area in (a)). Neurotransmitter(s) released from the facilitatory neuron activate second-messenger cascades in the sensory neuron. One of these is the cAMP cascade that leads to phosphorylation, and closure, of an S potassium channel, by a cAMP-dependent protein kinase (PKA). This, in turn, leads to delayed repolarizations of subsequent action potentials, hence increased Ca^{2+} influx and transmitter released per impulse. The synapse thus becomes facilitated. The phosphoryl moiety on the channel (or on a channel-regulating peptide) turns over rapidly, and short-term memory is possibly retained by enhanced phosphorylation. Other molecular processes are also initiated by the facilitatory transmitter(s), including activation of Ca^{2+} phospholipid-activated kinase(s) (PKC), leading to changes in Ca^{2+} handling and transmitter relase. In addition, inhibitory interneurons, transiently activated *in vivo* by the strong sensitizing stimulus, may counteract the action of facilitatory transmitters (e.g. dephosphorylate the S channel). The net degree of facilitation of the synapse is thus a function of an integrated network activity. (c): Classical conditioning is depicted as activity-dependent augmentation of presynaptic facilitation. Transmitter is the cellular code of the US, and Ca^{2+}, entering the cell on invasion of the sensory neuron terminal by the action potential, is the cellular code of the CS. A transmitter and Ca^{2+}-stimulated adenylate cyclase may be one of the molecular loci for CS–US convergence. Its coincident activation by transmitter and Ca^{2+} is hypothesized to lead to a larger increase in cAMP than that caused by transmitter or Ca^{2+} alone.

organism, the cellular level, and the molecular level. This classification is partly arbitrary; for example, one may further divide the 'cellular level' into the 'circuit level' and the 'single-cell level'. However, the three-level classification is parsimonious and adequately illustrates the use of the reductive approach in research on *Aplysia*. The transition from the whole organism to the cellular level is clear-cut, but the distinction between cellular and molecular levels is sometimes rather arbitrary; this will be further noted below. An additional clarification: in our discussion, a reductive step is considered to have been taken as soon as analysis shifts from a higher level to a lower level of organization of the biological structure, regardless of whether or not the analysis of the higher level is abandoned. For example, an experimenter recording from a neuron in a ganglion while at the same time still observing the behaviour of the organism, has taken a reductive step from the whole organism to the cellular level, although still clinging to the first level. And finally, as explained in Chapter 1, a simplifying step facilitates analysis without shifting levels, for example by lessening the amount of tissue in a preparation.

● Because the analysis of the GSW reflex is a prototype of a reductionist research programme, each of the reductive and simplifying steps performed in the course of studying the simplest reflex modification, namely habituation, is detailed and explained below. This is followed by an account of more complex forms of learning in the same reflex, analysed by a similar strategy. In this chapter we deal only with mechanisms of acquisition and short-term memory, lasting from minutes to hours. Long-term memory is discussed in Chapter 8.

Facets of reflex habituation: from behaviour to homosynaptic depression involving inactivation of Ca^{2+} channels and depletion of releasable transmitter

The behaviour

Habituation of the GSW reflex is detected in intact, freely moving *Aplysia*, following repetitive application of a mild tactile stimulus to the mantle or the edge of the siphon. However, repeated stimulation of freely moving *Aplysia* is not a convenient starting point for cellular analysis. Pinsker *et al.* (1970) have, therefore, immobilized *Aplysia* in a small aquarium with circulating sea water, pinned the edge of the mantle shelf to a substage, and delivered controlled, brief jets of water to the siphon (see Fig. 4.1(a)). Gill contractions were monitored by a photocell placed under the gill. The contractions diminished with repetitive monotonous stimulation.

Most of the parametric characteristics of habituation outlined by Thompson and Spencer (1966; see Chapter 2) were observed in the GSW reflex (Pinsker *et al.* 1970). In addition to response decrement with repetitive presentation of the constant stimulus, there was spontaneous recovery with rest, dishabituation by a strong tactile stimulus to another part of the animal, habituation of the dishabituatory stimulus with repeated presentations, and greater habituation with short interstimulus intervals and weak stimuli. Greater habituation with repeated periods of habituation and recovery, and prolongation of recovery by [*overlearning*] continued stimulation after the response has been abolished, were also observed (Carew *et al.* 1972; Carew and Kandel 1973). However, there was no significant generalization of habituation to a stimulus in another part of the receptive field.

The first reductive step: from behaviour to cells

A quick glance at a simplified scheme of the neuronal circuit that underlies the GSW reflex (Fig. 4.2(a)) reveals several potential cellular sites of habituation. Identification of those sites which play a role in the behavioural modification required electrophysiological manipulations. Kupfermann *et al.* (1970) have, therefore, modified the experimental set-up described above, to enable studies of cellular correlates of habituation in an almost intact, behaving *Aplysia*. Again the animal was immobilized in an aquarium with the siphon and gill exposed, but this time a slit was made in the neck, allowing the abdominal ganglion and its nerves to be externalized and pinned on a lucite stage. This preparation permitted both electrical stimulation of, and recording from, nerves and cells. Note the radical reduction in the level of analysis, from the whole organism ('the black-box approach') to neurons.

Several lines of evidence indicated that the habituation observed in this immobilized, cut-open preparation, under the stimulus regimen employed, was mainly due to processes [*Not sensory adaptation*] occurring in the central nervous system, upstream from the motoneurons' output. Sensory adaptation was excluded because recording from single afferents in peripheral nerves, and from individual sensory neurons in the exposed abdominal ganglion, while at the same time stimulating the skin at intervals that produced habituation in the intact animal, did not reveal changes in sensory responsiveness. Substantial alterations in the efficacy of moto- [*Not fatigue*] neurons, neuromuscular synapses, and muscles were excluded because repeated stimulation of identified motoneurons innervating the gill, at intervals that produce habituation, did not result in attenuation of muscle contraction.

In contrast, intracellular recordings from gill motoneurons during habituation did reveal a progressive decrease in the number and frequency of evoked action potentials in these motoneurons (Kupfermann *et al.* 1970). Moreover, recovery of the motoneurons' electrical discharge correlated with recovery of the behavioural response, following either rest or dishabituation. The motoneurons' threshold to constant current stimulation did not change, indicating a change in synaptic input. To unveil synaptic input, which is normally masked by spike activity in the motoneuron, the latter was hyperpolarized and excitatory post-synaptic potentials (EPSP) were monitored. Under these conditions, habituation was correlated with a progressive decrease in the amplitude of EPSP in the motoneuron, and rest or dishabituation were correlated with restoration of the EPSP (Castellucci *et al.* 1970).

A simplification of the experimental system: an isolated reflex

To facilitate further the analysis of the cellular mechanisms of habituation, the effector and affector organs, together with the abdominal ganglion, were removed from the animal and immobilized in a perfusion chamber (Carew *et al.* 1971; Byrne *et al.* 1974; Kandel 1976). Although *Aplysia* was reduced here from a whole mollusc to a piece of siphon skin connected to the gill via the appropriate nerves and the ganglion, the behavioural input and output remained very similar to those encountered in the intact animal: tactile stimuli were still delivered to the skin, and behavioural responses were still recorded by measuring gill contraction with a photocell (Fig. 4.1(b)). This preparation had two major advantages: first, it eliminated peripheral pathways between the siphon and the gill, thus simplifying the reflex to its central component; and second, it isolated the abdominal ganglion from the other ganglia, thus establishing that both the behavioural alterations and the cellular correlates of habituation could take place in the absence of the rest of the central nervous system (Kupfermann *et al.* 1971). The conclusion reached from the electrophysiological analysis of this preparation confirmed that the strength of synaptic input of indentified motoneurons correlated with the strength of behavioural response.

A further simplification of the experimental system: an isolated ganglion

Here only the abdominal ganglion was removed from the animal and pinned in a perfusion chamber (Castellucci *et al.* 1970; Kandel 1976). The afferent and efferent organs were no longer present. Instead, the siphon nerve was electrically stimulated to simulate pressure to the skin and elicit the reflex, and the output of an identified motoneuron was monitored to substitute for quantification of gill withdrawal. The isolated ganglion preparation permitted easier and better quantification of reflex input and output. Again, repeated electrical stimulation of the siphon nerve, at intervals producing habituation in the intact organism, caused a decrease in EPSP in the gill motoneuron; the EPSP recovered with rest.

A further simplification: monosynaptic component of the reflex

Reduced synaptic input to the gill motoneurons could be caused either by an increase in postsynaptic inhibition, or by a decrease in excitatory input. Both processes may involve the modulation of sensory neurons and interneurons. The following possibilities could be expected: transmission between the mechanosensory neuron and the motoneuron is either *monosynaptic*, hence direct, or *polysynaptic*, involving interneurons. The neuroanatomy of the abdominal ganglion reveals both kinds of sensory-motor connections (Fig. 4.2(a); Hawkins *et al.* 1981). In addition, an alteration in the effectiveness of the sensory-to-motor connection might be either *homosynaptic*, involving only the modulated synapse, or *heterosynaptic*, involving modulatory interneurons. Heterosynaptic inhibition of pathways was observed earlier in the abdominal ganglion (Tauc 1965).

A series of experiments was carried out to establish which of the above mechanisms contributes significantly to habituation of the GSW reflex. In some of these experiments, Castellucci *et al.* (1970; Castellucci and Kandel 1974) substituted electrical stimulation of the siphon nerve with electrical stimulation of individual sensory neurons in the isolated ganglion.

The electrical stimulus was adjusted to generate a single action potential in the sensory neuron. The first time the sensory neuron was caused to fire an action potential, it produced a large EPSP in the follower motoneuron, similar to that produced by localized stimulation of the skin. Transmission appeared monosynaptic due to the following. (1) There was a one-to-one relationship of the sensory neuron firing to the motoneuron EPSP. (2) The latency was short and constant. (3) Transmission was not abolished by high concentrations of divalent cations, known to block polysynaptic input. Subsequent action potentials, initiated in the sensory neuron at intervals that produced habituation in the intact animal, gave rise to progressively smaller EPSPs in the motoneuron. This depression was, again, not significantly affected by high concentrations of divalent cations, indicating a major homosynaptic component. Taken together, these results imply that homosynaptic depression of the excitatory efficacy of the monosynaptic sensory-to-motor connection accounts for much of the habituation.

• So far, several simplifying steps taken in the analysis of the cellular correlates of the GSW reflex have already restricted our attention to the central nervous system component of the reflex. An additional simplification was made at this stage. Although the circuit subserving the behaviour does involve both polysynaptic and monosynaptic components (Hawkins *et al.* 1981), attention was intentionally focused on the monosynaptic component (Fig. 4.2(a)). This component, clearly simpler to analyse than the polysynaptic components, was estimated to account for *c.* 60 per cent of the complex EPSP in the motoneuron (Byrne *et al.* 1978). It was also assumed that cellular and molecular processes in the monosynaptic component of the reflex are similar in nature to those in the polysynaptic components. From this stage on, unless otherwise indicated, we refer in our analysis of the GSW reflex only to the monosynaptic component of the reflex.

Note that although the electrophysiological studies described above implicated synaptic changes in habituation, the recordings were not made at synapses, but rather in the soma. The reason for this was the unfeasibility of recording from terminals in the compact, dense neuropil *in vivo*. An assumption was made that since the soma and the terminals are topographically close and electrotonically coupled, changes occurring in the latter are reflected in recordings from the former. Later experiments provided support for this assumption. Soma potential in *Aplysia* has been shown to regulate transmitter release (Shimahara and Peretz 1978); and recordings from growth cones of sensory neurons in culture revealed some conductance properties similar to those recorded in the soma (Belardetti *et al.* 1986).

The reconstructed monosynaptic component, or 'learning in a dish'

Recently, a novel, simplified system was introduced in the analysis of the GSW reflex. Following the demonstration that isolated *Aplysia* neurons can survive in primary cultures (Dagan and Levitan 1981; Schacher and Proshansky 1983), Rayport and Schacher (1986) succeeded in co-culturing identified sensory neurons and motoneurons, and reconstructing *in vitro* the equivalent of the most elementary GSW reflex module. A single spike elicited in a sensory neuron in culture resulted in an EPSP in a major gill motoneuron, onto which it synapsed in the dish. Repeated stimuli, at a rate that produced habituation in the intact animal, led to a gradual and marked decrement in the EPSP. The latter recovered after rest. Homosynaptic depression was thus reproduced in the isolated elementary circuit element, in the absence of other ganglion constituents.

A second reductive step: from the cellular to the molecular level

What is the mechanism of synaptic depression in habituation? In attempting to answer this question, we must explicitly address subsynaptic and molecular mechanisms. We therefore consider this shift in the level of analysis as a reductive step.

Two major possible mechanisms of synaptic depression could be envisaged: a presynaptic alteration in transmitter release or a postsynaptic alteration in receptor responsiveness. We consider here the analysis of subsynaptic mechanisms already within the realm of the molecular level. The classical method of distinguishing between these mechanisms is to perform quantal analysis of synaptic transmission (del Castillo and Katz 1954; Boyd and Martin 1956). The prevalent notion in cellular neurobiology is that a burst of transmitter is composed of elementary quanta, each corresponding to the molecular content of a presynaptic vesicle. Release is critically dependent on Ca^{2+} ions, which play a role in the interaction of the vesicles with the cytoskeleton and the plasma membrane (Smith and Augustine 1988). At a low level of transmitter release, usually obtained by bathing the preparation in low Ca^{2+} and high Mg^{2+}, one can determine the mean number (m) of transmitter quanta released by the action potential from the presynaptic terminal, and the magnitude of the postsynaptic potential produced by a single quantum (q). Since the number of transmitter molecules in each quantum does not change in a healthy, non-exhausted synapse, m can serve as an index of the total amount of transmitter released, and q can serve as an index of the sensitivity of the postsynaptic membrane. Measurements of m and q can, therefore, distinguish between pre- and postsynaptic processes.

Castellucci and Kandel (1974) performed quantal analysis in the sensory-to-motor neuron system in the isolated abdominal ganglion. Again, recordings were actually made in the soma. The evidence thus obtained suggested that the decrease in EPSP with synaptic depression is paralleled by a decrease in the number of transmitter quanta released, but not in the sensitivity of the postsynaptic membrane. This suggested that in the monosynaptic part of the reflex, habituation is at least largely due to events occurring in the presynaptic terminal of the sensory-to-motor neuron synapse, and leading to a progressive decrease in the amount of transmitter released per sensory signal.

And what causes the decrease in transmitter release? As noted above, a major factor controlling release is the intracellular concentration of Ca^{2+}. During the action potential, a depolarizing inward Ca^{2+} current turns on, leading to an increase in intracellular Ca^{2+}. One would wish, therefore, to compare the inward Ca^{2+} current during an action potential in a depressed (habituated) sensory neuron to the current in a non-depressed sensory neuron, and to determine whether changes in this current might account for the depression in transmitter release.

However, measuring individual conductances is not a straightforward procedure. Several types of currents flow concurrently across the membrane and mask each other. For example, during an action potential the Ca^{2+} current is normally masked by the inward Na^+ current and by the repolarizing K^+ current. Moreover, most ion channels are voltage dependent (see Chapter 3), so that the changes in the membrane potential caused by the opening or closing of a channel themselves alter the properties of this channel. Electrophysiologists circumvent the above difficulties by using a combination of two experimental tools: pharmacology and voltage clamping. Pharmacology involves substituting ions in the medium, and using channel-blocking drugs, to place the burden of conductance on certain channels and block other channels in a relatively specific way. Voltage clamping consists of monitoring the current flowing across the neuronal membrane, while maintaining the intracellular voltage at a desired value by an electronic feedback circuit. This makes it possible to study the conductance of a voltage-dependent ionic channel as a function of membrane potential while preventing the resultant changes in membrane current from influencing the channel (Smith *et al.* 1980). Together, these methods dissect the contribution of individual channel types to conductance at any given state of the membrane potential.

These two tools, pharmacology and voltage clamping, were employed to analyse the conductance changes in an habituated and non-habituated sensory neuron (Klein *et al.* 1980). The currents in a voltage-clamped mechanosensory neuron, and the EPSP in its follower motoneuron, were measured in the presence of the Na^+ channel blocker tetrodotoxin (to reduce the Na^+ component of the inward current) and of the K^+ channel blocker tetraethylammonium (TEA) (to reduce the outward K^+ current). Under these conditions, the inward current is expected to be an indicator of the Ca^{2+} conductance. Repeated stimulation of the sensory neurons resulted in a decrease in their inward current, paralleled by a depression of the postsynaptic potentials. The decrease in the inward current in the sensory neurons was furthermore detected when an additional inhibitor of K^+ current was added, Na^+ omitted, and Ba^{2+} (which passes only through the Ca^{2+} channel) substituted for Ca^{2+}. Taken together, the results suggested a cumulative inactivation of Ca^{2+} current with repeated firing.

Although the analysis of short-term habituation was reduced to the molecular level, explicitly implicating Ca^{2+} channels, molecular analysis was rudimentary. It was not systematically pursued, and detailed molecular processes were not elucidated. This is in contrast to sensitization and classical conditioning of the reflex (see below), where detailed molecular models have been proposed. Several possibilities could, nevertheless, be considered for the molecular mechanisms of homosynapatic depression. These include inactivation of Ca^{2+} channels by intracellular Ca^{2+} (Tilloston 1979), and by second-messenger cascades initiated by binding of the released transmitter to autoreceptors on the presynaptic terminal (illustrated in Cooper *et al.* 1986). Interestingly, anatomical analysis of habituated synapses revealed a decrease in transmitter vesicles presumably ready for release (Bailey and Chen 1988*b*). This indicates that inactivation of Ca^{2+} channels is not the sole manifestation of synaptic depression. Modelling the habituated synapse, using experimental electrophysiological data, also suggested that Ca^{2+} channel inactivation cannot exclusively explain the depression; additional processes, involving handling and depletion of releasable transmitter, should be considered (Gingrich and Byrne 1985). Moreover, although we have considered so far only a homosynaptic process in the monosynaptic component of the GSW reflex, heterosynapatic mechanisms clearly also take place in habituation *in situ*. Inhibitory interneurons and their aminergic and peptidergic transmitters have already been identified in the neuronal system subserving the reflex, but their relevance to short-term habituation is not known (Hawkins *et al.* 1981; Mackey *et al.* 1987; Piomelli *et al.* 1987; Sweatt *et al.* 1988).

- In conclusion, a series of reductive steps from one level of organization of the biological material to another, accompanied by simplifications of the experimental system at each level of organization, resulted in a partial mechanistic model for the monosynpatic component of short-term habituation. This model suggests that homosynaptic depression of the presynaptic terminal of the sensory-to-motor neuron synapse, resulting at least in part from inactivation of Ca^{2+} current and depletion of releasable transmitter, contributes significantly to the decrement in the behavioural response.

A note on dishabituation

In our general discussion of habituation in Chapter 2, we raised two potential explanations for dishabituation: dishabituation is removal of inhibition built up during habituation; or, alternatively, dishabituation is an independent process. Evidence from vertebrate systems indicated that dishabituation is independent of habituation. This is also the case in the GSW reflex in *Aplysia*. The reflex, we should remember, could be elicited by one or two anatomically distinct and functionally independent afferent pathways. One pathway leads from the siphon skin, the other from the mantle shelf. Habituation by repeated stimulation of one locus did not affect the responsiveness of the other locus, i.e. there was no generalization. If dishabituation were simply the removal of habituation, then the dishabituating stimulus should have facilitated only the habituated pathway. This was not the case (Carew *et al.* 1971). The dishabituating stimulus also facilitated the non-habituated pathway. Moreover, the response of the depressed pathway was often restored to levels exceeding the prehabituated level. The mechanism of dishabituation of the GSW reflex is, therefore, not the reversal of habituation.

Using a combination of behavioural and developmental studies, Rankin and Carew (1987) provided additional evidence for the independence of habituation and dishabituation, by showing that during ontogenesis, dishabituation emerges in *Aplysia* only after habituation.

The behavioural data suggest that dishabituation is a special case of sensitization. But are dishabituation and sensitization indeed the same thing? We shall defer the answer to a later stage in our discussion, following the analysis of cellular and molecular mechanisms of non-associative reflex intensification.

Facets of reflex dishabituation and sensitization: from behaviour to presynaptic facilitation mediated by phosphorylation cascades

Analysis of dishabituation and sensitization of the GSW reflex was performed along the same conceptual and methodological lines outlined above for habituation. Most reductive and simplifying steps need not, therefore, be explicitly spelled out, since they should now be self-explanatory. Instead, from now on we shall concentrate on the conclusions reached in each step. In general, the analysis of dishabituation and sensitization at the subcellular and molecular levels was much more extensive than that of habituation.

Again, at first came the behavioural observation. When a mechanical or electric shock is applied to the head or tail of *Aplysia*, the subsequent GSW reflex, elicited in response to a mild stimulus applied to the mantle shelf or siphon, is greatly enhanced. This could be demonstrated in a 'naïve' animal (sensitization), as well as in an habituated animal (dishabituation). At this stage of our discussion, we shall use the terms 'sensitization' and 'dishabituation' interchangeably; we will see later that this heuristic interchangeability does not hold as the analysis deepens.

Sensitization involves heterosynaptic facilitation

As in the case of habituation, attention in the study of sensitization was focused on the monosynaptic component of the reflex (Fig. 4.2(a)). Using the isolated abdominal ganglion preparation, Castellucci and Kandel (1976) first stimulated repeatedly and monotonously the sensory neuron, leading to synaptic depression and decrement of EPSP in the moto-neuron. Subsequently, they delivered a brief electrical stimulus to the connective from the head, simulating the sensitizing shock used in behavioural experiments. This rapidly facilitated the EPSP produced in the motoneuron on sensory stimulation. *Facilitation was considered to be a cellular correlate of behavioural dishabituation and sensitization, and was later used as a cellular analogue of these behavioural phenomena.* Facilitation lasted for many minutes, and followed a time course similar to that of behavioural dishabituation. The facilitating stimulus did not fire the sensory neuron, implying that facilitation was not post-tetanic facilitation resulting from repetitive activation of the sensory neurons, but rather an heterosynaptic process. A modulatory neuron had, therefore, to be added at this stage to the elementary neuronal module of the monosynaptic component of the reflex (Fig. 4.2(a)).

Facilitation is presynaptic

Facilitation could not be explained by the direct effect of the facilitatory stimulus on the motoneuron; directly firing the latter, even at high frequency, did not facilitate it (Castellucci *et al.* 1970). To determine further whether facilitation of the sensory-to-motoneuron synapse occurred presynaptically or postsynaptically, quantal analysis was again performed (Castellucci and Kandel 1976). The average size of a transmitter quantum did not vary significantly between pre- and post-facilitation measurements, whereas the number of quanta released per impulse did increase. This indicated that facilitation was presynaptic.

Presynaptic facilitation is associated with modulation of K^+ channels

Stimulation of the connective that carries the sensitizing input resulted in spike broadening in the sensory neurons (Klein and Kandel 1978). As mentioned above, during the action potential a depolarizing inward Ca^{2+} current turns on. Therefore, spike broadening means an increased Ca^{2+} influx, which is expected to enhance transmitter release. This increased Ca^{2+} influx could result from direct modulation and extended opening of Ca^{2+} channels or, indirectly, from a decrease in the repolarizing K^+ currents. To distinguish between these alternatives, the cell body of the sensory neuron was voltage clamped, and the various channels examined in isolation, by blocking other channels with pharmacological agents and manipulating the concentrations of the appropriate ions (Klein and Kandel 1980). These experiments showed that presynaptic facilitation is associated with a decrease in steady-state K^+ current, and in a transient K^+ current which could be elicited in the sensory neuron by brief depolarizing command steps. In other words, the results indicated that the facilitatory transmitter modulates the Ca^{2+} current not by directly affecting Ca^{2+} channels, but indirectly, by blocking K^+ conductance.

Neurons contain several types of K^+ channels (Hille 1984). *Aplysia* sensory neurons are known to contain at least four types of such channels. One of these is a chemically gated, serotonin-regulated channel, named the S channel. The S channel is active at the resting potential, is only moderately dependent on membrane potential, and its gating is affected by intracellular Ca^{2+}. Serotonin indirectly closes the channel. This closure contributes to the spike broadening which subserves short-term presynaptic facilitation (Klein *et al.* 1982; Siegelbaum *et al.* 1982).

Actually, in a physiological medium, spike broadening observed in the sensory neuron during facilitation is rather small (10–30 per cent). It can be artificially made quite dramatic by adding TEA to the medium (Klein and Kandel 1978). This is because at concentrations at which TEA blocks several other K^+ channels it does not block the S channel. TEA thus places the burden of repolarizing the action potential on the modulated S channel. Note that TEA was used here in a different way than in the study of habituation. Whereas in synaptic depression TEA permitted a crude measure of Ca^{2+} influx, in synaptic facilitation it aided in indirectly measuring the change in the relevant K^+ current.

Is the relatively small broadening observed in the absence of TEA sufficient to account for a substantial increase in transmitter release? This was investigated in the following way: spike broadening can be simulated in a controlled manner by depolarizing voltage–clamp steps in the sensory neuron. The effect of these depolarizing steps on transmitter release can be quantified by measuring the EPSP in the motoneuron. Such analysis indeed showed that even small prolongations in depolarizing steps in sensory neurons, *in situ* and in culture, have a profound effect on transmitter release (Hochner *et al.* 1986*a*). Moreover, preventing the relatively small spike broadening by voltage clamping the presynaptic neuron in culture, substantially reduced or blocked facilitation (Hochner *et al.* 1986*a*).

In addition to its effects on Ca^{2+} influx, and hence indirectly on transmitter release, closure of the S channel in the sensory neuron was also proposed to increase excitability by (1) decreasing the threshold for spike initiation; and (2) reducing accommodation, which is the decrease in response to a prolonged constant stimulus. This 'anti–accommodation' effect results in increased firing during prolonged stimuli (Klein *et al.* 1986).

A molecular model for non–associative learning

Prolonged inhibition of protein synthesis did not affect short-term sensitization of the GSW reflex (Schwartz *et al.* 1971). It was, therefore, hypothesized that short-term pre-synaptic facilitation is caused by alteration in the level or function of existing molecules. Second messengers and their associated molecular cascades seemed to be reasonable candidates. At that time, much attention was focused on the role of cyclic adenosine mono-phosphate (cAMP) as a second messenger and mediator of neuronal plasticity (reviewed in Nestler and Greengard 1984; Dudai 1987). A search for alterations in cAMP was, therefore, a natural choice.

Again, the isolated ganglion system was used. Prolonged electrical stimulation of the pleural–abdominal connective was indeed accompanied by an increase in the level of cAMP in the abdominal ganglion (Cedar *et al.* 1972). A biogenic amine, serotonin, which activates *Aplysia* adenylate cyclase and hence increases the cAMP level, facilitated synaptic trans-mission between the sensory neuron and the motoneuron (Brunelli *et al.* 1976). Refinement of microbiochemical techniques made it possible later to show that serotonin increases the cAMP level in single sensory neurons (Bernier *et al.* 1982).

cAMP-dependent phosphorylation and S channel closure

On the basis of the afore-mentioned results, a molecular model could already be proposed for short-term sensitization (see Fig. 4.2(b)). According to this model, the sensitizing stimulus excites facilitatory interneurons, leading to the release of modulatory transmitter(s), such as serotonin, other biogenic amines, or neuropeptides (Abrams *et al.* 1984). These transmitter(s) activate adenylate cyclase. The level of cAMP rises in the sensory neuron. cAMP exerts its known effects in eukaryotic cells by binding to the regulatory subunits of cAMP-dependent protein kinease (PKA). This binding dissociates the regulatory subunits from the catalytic subunits and activates the latter. The kinase then phosphorylates substrate protein(s), changing their properties and hence altering their cellular activity (Nestler and Greengard 1984; Dudai 1987). In view of what has been said above about the physiological correlates of facilitation in the sensory neuron, one might expect phosphorylation to alter the S channel, either directly or indirectly.

A series of electrophysiological, pharmacological, and neurochemical experiments strengthened the evidence that the S channel is indeed closed, either directly or indirectly, by cAMP-mediated phosphorylation. Furthermore, memory of facilitation is retained in persistent activation of the cAMP cascade within the first minutes after training. Injection of the catalytic subunit of PKA into the sensory neuron simulated facilitation, measured by spike broadening and enhanced EPSP in the follower motoneuron (Castellucci *et al.* 1980). Injection of a specific protein inhibitor of PKA blocked facilitation (Castellucci *et al.* 1982). Injection of GDPβS, an inhibitor of adenylate cyclase, terminated spike broadening (Schwartz *et al.* 1983).

The latter results were taken to indicate not only that the cAMP cascade is necessary for short-term memory, but, more specifically, that memory during the first minutes of facilitation resides in persistant activation of adenylate cyclase. This conclusion was, however, later questioned, since *in vitro* no conditions were found under which adenylate cyclase is persistently activated following a transmitter pulse (Yovell *et al.* 1987). Since in the above cyclase-blockade experiments the sensory neuron was continuously stimulated, the blockade by GDPβS might have merely reflected interruption of transmitter stimulation. An alternative possibility is that memory is retained by persistent activation of the PKA. This could be achieved, for example, by degradation of regulatory subunits, so that fewer of them are available to inhibit the catalytic subunit; or by decreasing the affinity of the regulatory to the catalytic subunit by post-translational modification of the former (Schwartz and Greenberg 1987; Buxbaum and Dudai 1989; see also Chapter 8).

More direct evidence concerning the relationship between cAMP-mediated phosphorylation and the activity of the potassium S channel, emerged from single-channel recordings. These were performed by the patch-clamp technique, in which a membrane patch is sealed on the tip of a micropipette so that a current flowing through single channels can be controlled and recorded (Sackmann and Neher 1983). Recordings were made first in membrane patches still attached to the whole cell (Siegelbaum *et al.* 1982), and later in membrane patches detached from the cell (Shuster *et al.* 1985). In cell-attached membrane patches, the application of serotonin to the cell body, or the intracellular injection of cAMP, caused prolonged and complete closure of the S channel. In inside-out cell-free membrane patches, prepared from the sensory neurons, the purified catalytic subunit of mammalian PKA produced all-or-none closures of individual S channels. Under these conditions, fewer channels were closed for shorter periods than in cell-attached membranes treated extracellularly with serotonin or intracellularly with cAMP. The differences between cell-attached and cell-free patches were attributed to the lack, in the latter, of cytoplasmic inhibitors of phosphoprotein phosphatases, resulting in frequent dephosphorylation and opening of the channel. The difference between the behaviour of S channels in cell-attached and cell-free patches may also result from different activity in these preparations of other second-messenger cascades which regulate the channel (Belardetti and Siegelbaum 1988).

An additional process, independent of S-channel closure, contributes to presynaptic facilitation

Closure of the S channel is not, however, the exclusive mechanism responsible for presynaptic facilitation in the sensory-to-motor monosynaptic component of the GSW reflex. Boyle *et al.* (1984) used a Ca^{2+}-sensitive dye, arsenazo III, to monitor intracellular free Ca^{2+} transients in cell bodies of voltage-clamped sensory neurons during long depolarizing steps. Ca^{2+} transients elicited by depolarizing steps markedly increased in amplitude following the application of serotonin. Since the cells were voltage clamped, the increase in free Ca^{2+} could not have been due to serotonin-induced alterations in the S current. It was suggested that since transmitter release depends on Ca^{2+}, the S channel-independent action of serotonin on Ca^{2+} handling increases the availability of releasable transmitter. The Ca^{2+} phospholipid-dependent protein kinase, PKC, was implicated in this Ca^{2+} mobilization process, because phorbol ester, a PKC activator, produced facilitation without a change in the S current in sensory neurons (Hochner *et al.* 1986*c*). Serotonin, which mimics the effect of the facilitatory transmitter in physiological experiments, can indeed activate PKC in *Aplysia* neurons (Sacktor *et al.* 1986).

The contribution to presynaptic facilitation of each of the two serotonin-stimulated mechanisms depends on the cell's history. As synaptic transmission becomes depressed with repeated stimulation of the sensory neuron, changes in duration of the action potential have a progressively weaker influence on transmitter release. Once homosynaptic depression declines, the sensitivity of transmitter release to spike duration is restored. It was, therefore, suggested that prolongation of the action potential by closure of S channels mediates the effects of sensitization, whereas the S channel-independent mechanism mediates the effects of dishabituation (Hochner *et al.* 1986*b*).

• We are now, thus, in a position to answer a question posed in the previous section: are dishabituation and sensitization identical processes? Molecular and cellular analysis indicates that they are not. A similar conclusion emerged from a comprehensive behavioural analysis of dishabituation and sensitization, recently performed by Tom Carew and his colleagues (Marcus *et al.* 1988; Rankin and Carew 1988). They found that dishabituation and sensitization of the GSW reflex can be dissociated in three ways. First, dishabituation appears immediately after the delivery of the dishabituating stimulus, whereas sensitization has a delayed onset. Second, dishabituation is preferentially produced by relatively weak stimuli, whereas sensitization is produced only by strong stimuli. And third, dishabituation and sensitization differ in their ontogenesis: very young *Aplysia* can dishabituate but cannot be sensitized. The distinction between dishabituation and sensitization was not appreciated in the early studies on *Aplysia*, and became apparent only when the analysis deepened. A question that must now be addressed is which additional molecular and cellular mechanisms differentiate behavioural sensitization from dishabituation. For example, which mono- and polysynaptic processes account for the delayed onset of sensitization? One possibility is that the strong sensitizing stimulus activates *in vivo* inhibitory network components, which transiently mask facilitation (Mackey *et al.* 1987, 1988).

Facets of classical conditioning of the reflex: activity–dependent presynaptic facilitation

Early attempts to demonstrate associative learning in *Aplysia* encountered difficulties. However, the great experimental advantages offered by the GSW reflex, and the potential for comparing the mechanism of non-associative and associative learning in the same neuronal system, prompted researchers to establish that *Aplysia* can indeed learn to associate the reflexive response with a relatively neutral stimulus (Carew *et al.* 1981; Lukowiak and Sahley 1981). In the most studied version of classical conditioning of the GSW reflex, a mild tactile stimulus to the siphon, which normally produces only weak withdrawal, is the conditioned stimulus (CS). A strong electric shock to the tail, which produces strong withdrawal, is the unconditioned stimulus (US). Specific temporal pairing of the CS and US markedly enhances the subsequent GSW reflex in response to the CS (Fig 4.3). This is, thus, a case of conditioning which is traditionally termed alpha conditioning (see p. 26).

As in many other cases of classical conditioning, here, too, successful conditioning depended on temporal pairing of the CS and US, and the order of their presentation. Optimal conditioning occurred when the US began *c.* 500 ms after the onset of the CS. Increasing the delay to 2 s prevented effective pairing. Backward conditioning was not effective (Hawkins *et al.* 1983*a*).

Moreover, the GSW reflex could be differentially conditioned, by using two CSs as discriminating stimuli (Carew *et al.* 1983). These CSs were two similar mechanical stimuli, applied to two different sites on the skin (e.g the mantle shelf and the siphon), or to two

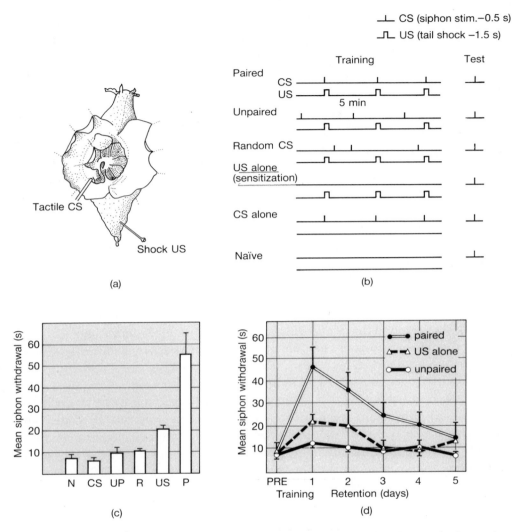

Fig. 4.3. Classical conditioning of the GSW reflex. (a): A dorsal view of *Aplysia*, with the respiratory chamber exposed, illustrating the CS and US. (b): The training (paired) and control procedures. (c): The siphon withdrawal response, expressed in withdrawal time, measured 24 hours after 31 training trials. (d): Time course of retention of the conditioned response. (After Abrams 1985.)

different sites in the field of innervation of a single cluster of sensory neurons. One CS (denoted CS^+) was paired with the US, whereas the other (denoted CS^-) was unpaired. Conditioning was measured by comparing the responses to CS^+ and CS^- in the same animal after training. CS^+ was much more effective in eliciting siphon withdrawal.

The noxious stimulus used in this paradigm was similar to the one used in the studies of reflex sensitization. It should, therefore, be expected to activate the same facilitatory neurons involved in sensitization. A plausible working hypothesis might thus be that classical conditioning also involves presynaptic facilitation as a mechanism for strengthening the CS pathway. A difference should, however, exist between the mechanisms of sensitization and classical conditioning, since in the latter there is a requirement for an appropriate CS–US contiguity.

Activity-dependent facilitation contributes to classical conditioning of the GSW reflex

To find out what cellular mechanism(s) provide(s) the animal with the ability to associate specifically the CS and US, reductive steps and simplified experimental preparations, similar to those described above, were used. Advantage was also taken of the phenomenon of differential conditioning, which provides the experimenter with an 'internal control' in the same animal (see below). The preparation most widely used was the isolated central ganglia of *Aplysia*, attached to the tail via the posterior pedal nerves (see Fig. 4.1(c); Hawkins *et al.* 1983*b*). The amplitudes of the monosynaptic EPSPs in a motoneuron were measured following intracellular stimulation of each of two siphon mechanosensory neurons, chosen arbitrarily from an apparently homogeneous cluster of such neurons. Stimuluation of each of these cells thus replaced the tactile stimulus used as the CS in the intact animal.

During training, stimulation of one of the 'paired' sensory neurons immediately preceded the electric shock applied to the tail (or to the posterior pedal nerve transmitting the stimulus from the tail; see Fig. 4.1(c)). Stimulation of the other, 'unpaired' sensory neuron followed tail shock by a few minutes. In both cases, the EPSP in the follower motoneuron, resulting from stimulation of the sensory neuron, was facilitated during training. This was expected: tail shock sensitizes the sensory pathway. However, facilitation of the EPSP resulting from activity in the paired neuron was much greater than that of the EPSP resulting from activity in the unpaired neuron. Moreover, the EPSP resulting from activity in the unpaired neuron was facilitated only early in training, and decreased later, probably due to habituation of the pathway by repeated firing. In other words, the paired pathway showed activity-dependent amplification of facilitation, when compared to the unpaired pathway, which served as an internal control in the same animal. The phenomenon of differential conditioning was thus reduced to a cellular analogue.

Is the activity-dependent amplification of facilitation subserved by presynaptic mechanisms in the sensory-to-motoneuron synapse similar to the mechanisms of sensitization? To answer this question, the duration of the action potentials in the paired sensory neuron was measured, and compared to that of the unpaired sensory neuron. TEA was used, as explained above, to augment alterations in spike duration. The results of these experiments (Hawkins *et al.* 1983*b*), combined with voltage-clamp analysis to resolve the individual currents in the modulated sensory neuron (Hawkins and Abrams 1984), showed a differential increase in spike duration in the paired sensory neuron, and indicated that this

was due to blockade of the S channel. Thus, in classical conditioning the sensory neuron ✳
seemed to be facilitated by ionic mechanisms similar to those employed in sensitization, but
in classical conditioning spike activity in the sensory neuron within the appropriate time
window somehow enhanced the ability of the same neuron to respond to the facilitatory
input.

Experimental tests of theoretical assumptions on synaptic facilitation

The research groups investigating classical conditioning in *Aplysia* attempted, at this stage,
to suggest a molecular model for the interaction between spike activity in the sensory
neuron (the presumptive cellular code of the CS) and facilitation (the US). However, before
describing the suggested model, we should be aware of at least two other types of
mechanisms which could account for conditioning of the reflex in the present system (Fig.
4.4). First, the CS and US pathways might converge on the same facilitator neuron(s), so that
paired presentation of both stimuli increases firing of the facilitator(s), as compared to non-
paired presentation (Fig. 4.4(a)). However, this is not a likely explanation. Differential
facilitation was observed in arbitrarily chosen pairs of sensory neurons; if the CS and US do
converge on facilitators, this would imply a separate facilitator for each sensory neuron, a
strikingly non-parsimonious mechanism.

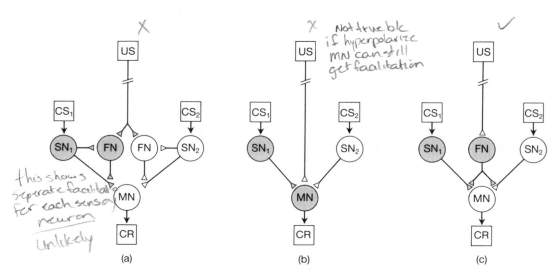

Fig. 4.4. Possible molecular mechanisms for synaptic facilitation during classical conditioning in the
monosynaptic sensory-to-motor component of the neuronal system subserving the GSW reflex. In each case
acquisition of facilitation requires paired spike activity in the shaded neurons and terminals. (a): Summation
of the US and CS on common facilitators. (b): A Hebbian synapse, requiring paired activity of the presynaptic
(sensory) and postsynaptic (motor) neurons. (c): Activity-dependent presynaptic facilitation. The experimental
data supported the latter mechanism. Abbreviations as in Fig. 4.2. (Adapted from Hawkins *et al.* 1983*b*.)

Another possible mechanism is of great interest from both a theoretical and a historical point of view. In his book *The organization of behaviour*, Hebb (1949) formulated a Darwinian neurophysiological postulate of learning, which has since come to exert great influence on the field of learning research. He postulated that: 'When an axon of cell A is near enough to excite a cell B and repeatedly or persistently takes part in firing it, some growth process or metabolic change takes place in one or both cells such that A's efficiency, as one of the cells firing B, is increased' (Fig. 4.4(b)). Neuroscientists have repeatedly and persistently tried to integrate Hebb's postulate into cellular models of learning (Stent 1973, Rauschecker and Singer 1981, and Heidmann and Changeux 1982, are just three examples). Associative modification of the GSW reflex provides an opportunity to put the postulate to an experimental test.

The contribution of Hebb's postulated mechanism to associative modification of the GSW reflex could not be a priori excluded. This is because a shock to the tail (i.e. the US), in addition to causing presynaptic facilitation, also causes the postsynaptic motoneuron to fire via polysynaptic pathways (Walters *et al.* 1983*b*). Experiments by Carew *et al.* (1984) showed, however, that the mechanism postulated by Hebb is neither necessary nor sufficient for associative modification of the GSW reflex. In one set of experiments, the postsynaptic motoneuron was held hyperpolarized and thus prevented from firing, yet activity-dependent facilitation did occur. This contradicted Hebb's prediction that the postsynaptic cell must fire in response to its presynaptic partner for the synapse to become facilitated. Hebb's mechanism is, therefore, not necessary for facilitation in this synapse. In another set of experiments, intracellularly stimulated spike trains in the motoneurons were paired with spike trains in the sensory neurons. The synaptic connection was not facilitated. Hebb's mechanism is therefore not sufficient in this system.

To do some justice to Hebb, it should be noted that he himself did not consider his hypothesis applicable to sensory–motor connections in the adult animal, but rather to sets of coactive neurons subserving higher brain functions (Hebb 1949). It should also be noted that the experiments by Carew *et al.* (1984) did not exclude the possibility that a Hebbian mechanism plays a role in modifying the connection between the facilitatory neuron and the presynaptic sensory neuron (Sahley 1985).

A molecular model for associative learning

The results described so far led to the conclusion that spike activity in the presynaptic sensory neuron enhances the ability of this neuron to respond to the facilitatory transmitter. We also know from our discussion of sensitization that at least one action of the facilitatory transmitter is activation of the cAMP cascade in the sensory neuron. The level of cAMP in sensory neurons that had received a puff of serotonin immediately after a brief train of action potentials, was higher than in sensory neurons that had received a puff of serotonin alone (Abrams 1985). The assumption was, therefore, made that the action potential in the sensory neuron interacts with the cAMP cascade, and the question was posed: which component of the action potential interacts with which component of the cAMP cascade?

Four elements of the action potential may be considered as potential contributors to the augmentation of the facilitatory transmitter response: (1) membrane depolarization; (2) Na^+ influx; (3) K^+ efflux; and (4) Ca^{2+} influx. Ca^{2+} was a plausible candidate, since it is well known to interact with the cAMP cascade (reviewed in Dudai 1987). Depleting most of the free Ca^{2+} from the medium prevented the augmentation of spike broadening by spike activity in the sensory neurons (Abrams 1985). This suggested that Ca^{2+} was crucial to the cellular code of the CS.

A candidate convergence locus for molecular associations

How does Ca^{2+}, a presumptive cellular code of the CS, interact with the facilitatory transmitter, the presumptive cellular code of the US? One plausible convergence site is the adenylate cyclase complex, which can be stimulated by both transmitter and Ca^{2+} (Abrams 1985; Eliot *et al.* 1986; Abrams and Kandel 1988). If convergence of the cation and the transmitter on the cyclase indeed plays a role in associative learning, then one might expect the optimal order of interaction of these stimuli with the enzyme complex to correspond to the temporal constraints of associative learning. In other words, the interaction of Ca^{2+} and transmitter should stimulate the cyclase more efficiently if Ca^{2+} precedes the transmitter (forward conditioning), and less efficiently if the transmitter precedes the Ca^{2+} (backward conditioning). To test whether this is the case, Yovell and Abrams (1988) pulsed adenylate cyclase from *Aplysia* sensory neurons *in vitro* with transients of Ca^{2+} and serotonin. They found that under certain experimental protocols, activation of the enzyme was somewhat larger when Ca^{2+} preceded the transmitter, than when the transmitter preceded Ca^{2+}. This suggests that a basic property of classical conditioning, namely the potency of forward conditioning, can be translated into molecular terms, without invoking emerging network properties. Note, however, that it is not at all clear whether the rather weak order dependency which is revealed in cyclase activation under limited *in vitro* conditions, is a property of the enzyme complex or of certain experimental constraints. Also, there is no proof that such a mechanism, even if it is physiologically meaningful, indeed subserves the temporal specificity which is detected on the behavioural level.

• In conclusion, the following molecular model has been offered for short-term associative conditioning in the monosynaptic component of the GSW reflex in *Aplysia* (see Fig. 4.2(c)): the action potential, invading the sensory neuron terminal on application of the CS, increases Ca^{2+} influx. This in turn activates a transmitter-sensitive, Ca^{2+}/calmodulin-sensitive adenylate cyclase. The facilitatory neuron, firing on application of the US, releases transmitter(s) that also activate(s) the same adenylate cyclase. The activation of the cyclase by Ca^{2+} and transmitter within the appropriate time window and order, results in a larger rise in cAMP than that obtained by activation of the cyclase by each of the ligands independently. This rise in cAMP activates PKA, leading to closure of S channels and enhanced transmitter release by future impulses.

Note again that the model is rudimentary. Moreover, even if Ca^{2+}/calmodulin-stimulated adenylate cyclase is a site for stimulus convergence during associative modification of the GSW reflex, it is almost certainly not the exclusive site. Additional molecular loci of cross-talk between the Ca^{2+} and the cAMP cascades are known, including receptors for neurotransmitters and phosphoproteins involved in transmitter release (reviewed in Dudai 1987).

Another defensive withdrawal reflex in *Aplysia* shares mechanisms with the GSW reflex

A mechanical or electrical stimulus applied to the tail of *Aplysia* elicits quick withdrawal of this organ (Walters *et al.* 1983*a*). Modification by experience of the tail withdrawal reflex has been extensively studied. The tail withdrawal response is mediated via neurons in the pleural and pedal ganglion. The response to a weak test stimulus to the tail can be sensitized by a tail shock, and sensitization correlated with heterosynaptic facilitation of the synapses between the sensory neurons and the motoneurons (Walters *et al.* 1983*b*). A cellular analogue of classical conditioning has also been demonstrated in this system: pairing a shock to the tail (US) with current injection (CS) into tail sensory neurons innervating a receptive field not activated by the shock, resulted in enhanced facilitation of the monosynaptic connections of the sensory neurons to the tail motoneurons (Walters and Byrne 1983).

In recent years, Byrne and his colleagues have investigated the cellular and molecular mechanisms underlying sensitization and classical conditioning of the tail withdrawal reflex (Byrne 1985). Their strategy was similar to that employed in studying the GSW reflex, and insight into the cellular and molecular correlates of both defensive reflexes was often obtained concurrently. Activation of the cAMP cascade by the sensitizing stimulus, and phosphorylation-induced closure of K^+ channels, were suggested to contribute to the presynaptic facilitation underlying sensitization (Walsh and Byrne 1984; Ocorr and Byrne 1985). Serotonin, which mimics the effect of the sensitizing stimulus, modulates several K^+ channels, including the S channel, a Ca^{2+}-dependent K^+ channel, and a voltage-dependent K^+ channel (Baxter and Byrne 1986). Classical conditioning was suggested to result from activity-dependent neuromodulation, in which spike activity and the resulting Ca^{2+} influx enhance the synthesis of cAMP in response to the neuromodulator released by the US. Here, too, Ca^{2+}/calmodulin-sensitive adenylate cyclase was proposed as a molecular site for stimulus convergence (Walters and Byrne 1983; Byrne 1985; Ocorr *et al.* 1985). It appears, therefore, that cellular and molecular mechanisms that contribute to non-associative and associative plasticity are shared by modifiable defensive reflexes in *Aplysia*.

What is still missing?

Interdisciplinary analysis of the GSW reflex and the tail withdrawal reflex in *Aplysia* has led us from behaviour to cells and molecules. Two shortcomings should, however, be noted.

The picture is still fragmentary

Several systematic simplifications were intentionally made in the process of analysing the GSW reflex. Here lies the great power, but also the weakness, of the approach. At each stage of the analysis, pieces of the picture were abandoned for the sake of focusing on other pieces. The models depicted above refer to a cellular module representing a monosynaptic part of a central component of the reflex. Yet, we already know that *in vivo* the GSW reflex may be executed by a concerted action of central and peripheral neuronal systems (Kupfermann *et al.* 1971; Lukowiak and Jacklet 1972; Jacklet *et al.* 1975; Peretz and Lukowiak 1975; Lukowiak 1979). The neuromuscular junction itself shows plasticity in non-associative modification of the reflex (Jacklet and Rine 1977). In the central nervous system, only about 30 per cent of the total response to a mild-to-moderate stimulus is contributed by the major gill motoneuron commonly studied (Kandel *et al.* 1983). As previously noted, only about 60 per cent of the complex EPSP in the motoneuron is attributed to monosynaptic input (Byrne *et al.* 1978).

Scores of neurons, only part of which are represented in the above-mentioned sensory-to-motor monosynaptic module, were implicated by neuroanatomical and electrophysiological analysis of the central components of the reflex (Kandel 1979; Hawkins *et al.* 1981). Simultaneous recording of activity from many neurons in the abdominal ganglion, made possible by real-time optical measurements of voltage-sensitive dyes, revealed activity in 150 neurons during a withdrawal reflex, and suggested that altogether up to about 400 neurons are active when the reflex is executed (Wu *et al.* 1987). Some of these neurons may be causally related to the reflex, and in some of them the mechanisms of plasticity may differ from those depicted above. For example, postsynaptic facilitatory changes are already known to take place in certain motoneurons during sensitization (Frost *et al.* 1985*b*, 1988). Presynaptic inhibition, which attenuates facilitation during sensitization, should also be considered (Mackey *et al.* 1987; Frost *et al.* 1988; Sweatt *et al.* 1988). The contribution of such additional cellular loci and mechanisms in the central and peripheral nervous systems, and in the monosynaptic and polysynaptic components of the central neuronal network, still awaits analysis (Frost and Kandel 1984; Frost *et al.* 1985*a,b*, 1988; Cleary and Byrne 1986; Mackey *et al.* 1987).

Many gaps also exist in our knowledge of those molecular mechanisms which are already included in cellular and molecular models of learning in *Aplysia* defensive reflexes. One of the most important pieces of missing information is the identity of the transmitter(s). The facilitatory neurons were at first thought to be serotonergic, since the application of serotonin mimics their effect, but this monoamine is absent from some facilitators (Kistler *et al.* 1985). Facilitatory transmitter(s) may be aminergic and/or peptidergic, like small cardioactive peptides (SCP) (Abrams *et al.* 1984).

The identity of the second-messenger cascades activated by the facilitatory transmitter(s) is of great importance. The cAMP cascade is only one of the second-messenger systems activated by transmitters, and the same transmitter often interacts with different receptors, coupled to different second-messenger cascades (Nestler and Greengard 1984; Miller 1986; Dudai 1987; Kaczmarek 1987). Identification of the relevant second-messenger systems should complement the existing picture of what happens in the sensory neuron when modulated by the facilitatory transmitter. Such information may also lead to reinterpretation of the data. For example, pharmacological experiments that have implicated G-regulatory subunit(s) in facilitation have assumed that these subunits regulate adenylate cyclase activity. However, it is now clear that transmitters may exert their cellular effect via a G-dependent, adenylate cyclase-independent mechanism (Bourne 1986; Neer and Clapham 1988).

Also still unidentified are the molecular mechanisms by which the facilitatory transmitter(s) exert(s) the S channel-independent effects on transmitter release (Hochner *et al.* 1986*b*). Is mobilization of the transmitter involved, and if so, how? An what is the role of lipids (Piomelli *et al.* 1987), and of PKC, known to regulate release (Miller 1986; Kaczmarek 1987)?

Moreover, as noted above, it is now clear that inhibitory transmitters and modulators, such as the peptide Phe-Met-Arg-Phe-NH$_2$, counteract the effect of facilitatory transmitters such as serotonin on the sensory neurons. The action of inhibitory transmitter(s) involves activation of lipid cascades (Piomelli *et al.* 1987), and at least one facet of their effect is reopening of the S channel (Belardetti and Siegelbaum 1988). How this is done is not yet clear. Does it involve activation of a protein phosphatase or, alternatively, inhibition of protein kinase (Sweatt *et al.* 1988)?

In other words, the picture becomes more and more complex and, admittedly, more interesting, as the analysis deepens. It appears that the richness of molecular and cellular mechanisms involved in a relatively simple learning process is much greater than was thought only a few years ago. And we may be seeing only tips of the iceberg.

Necessary mechanisms are not necessarily sufficient, and sufficient mechanisms are not necessarily exclusive

Cellular and molecular correlates of learning, and molecular mechanisms necessary for processes which are considered as cellular analogues of learning, have been identified. However, whether these mechanisms are sufficient for learning has not yet been established. To demonstrate that a given process is sufficient for learning *in situ*, the appropriate cellular alteration must be induced in the intact (or relatively intact) organism, and the resulting behavioural change observed. Note that even if this very difficult procedure is followed, and mechanisms sufficient for learning are identified, it does not prove that these mechanisms indeed function *in situ*; and even if they do function *in situ*, they may not be exclusive. There might be alternative pathways for modifying a given behaviour.

Tentative conclusions: synaptic plasticity, multiple loci and mechanisms, and ubiquitous molecular cascades

In spite of the above shortcomings, several conclusions can be extracted from the extensive studies of modifiable defensive reflexes in *Aplysia*.

Cellular and molecular correlates of learning can be identified

Although the *Aplysia* results do not exclude the contribution of molar, emergent network properties to learning, they do show that processes correlating with learning are identifiable at distinct cellular and molecular loci. These latter processes can provide partial explanations for changes in the behaviour.

Short-term learning involves modifications in the network that has subserved the behaviour beforehand

No change has to be invoked in the wiring, at least in the short term. This might be expected in alpha conditioning. We cannot, however, rule out the possibility that 'dormant' networks, which do not subserve the behaviour in the naïve animal, are recruited in controlling the modified behaviour.

Changes take place in synapses

The alteration in the efficacy of information transfer, occurring during learning in the monosynaptic component of the network subserving the behaviour, is due to alteration in the efficacy of connections between the network elements, resulting primarily from the modulation of transmitter release. Referring to the generalized neuron (see Chapter 3), plasticity is revealed here in synaptic integration and neuronal output. The contribution of additional facets of neuronal integration, and of responsiveness to input, still awaits clarification.

Complex forms of learning share mechanisms with simpler forms

In the GSW reflex, classical conditioning is depicted as an elaboration of sensitization.

There are multiple cellular loci of learning even in a simple system

Scores, possibly hundreds of cells change their activity when the GSW reflex is modified by experience.

Multiple molecular mechanisms operate in cellular loci of plasticity

Some of these mechanisms probably operate in parallel, with a cross-talk between them, for example the cAMP, Ca^{2+}, and phospholipid intracellular cascades.

Ubiquitous molecular cascades are recruited in learning

- The same types of ubiquitous signal-mediating systems that function in many types of cells, also function in learning. The capability of storing changes in activity over seconds to hours is probably a universal property of intracellular signal transduction cascades, and is recruited by the nervous system for the purpose of altering representations. It is the macro-circuit context, not the micromechanisms, that endows the cellular change with a specific behavioural meaning.

To conclude, even a relatively simple learning system is more complex than might have been expected. Multiple spatial and temporal processes, in and among cells, take place when *Aplysia* learns to modify the withdrawal of its gill. On the other hand, so far no new types of molecular mechanisms had to be invoked to explain neuronal changes which subserve learning. We now proceed to enquire how general are the conclusions emerging from the *Aplysia* studies.

5 Cellular mnemonic devices detected in relatively simple nervous systems: the case of Hermissenda

Aplysia is but one invertebrate species in which cellular correlates of simple associative learning have been extensively investigated. Other selected examples are conditioning of food preference in the garden slug, *Limax maximus* (Gelperin 1975; Chang and Gelperin 1980; Sahley *et al.* 1981a,b; Gelperin *et al.* 1985), and in the marine mollusc, *Pleurobranchaea californica* (Mpitsos and Davis 1973; Davis *et al.* 1980; London and Gillette 1986; Morielli *et al.* 1986); conditioning of phototaxis in another marine mollusc, *Hermissenda crassicornis* (see below); and conditioning of leg position in arthropods (Horridge 1962; Hoyle 1980; Forman 1984). Information is also available on cellular components involved in the experience-dependent modification of reflexes in other invertebrates (Selverston 1985; Carew and Sahley 1986; Byrne 1987).

The research strategy pursued in the above studies was, as described for *Aplysia*, constitutive and partial explanatory reductionism practised on simplified preparations. In this section, we address only one additional example, that of conditioned phototaxis in *Hermissenda*. Substantial information is available in this case on the neuronal system that subserves the behaviour, and cellular and molecular mechanisms have been proposed to account for the modification of this neuronal system by experience. Although some of the conclusions reached in the *Hermissenda* studies are basically similar to those emerging from the *Aplysia* studies, the differences do merit attention.

Hermissenda crassicornis is a colourful carnivorous marine snail with no shell or mantle. It has no gill either, and gas exchange is performed in plumlike appendages (branchiae) covering the back, and through the body wall. *Hermissenda* normally moves towards light. This positive phototaxis has probably evolved to direct the slug to the plankton at the water surface, on which it feeds. However, in turbulent water, phototaxis and locomotion are depressed, and the animal adheres to underwater rocks, apparently to avoid injury. Phototaxis and response to turbulence are guided by two major sensory systems in the head: the visual and the vestibular systems respectively (Fig. 5.1(a)).

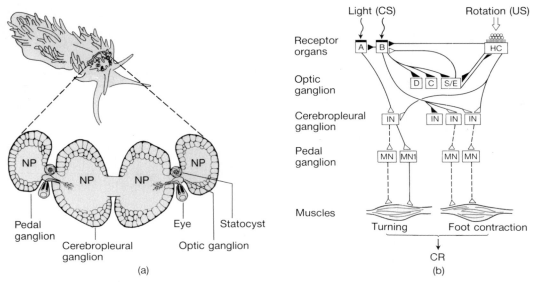

Fig. 5.1. The sensory and neuronal systems subserving *Hermissenda's* phototaxis and its modification by graviceptive stimuli. (a): *Hermissenda*, and a diagram of the dorsal surface of the circumesophageal nervous system. The eyes and statocysts are juxtaposed between the cerebropleural and pedal ganglia, on both sides of the head. NP, neuropil. (Modified from Crow *et al.* 1979; Alkon 1984.) (b): A highly simplified diagram of the pathways that process information about light and rotation. Loci of convergence for visual and graviceptive information exist at various levels, including the sensory receptors, and interneurons at the optic and cerebropleural ganglia. Open triangles, excitatory synapses; closed triangles, inhibitory synapses; broken lines, postulated, not yet identified pathways. For simplicity, the heterogeneity in A and B photoreceptors, and in various types of interneurons (IN) and motoneurons (MN), is omitted. C, D, S/E, optic ganglion cells; HC, hair cells. One model proposes that following conditioning, network interactions convert an enhanced excitability of B photoreceptors into a depressed phototactic reponse. Note, for example, that enhanced firing of B would inhibit A and MN1, leading to depression of muscle contraction. This may, in turn, result in attenuated orientation towards light. The behavioural response, however, depends on the recruitment of several motor systems, of which the two major ones, subserving orientation and clinging, are schematically depicted. For example, increased firing of B may also enhance contraction of muscles responsible for clinging. Another model suggests that attenuation of B responses to sustained illumination contributes to the experience-dependent suppression in behaviour. For further details, see text. (After Lederhendler *et al.* 1986.)

Modification of phototaxis by rotation, or *Hermissenda* on a turntable

Dan Alkon and his co-workers investigated the neuroanatomy and cellular physiology of the visual and vestibular pathways in *Hermissenda*. They decided to explore the possibility of using this experimental system for analysing the cellular and network mechanisms of associative learning. The idea was to teach the animal to associate stimuli processed by the two independent yet converging central sensory pathways. A conditioning paradigm was developed, in which light was the CS, rotation the US, and a depressed phototaxis the CR (Crow and Alkon 1978; Alkon 1983).

The following procedure was used: *Hermissenda* were individually placed in glass tubes filled with sea water, arranged radially on a turntable (Fig. 5.2). A removable plug confined each animal at the distal end of the tube. A light spot was projected onto the centre of the turntable, illuminating a circular area with photocells on its circumference. On removal of the plug, the animals locomoted towards the illuminated area, interrupting the light between the source and the photocell, and thus activating an event marker. Stimulation of the vestibular organ was produced by rotating the turntable at about 95 rpm. Each training trial consisted of a 30 s session of light, either paired or non-paired with rotation, depending on the experimental group. Three days of training, consisting of 50 trials per day,

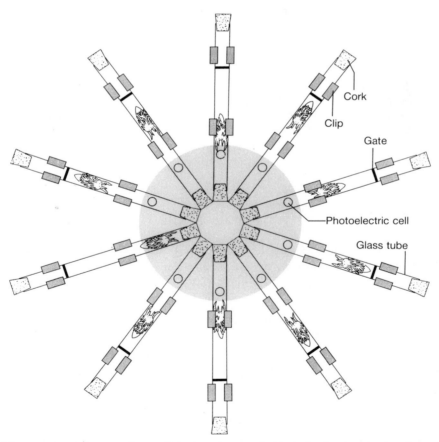

Fig. 5.2. The experimental system for conditioning *Hermissenda*. The animals are trained and tested in glass tubes mounted on a turntable. Light is the CS and rotation the US. Pairing both stimuli results in a decreased phototaxis, monitored by photoelectric cells that detect when each individual reaches the spot of light (depicted as a grey circle at the centre). (After Alkon 1983.)

resulted in a significant depression of phototaxis in animals that had received paired presentations of light and rotation, compared to non-paired controls (Crow and Alkon 1978). The depressed response to light persisted for about 3 days. The behavioural modification was stimulus-specific with regard to both the CS and US (Farley and Alkon 1980, 1982; Crow 1985*a*). Detailed analysis of the effect of the training schedule on acquisition and retention later showed that brief training leads to a non-associative behavioural modification, which builds up and decreases within approximately 1 hour; more intensive training was required to bring about long-lasting associative conditioning (Crow 1983).

Hermissenda's response to light is complex and multiphasic, and so is its modification by experience. Whereas in *Aplysia* the CR was a relatively simple motor response, such as gill or tail contraction, here the CR was a chain of reactions, including arousal, initiation of locomotion, orientation, approach, foot shortening, and more (Lederhendler *et al.* 1986). Only part of these reactions was systematically quantified, and their causal role in the overall behavioural depression still awaits analysis. For example, animals tested more than 1 day after training were much slower in initiating locomotion, but reports differed with regard to the effect of learning, if at all, on the velocity of locomotion after its initiation (Farley and Alkon 1982; Crow and Offenbach 1983). One factor contributing to the initial delay in phototaxis may be a shortening of the foot, a response involved in clinging to substrates. This response is normally evoked by rotation, but after training it is also elicited by light (Lederhendler *et al.* 1986). As we shall see below, the complexity of the phototactic response complicates interpretation of data on possible cellular correlates of learning in this case.

Cellular loci of plasticity: the role of the B photoreceptors

How does pairing of rotation with light suppress positive phototaxis in *Hermissenda* for days? Additional information on the sensory systems involved is necessary at this point. The CS is mediated via the eyes, situated on the dorsal surface of the circumesophageal nervous system, between the cerebropleural and pedal ganglia (Alkon 1983; Crow 1984; see Fig. 5.1(a)). Each of the two eyes contains five photoreceptors, classified on the basis of their morphological and electrophysiological properties as type A (two cells) and type B (three cells). Type A and type B photoreceptors are further classified on the basis of their position within the eye (e.g. medial, lateral). Photoreceptors at different positions are not functionally equivalent. Much attention was focused on the type B photoreceptors, and mainly on the medial type B cells. Type B photoreceptors are spontaneously active in the dark, and are responsive to dim illumination after dark adaptation.

The US is mediated via the statocysts, which are gravity-detecting organs, situated juxtaposed to the eyes between the cerebropleural and pedal ganglia (Alkon 1983; Crow 1984; see Fig. 5.1(a)). Each statocyst is a hollow sphere containing statolymph and 150–200 small stones, called statoconia. As the animal orients in water, the statoconia fall through the statolymph and deflect the cilia of one or more of the 13 hair cells that line the inner statocyst wall. This provides the central nervous system of *Hermissenda* with information on its spatial position. *Hermissenda* with defective statocysts cannot be effectively conditioned to associate light and rotation, indicating that the statocysts indeed mediate at least the majority of the US (Alkon 1983; Crow 1984).

Synaptic interactions between the visual and vestibular pathways, and some interactions of these systems with motor systems, have been characterized (Alkon 1983, 1984; Goh and Alkon 1984; see Fig. 5.1(b)). The network is complex and only partly understood. It has been suggested that medial type A photoreceptors excite, via interneurons in the cerebropleural ganglia, identified motoneurons in the pedal ganglia. These motoneurons (denoted MN1) contribute to turning of the animal's foot in the direction of the light. The same motoneurons can also be excited by hair cells, via interneurons. Medial and intermediate type B photoreceptors were suggested to inhibit medial type A photoreceptors. Type B photoreceptors, as well as caudal hair cells in the statocysts, inhibit cells in the optic ganglia. The latter ganglia cells (denoted S/E) excite type B cells and inhibit caudal hair cells. Additional excitatory and inhibitory interconnections were identified, which may also serve as potential convergence loci between light and rotation.

Electrophysiological studies soon revealed that conditioning is correlated with alterations in the electrical properties of type B photoreceptors. These alterations were, however, fairly complex, and depended on the time after the onset and cessation of light.

Early changes in photoreceptor properties

To facilitate the following brief discussion, we use the term 'paired animals' for animals subjected to paired light and rotation during training, and 'control animals' for animals subjected to random pairing of light and rotation, or to rotation alone. The spontaneous activity of dark-adapted B photoreceptors in an isolated circumesophageal nervous system preparation, measured immediately after the last training session, was higher in nervous systems removed from paired animals than in those from control animals (Crow and Alkon 1980). The resting potential of the B photoreceptors was more depolarized in the paired animals, and their input resistance was increased. In other words, the excitability of the 'paired' B photoreceptors was enhanced. These photoreceptors responded with increased impulse frequencies to light pulses of various intensities. In addition, they displayed enhancement of a postillumination 'tail' of depolarization. This 'tail' is observed in the B photoreceptors of naïve animals following the cessation of light; it became larger, and lasted longer, with training (Farley and Alkon 1982). The enhanced spontaneous activity and the depolarized resting potential lasted less than 24–48 hours after training, but the elevated input resistance and the enhanced responses to light persisted longer.

The modifications in the B photoreceptors could, in principle, result either from alterations intrinsic to the photoreceptors, or from long-lasting modifications in the input to these cells. These alternatives could be tested, because in the photoreceptors the sensory and synaptic input can be surgically separated. This is because phototransduction takes place in the microvilli of the rhabdomeres near the cell body, whereas the major synaptic input is near the distal end of the axon. Axotomized photoreceptors are thus deprived of their synaptic input, but can still generate light-evoked receptor potentials. Crow and Alkon (1980) axotomized the B photoreceptors after training, and discovered that the photo-receptors from the trained animals retained their modified membrane properties, including the enhanced tail of postillumination depolarization. This demonstrated that experience-dependent modifications in B photoreceptors were intrinsic to these cells, and that retention did not require sustained input from other circuit components.

Is this intrinsic modification in the properties of B photoreceptors *causally related* to the behavioural change, or only *correlated* with it? To answer this, *Hermissenda* were restrained, and a B cell impaled with a microelectrode. The cell was then subjected to paired presentations of 30 s light and 30 s depolarizing current. Control animals were presented with unpaired stimuli, or impaled with an intracellular electrode but without stimulation. Pairing of light and depolarization led to an increase in depolarization and input resistance in the B photoreceptor. The recovered animals were then subjected to behavioural tests. Admittedly, the slugs did not like the treatment too much: only about 40 per cent of them displayed essentially normal locomotion after the surgical and electrophysiological manipulations, and all the recovered animals exhibited phototactic latencies longer than the initial baseline level. Nevertheless, the phototactic latencies in the paired animals were longer than in the controls (Farley *et al.* 1983). This suggested that alterations in the properties of B cells are indeed causally related to conditioning. Note, however, that these experiments did not determine whether the treatment of B cells led to alterations in other cells as well, and hence whether the modification in the properties of the B cells was necessary and sufficient to produce the behavioural depression.

Altered responses to sustained illumination

In the experiments described above, B photoreceptors were analysed during the early phase of their response to light, or immediately after termination of illumination. As mentioned above, some changes in the behaviour of conditioned *Hermissenda* can, indeed, be detected within seconds of the initial exposure to light; but major behavioural differences between the trained and naïve individuals are detected only after a sustained exposure to light. Terry Crow has, therefore, reasoned that one should also measure and pay attention to the properties of B cells in trained animals following several minutes of light adaptation, and not only immediately after the onset or cessation of light (Crow 1985*b*). He settled on 5 minutes of light adaptation, because in behavioural experiments more than 70 per cent of the conditioned animals started moving towards the light after 5 minutes of illumination.

An interesting complexity in the response of B cells was revealed in these experiments. In contrast with the enhanced excitability observed in all the type B cells in the early part of the light response (see above), the response of part of the B cells to sustained illumination became attenuated after training (Crow 1985*b*; Bridge and Crow 1986). This was expressed in a reduced discharge rate, and, in B cells surgically deprived of their synaptic input, in a smaller, light-adapted receptor potential. Thus, conditioning-correlated alterations in intrinsic properties of type B photoreceptors led to either increased or decreased excitability, depending on the subtype of the cell and on the time course of its exposure to light.

The B photoreceptors are clearly not the only site of plasticity correlated with reduced phototaxis. A persistent, pairing-specific decrease in steady-state receptor potentials was also observed following conditioning in the type A photoreceptors (Richards and Farley 1984). Modifications in second-order sensory neurons have not yet been systematically explored.

Molecular mechanisms of plasticity: the role of Ca²⁺

Whereas all the reports agree that plastic changes take place in the primary photoreceptors during learning in *Hermissenda*, opinions differ on the behavioural relevance of identified cellular and molecular correlates of conditioning. All current models of learning in *Hermissenda* do assign, however, a critical role to experience-dependent alterations in intracellular Ca^{2+}.

Crow and Alkon (1980) suggested that the modifications observed in the membrane properties of B photoreceptors could be explained by a persistent decrease in a resting, voltage-dependent K^+ conductance across the cell membrane. They proceeded to identify the nature of the modified current(s). As expected, *Hermissenda* neurons are equipped with a variety of voltage-gated and chemically gated K^+ channels. A series of electrophysiological studies, utilizing methods and rationales similar to those described earlier for *Aplysia*, has led to the conclusion that conditioning in *Hermissenda* is correlated with a reduction and rapid inactivation of a rapidly rising, rapidly inactivating K^+ current, I_A, and of a Ca^{2+}-dependent K^+ current, $I_{K(Ca)}$ (Alkon *et al.* 1982, 1987*a*).

These K^+ channels are inhibited, either directly or indirectly, by phosphorylation. Intracellular iontophoretic injection of the catalytic subunit of mammalian cAMP-dependent protein kinase (PKA) decreased I_A, and also a delayed K^+ current, denoted at that time I_B and later found to be partially contributed by $I_{K(Ca)}$. The effect on the delayed current was, however, larger than that on the early current (Alkon *et al.* 1983). This did not mimic the physiological situation, in which the early current was also markedly modified. Note that at high concentrations, PKA readily phosphorylates unnatural substrates (Nestler and Greengard 1984; Dudai 1987), and iontophoretic injection introduces very large amounts of the catalytic subunit into the cell (Levitan 1985). The physiological relevance of the cAMP-dependent phosphorylation was not, therefore, clear. Injection of various types of Ca^{2+}/calmodulin-dependent kinases (protein kinase B, PKB) had a larger effect on reduction of the early K^+ current, and also reduced the late current (Acosta-Urquidi *et al.* 1984; Sakakibara *et al.* 1985). Both PKA and PKB injections increased input resistance and enhanced the post-illumination depolarization tail. Intracellular injection of another Ca^{2+}-dependent kinase, the Ca^{2+}- and phospholipid-dependent kinase (PKC), reduced I_A and $I_{K(Ca)}$ in B photoreceptors (Farley and Auerbach 1986). Phorbol ester, an activator of PKC, had similar effects. Taken together, the results did not establish which protein kinase(s) act(s) *in vivo*, but clearly demonstrated a role for Ca^{2+}-dependent phosphorylation cascades.

Although much attention has been focused on K^+ channels, other currents also change with training, including a voltage-dependent Ca^{2+} current (Farley *et al.* 1984) and a light-evoked current (Crow 1985*c*). The Ca^{2+} current, similarly to the K^+ currents, was enhanced in B cells by injection of PKC and by phorbol ester (Farley and Auerbach 1986).

Cellular and molecular models of conditioning in *Hermissenda*

Since conditioning of phototaxis in *Hermissenda* is correlated with a variety of behavioural, cellular, and molecular changes, it is not surprising that different mechanistic models were offered for acquisition and retention in this system, and that each of these models accommodates only part of the experimental observations. Alkon and his colleagues have suggested a model which assumes that enhanced excitability of type B photoreceptors, detected in trained animals after light onset (see above), plays a causative role in the behavioural modification (Alkon 1983, 1984; Lederhendler *et al.* 1986; Alkon *et al.* 1987*b*; see Fig. 5.1(b)). This assumption necessitates taking into account rather complex interactions in the network that processes and converges the visual and vestibular input, in order to explain how an enhanced photoresponse leads to reduced phototaxis. Additional ingredients in the model are the increased intracellular Ca^{2+} detected in B photoreceptors during acquisition (Connor and Alkon 1984), and the ability of Ca^{2+}-activated kinases to reduce I_A and $I_{K(Ca)}$.

According to this model, the paired presentation of light (CS) and rotation (US) results in cumulative depolarization in the B photoreceptors. This stems from integrated interactions of B photoreceptors, statocysts' hair cells, and visual ganglion cells, including S/E cells (see Fig. 5.1(b)). The following is a hypothesized *acquisition* process: light depolarizes the B photoreceptors, and its cessation generates a depolarizing tail of the receptor potential. The B cells are inhibited by the caudal statocysts' hair cells. During rotation, inhibition of the B cells by the hair cells increases. However, when rotation stops, the hair cell activity is reduced below its tonic level and as a result there is rebound excitation in the B cells. In addition, the B cells as well as the hair cells send inhibitory input to the S/E cells, whereas the latter send excitatory input to the B cells. During paired light and rotation, S/E cells are even more inhibited, but at the cessation of pairing, S/E cells rebound from inhibition. This causes further excitation of the B cells. All in all, the hypothesis in this part of the model is that coincident cessation of both light and rotation enhances depolarization of B photoreceptors. This is a somewhat unconventional contiguity constraint. (Comparison of the effectiveness of temporal contiguity schedules indeed showed that simultaneous conditioning, in which light and rotation stopped together, was superior to delay conditioning

(Farley 1986), but the role in learning proposed for coincident termination of the CS and the US requires further examination.)

Membrane depolarization enhances Ca^{2+} influx. This leads to Ca^{2+}-dependent phosphorylation (via PKB and/or PKC) of the K^+ channels. Closure of these channels increases the input resistance and excitability. This, in turn, again enhances Ca^{2+} influx and kinase activity. Upon repeated training, the process establishes enduring phosphorylation and closure of the channels carrying I_A and $I_{K(Ca)}$.

According to the above hypothesis, the CS and US are coded by depolarizations and ultimately by Ca^{2+} surges, which converge on Ca^{2+}-activated protein kinase(s) (Fig. 5.3). Convergence is thus the additivity (or synergy) of the Ca^{2+} effect. However, additional possibilities exist. The US may be recoded, at least partially, into a neurotransmitter-activated, second-messenger cascade which utilizes cAMP, Ca^{2+}, and/or lipids as intracellular signals; and convergence may be based on a cross-talk between different second-messenger cascades, for example cAMP and Ca^{2+}. Several independent pieces of experimental evidence accord with the latter type of mechanism, which implicates transmitter-activated, second-messenger cascades, and not only depolarizations, in conditioning. Thus, α-adrenergic agonists regulate the membrane properties and the response to light of B photoreceptors (Sakakibara *et al.* 1987). Pairing light with the direct application of serotonin to the exposed circumesophageal nervous system reduces phototaxis (Crow and Forrester 1986) and alters the light response of identified B photoreceptors (Forrester and Crow 1987).

The closure of K^+ channels in the B photoreceptors contributes to *retention*: if the channels are closed, and the repolarizing K^+ currents diminished, the CS (light) results in an enhanced receptor potential. Here enters a critical prediction of the model: facilitation of

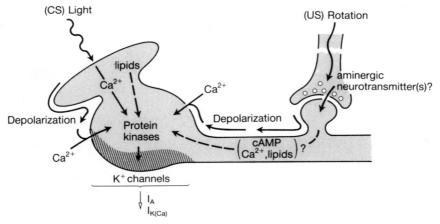

Fig. 5.3. A sketch of a type B photoreceptor, depicting the role of Ca^{2+} in the cellular coding and convergence of the US and CS in part of the neuronal system subserving conditioned phototaxis in *Hermissenda*. Additional ion channels and the cross-talk between second-messenger cascades are omitted for simplicity. (Modified from Alkon *et al.* 1987*b*.)

the B photoreceptors leads to suppression of phototaxis, because B photoreceptors inhibit medial type A photoreceptors, which in turn excite muscles subserving positive phototaxis (see Fig. 5.1(b)).

An alternative cellular model, suggested by Crow (1985b), assumes that the reduced excitability, detected in part of the B cells following sustained illumination, is relevant to the depressed phototaxis. According to this model, positive phototaxis is suppressed because the photoreceptors are less responsive to light; therefore, there is no need to invoke complex interactions between network components, in order to convert the enhanced excitability of B photoreceptors into a depressed behaviour. The postulated network of the phototactic response contains elements that can account for a direct translation of suppressed photo-receptor response to a behavioural depression (see Fig. 5.1(b)). The attenuated responsiveness of B cells may be contributed to, according to this model, by an increase in light-evoked receptor desensitization as well as by a delayed recovery of sensitivity following light adaptation. It is not yet clear how modifications in intracellular Ca^{2+} and in K^+ channels are involved in these desensitization processes.

Actually, the above two models are not mutually exclusive. Since the behavioural response of conditioned *Hermissenda* to light involves several phases of network activity, different types of B cells may contribute differentially to different pieces of the behavioural response. A crucial test of all the models is their ability to explain how changes in the responsiveness to the CS relate to the modified motor behaviour. Two points must be clarified: (a) the detailed experience-dependent modifications in light responses in individual subtypes of B photoreceptors, as a function of time after training and of light intensity (this is critical, because the response of B cells to sustained illumination appears to depend on the subtype of these cells) and (b) the identity of the motor systems involved in the phototactic behaviour, their interrelationship, and their connections to the CS path-ways (Schuman *et al.* 1986; Hodgson and Crow 1987). Future models should also incorporate additional cellular sites of plasticity, in the interneurons and possibly in the motor systems.

Differences and similarities between the *Aplysia* and *Hermissenda* models: synaptic vs. somatic changes, second-messenger cascades, and modulation of ion channels

The models proposed for associative learning in *Hermissenda* differ in several aspects from those proposed for associative learning in *Aplysia*, and yet the two systems do have common denominators.

Localization

In both *Aplysia* and *Hermissenda*, a major cellular site of plasticity has been identified in the primary sensory neuron responding to the CS. In both systems, multiple sites of plasticity exist. In *Aplysia* attention has been focused on plastic changes in *synapses*; in *Hermissenda*, on plastic in the neuronal *soma*. However, persistent somatic alterations in *Aplysia* do take place, and persistent synaptic alterations in *Hermissenda* may also occur.

Network modification

Our knowledge of the networks involved in the modifiable behaviours in both *Aplysia* and *Hermissenda*, is fragmentary, probably more so in the case of *Hermissenda*. In *Aplysia*, learning is a modification in the probability of a simple reflex, resulting, at least in the short-term, from alterations in pre-existing, functionally active components of the neuronal network. In *Hermissenda*, the measured behavioural response is more complex, and it is not yet clear how much of it is contributed by the recruitment of dormant connections subserving the emergence of apparently novel responses to the CS, such as foot shortening.

Cellular mechanisms of acquisition and retention

• In both *Aplysia* and *Hermissenda*, plastic changes consist of alterations in neuronal membrane properties. In *Aplysia*, attention has been primarily focused on an alteration in the efficacy of the neuronal output device. In *Hermissenda*, attention has been focused on alterations in the integrative properties of the neuron.

In both *Aplysia* and *Hermissenda*, the closure of K^+ channels contributes to experience-dependent enhanced excitability. In *Aplysia*, a major role is played by the S channel. In *Hermissenda*, major roles are played by the channels mediating the $I_{K(Ca)}$ and I_A conductances. In both systems, the K^+ channels are closed by phosphorylation, mediated by ubiquitous second-messenger cascades. In both systems, Ca^{2+} plays a central role in the cellular coding of the CS and in the convergence of the US and CS. In both systems, convergence is suggested to take place on second-messenger cascades. In *Aplysia* much attention has been focused on the cAMP cascade while in *Hermissenda* it has been focused on the Ca^{2+}-and phospholipid-dependent cascades. In both cases, several second-messenger cascades operate in concert.

Together, *Aplysia* and *Hermissenda* provide us with insight into the richness of cellular and molecular correlates of learning and short-term memory in relatively simple invertebrate systems. Can molecular putative mnemonic devices be identified in the vertebrate brain, too? Yes, they can; and an example is provided in the next chapter.

6   *A cellular mnemonic device in the mammalian brain: long-term potentiation*

Instead of analysing the ganglia of relatively simple invertebrates, one may search for cellular and molecular mechanisms of learning in subsystems of complex brains. This could be done using two approaches. The first is to choose a conditioned behaviour, identify its neural substrate, and elucidate alterations correlated with conditioning and necessary for it. This is similar to the approach used in *Aplysia* or *Hermissenda*, except that the experiments and their interpretation are often more complex. Examples of conditioned reflexes in vertebrates, which yielded some information on cellular correlates of learning, are provided in Chapter 11.

A second approach is to start with a cellular phenomenon suspected of involvement in learning, rather than with a behaviour. This phenomenon is then investigated in neuronal preparations amenable to electrophysiological, neurochemical, and/or neuroanatomical manipulations, irrespective of whether or not the preparations are suitable for the analysis of specific learning tasks. This is, in fact, the bona fide, ultimate bottom–up approach (see Chapter 3). In this chapter we illustrate the latter approach by addressing long-term potentiation (LTP), a widespread neuronal phenomenon suggested to be a mnemonic device in the mammalian brain. LTP could be described, at the time of writing, as a cellular device in search of a behaviour. We will, however, discuss evidence suggesting that it is indeed involved in neuronal and behavioural plasticity.

How long is long-term potentiation?

In many preparations, repetitive neuronal stimulation leads to alterations in synaptic efficacy that outlast the stimulation. These alterations may, therefore, be regarded as 'cellular memories'. They have been classified according to their time course (Magleby and Zengel 1982). An increase in transmitter release lasting hundreds of milliseconds is termed *facilitation*. Appropriate protocols of repetitive stimulation may lead to *augmentation*, lasting seconds, and to *potentiation*, which lasts much longer.

The term 'potentiation', combined with the appropriate qualifiers, is used in the physiological literature to describe changes in synaptic efficacy with different time courses and different underlying mechanisms. Often, potentiation after tetanic stimulation persists for only a few minutes. It is termed *post-tetanic potentiation (PTP)*. In some preparations, a stable increase, lasting from hours to weeks, is observed in the magnitude of postsynaptic responses to a constant afferent volley, following brief tetanic stimulation of the same afferents. This enduring change is termed *long-term potentiation (LTP)* (Andersen 1987; Collingridge and Bliss 1987; Teyler and DiScenna 1987).

Facilitation, augmentation, and short-term potentiation have been studied mainly at the neuromuscular junction, and have been shown to result from increased availability of Ca^{2+} required for transmitter release (Katz and Miledi 1968; Lev-Tov and Rahamimoff 1980). Phenomena with similar time courses have also been studied in central synapses (McNaughton 1982). LTP has been studied mainly in the hippocampal formation of the mammalian brain, but has also been demonstrated in other brain regions (Artola and Singer 1987), peripheral neurons (Dolphin 1985), and invertebrates (Walters and Byrne 1985; Miller *et al.* 1987).

LTP has long gained the attention of students of memory, since it is experience dependent and enduring, and displays additional characteristics which suggest a possible role in relatively long-term information encoding. It was first systematically studied in the hippocampus by Bliss *et al.* (Bliss and Gardner-Medwin 1973; Bliss and Lomo 1973), following earlier reports of persistent effects of brief tetani in brain and ganglia (Larrabee and Bronk 1947; Lomo 1966).

In their now classical experiment, Tim Bliss and Terje Lomo (1973) exposed the dorsal hippocampus of an anaesthetized rabbit, placed extracellular recording microelectrodes in the dentate gyrus, and stimulated with a 'conditioning' train the perforant path fibres. These fibres originate in the entorhinal cortex and synapse onto apical dendrites of granule cells in the dentate molecular layer (Fig. 6.1). The conditioning trains were delivered at a rate of 10–20 Hz for 10–15 s, or 100 Hz for 3–4 s. Bliss and Lomo measured the population response in the dentate by using three parameters: the amplitude of the population EPSP and the spike, and the latency of the latter. In the majority of the rabbits, one or a few trains sufficed to potentiate the response to single perforant path volleys for periods ranging from 30 minutes to 10 hours. A similar enduring increase in the efficacy of transmission at the tetanized path synapses was obtained in unanaesthetized rabbits, chronically implanted with stimulating and recording electrodes and subjected to the same 'conditioning' procedure.

These observations were soon confirmed and extended in many laboratories, using multi- and single-unit recordings, both *in vivo* and in brain slices, in a variety of species. It is not yet clear how similar LTP-like processes are in different neuronal systems. The hippocampus still remains the favourite preparation for studying the phenomenon.

LTP in the hippocampus of an intact animal may persist for days to weeks (Bliss and Gardner-Medwin 1973; de Jonge and Racine 1985). Its decay kinetics indicate a multiphasic process (Racine *et al.* 1983). In hippocampal slices, which are fairly labile, LTP can be monitored for tens of minutes up to several hours, practically as long as the slice survives.

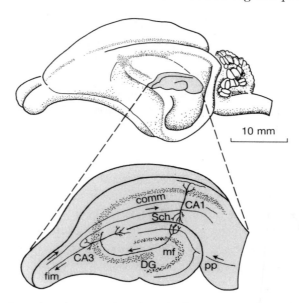

Fig. 6.1. Most studies of LTP have been conducted on the hippocampus *in vivo*, or on hippocampal slices. The diagram shows the location of the hippocampus in the rabbit brain, and a scheme of a slice, with the major areas and pathways marked. All the principal excitatory pathways of the hippocampus can sustain LTP. CA1, CA3, pyramidal cell fields of the hippocampus; comm, commissural projection to area CA1 (projection to CA3 not shown); DG, dentate gyrus; fim, fimbria; mf, mossy fibres; pp, perforant path; Sch, Schaffer collaterals. (Modified from Andersen *et al.* 1971.)

Specificity and associations in long-term potentiation

In addition to endurance, the following four characteristics of LTP make it attractive as a putative mnemonic mechanism. (1) It is generated by a very brief stimulus, and is thus potentially adapted to encode transient physiological events. (2) It is input specific, which is essential for any meaningful learning system. (3) A certain stimulus threshold is required, which confers a certain immunity to noise. (4) It shows associativity, which means that the system is adapted to encode pairing of independent events. Note that none of these characteristics provides a proof that LTP is indeed involved in learning.

1. Initiation by a brief stimulus

A conditioning stimulus of only a few seconds is sufficient to induce LTP for many hours (see above).

2. Input specificity

LTP is restricted to the conditioning path. This is best illustrated by experiments conducted on hippocampal slices, in which the circuitry is more easily defined and manipulated (Schwartzkroin and Wester 1975; Andersen *et al.* 1977, 1980; Lynch *et al.* 1977; Dunwiddie and Lynch 1978; Barrionuevo and Brown 1983).

In a representative set of studies, Per Andersen and his colleagues (Andersen *et al.* 1980) delivered brief tetanic stimuli to one of two independent afferent pathways to the CA1 pyramidal cells (see Fig. 6.1): one at the layer of the basal dendrites (stratum oriens), the other at the layer of the apical dendrites (stratum radiatum). Population EPSPs were recorded from the same layers and population spikes were recorded from the pyramidal cell body layer. In some experiments, extra- or intracellular recordings were also made from single pyramidal cells. A brief tetanus to either of the two input paths produced LTP, as revealed by increased population EPSP and population spike and a reduced latency of the latter. The LTP was detected only in response to stimulation of the tetanized path. No changes were detected in soma membrane potential, resistance, and excitability, implying that LTP is a local process, specific to the tetanized synapses.

The issue of specificity deserves a remark. First, tetanization can also produce a generalized depression of the postsynaptic cell (Bliss and Lomo 1973; Lynch *et al.* 1977; Andersen *et al.* 1980; Abraham *et al.* 1985). In these experiments, LTP was, therefore, described as a homosynaptic potentiation superimposed on a heterosynaptic depression. (The depression, which may be manifested as long-term depression (LTD) of pathways which converge on the postsynaptic cell but are inactive during tetanization of the potentiated pathway, will not be further discussed here, because its cellular basis is less understood than that of LTP.) Second, as explained below, under certain conditions tetanization may facilitate LTP in converging, active pathways (see 'Associativity' below).

3. Co-operativity

LTP has a stimulus intensity threshold, below which only PTP may develop (McNaughton *et al.* 1978). Above this threshold, potentiation is a function of the number of fibres activated in the pathway (McNaughton *et al.* 1978; Lee 1983).

4. Associativity

Since co-operativity indicates that LTP is a function of the number of activated fibres, the question arises whether two separate yet converging pathways can associate to establish and/ or increase LTP in one of these pathways. This is indeed the case: activation of adjacent, convergent afferents can yield greater LTP in one of these afferents (Lee 1983). Moreover, a weak input, incapable of sustaining LTP, can sustain it if activated concurrently with a strong stimulus to another, converging pathway. This phenomenon is termed *associative LTP*, and has been demonstrated *in vivo* (e.g. Levy and Steward 1979) and in slices (e.g. Barrionuevo and Brown 1983).

A set of *in vivo* experiments by Levy, Steward, and co-workers on LTP in the pathways from the entorhinal cortex to the dentate gyrus of the rat, illustrates several interesting facets of associativity (Levy and Steward 1979, 1983; White *et al.* 1988). The studies took advantage of the fact that independent temporodentate (entorhinal-dentate) pathways converge on different dendritic domains of the dentate granule cells. The relevant features of the topographical organization are the following (Fig. 6.2): first, the temporodentate pathways have both ipsilateral and crossed (contralateral) components. The ipsilateral input (which is the perforant path) is strong, comprising more than 90 per cent of the synapses at the entorhinal terminal zone. In contrast, the crossed input is weak, comprising less than 10 per cent of the synapses. Second, in the ipsilateral and the crossed temporodentate paths, projections from the lateral entorhinal cortex terminate on the distal dendrites of the dentate granule cells, whereas projections from the medial entorhinal cortex terminate on the dendrites more proximal to the soma.

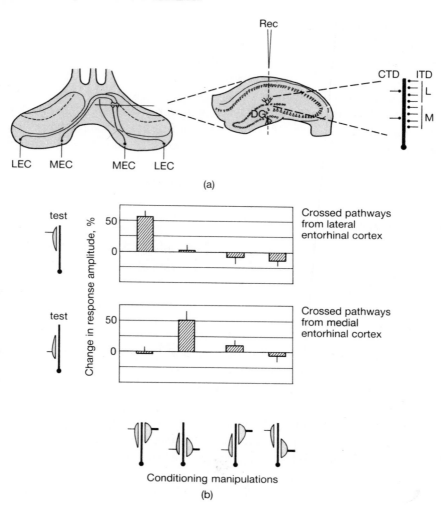

Tetanization of the strong, ipsilateral pathway in anaesthetized rats results in LTP of this pathway. In contrast, tetanization of the weak, crossed pathway does not result in LTP, or at most results in only a weak LTP in this pathway. However, marked LTP was induced in the weak pathway on co-tetanization with the strong pathway (Levy and Steward 1983). Moreover, LTP in the weak pathway resulted only if its terminals on the granule cell dendrites were in the same local dendritic domain on which the terminals of the strong pathway synapse. In other words, LTP developed only if the lateral ipsilateral pathway was co-conditioned with the lateral crossed pathways, or the medial ipsilateral co-conditioned with the medial crossed pathways (White *et al.* 1988; see Fig. 6.2). In addition to demonstrating associativity, these experiments provide further evidence of input specificity (see above). Furthermore, they show that LTP can be compartmentalized in local dendritic domains of a single cell.

Even more intriguing, order dependency, a phenomenon so often characteristic of associative learning, was reported to be crucial for successful associative LTP. In the weak, crossed temporodentate pathway LTP was induced only if stimulation of this pathway was concurrent with, or immediately preceded, the strong, ipsilateral input (Levy and Steward 1983). If the interval between the two activities became too long (200 ms), or if the order was reversed and activity in the ipsilateral pathway preceded activity in the crossed pathway, the responses evoked in the latter were depressed.

Temporal specificity of associative LTP was also detected *in vitro* in hippocampal slices. Kelso and Brown (1986) paired weak and strong activations of the Schaffer collateral and

---

Fig. 6.2. Associative interactions in LTP. (a): Schematic representation of the topographical characteristics of pathways from the entorhinal cortex to the dentate gyrus. *Left:* A dorsal view of the ipsilateral and crossed (contralateral) components of the pathways. The ipsilateral pathways contribute more than 90 per cent of the synapses in the entorhinal terminal zone in the dentate. In both the ipsilateral and crossed pathways there are projections from the lateral and from the medial entorhinal cortex. The lateral projections synapse onto the distal dendrites, and the medial projections synapse onto the proximal dendrites of the granule cells. Also shown are the positions of stimulating electrodes placed by White *et al.* (1988) in their study of microscopic associative interaction during the induction of LTP in this system. LEC, electrodes in lateral entorhinal cortex; MEC, electrodes in medial entorhinal cortex. *Centre:* Cross-section through the hippocampus. Broken line, the path in which the recording electrode was advanced through the dentate gyrus. *Right:* Diagram of the arrangements along the granule cell dendritic tree of the synapses activated by the lateral and medial stimulating electrodes. CTD, crossed temporodentate pathway; ITD, ipsilateral temporodentate pathway; L, lateral projections; M, medial projections. The circle represents the cell soma, and the shaft the dendritic tree. (b): Changes in the amplitude of response in the lateral and medial weak crossed pathways after co-conditioning with the strong ipsilateral pathways. In this system, tetanization of the ipsilateral pathways alone results in LTP in these pathways, but tetanization of the crossed pathways does not result in LTP in these pathways. *Top:* The post-conditioning changes in response amplitude in the crossed pathways from the lateral entorhinal cortex. *Centre:* The post-conditioning changes in the crossed pathways from the medial entorhinal cortex. *Bottom:* Data depicting the corresponding co-conditioning arrangements. Note that only co-conditioning of two pathways that converge on the same local dendritic domain resulted in potentiation of the crossed pathways. The data thus show (1) associative LTP; (2) multiple units of integration for synaptic modification by LTP exist in a single postsynaptic cell. (Adapted from White *et al.* 1988.)

commissural projections to area CA1 in rat hippocampal slices. Three stimulating electrodes were introduced in three different loci, one to deliver a strong input, the others to deliver weaker stimuli. The strong stimulus alone, or pairing of the two weak stimuli, did not produce LTP in either of the two weakly activated inputs. When the weak stimulus preceded the strong one by 0.2 s and terminated with it, persistent potentiation was elicited. In contrast, when the weak stimulus began 0.6 s after the strong stimulus, no lasting potentiation was detected.

Naturally, as noted above, the phenomenon of order dependency is appealing because of its analogy to a central behavioural parameter of associative conditioning. This analogy should not, however, be overstretched for the following reasons. (1) It is not known whether order dependency is an inherent characteristic of LTP, or only a by-product of particular experimental situations. For example, in the case of the temporodentate pathways, parsimonious interpretation of the data takes into account only the ipsilateral and crossed paths, but activation of each of these may also activate additional hippocampal paths, such as inhibitory commissural connections. The integrated effect of these additional network components on the temporal constraints of LTP induction must be clarified. In other words, order dependency might be due to properties of the particular network and not of LTP. (2) It is not clear whether the temporal constraints of order dependency in LTP parallel those of the supposedly analagous behavioural phenomenon. (3) It is not known whether the strong and weak inputs can represent a US and a CS respectively; therefore, interpretating the physiological phenomenon in terms of a concrete behavioural phenomenon is premature. This point, concerning the lack of information on cellular and molecular coding of potential conditioning stimuli in systems used to study LTP, will be re-emphazised below, in evaluating the relevance of LTP to learning in general.

The cellular basis of co-operativity and associativity was interpreted in the following way: above-threshold tetanization of a strong input, capable of sustaining LTP by itself, depolarizes the postsynaptic cell. A weak input cannot sufficiently depolarize the postsynaptic cell, but in associativity, massive depolarization is provided by the strong, co-conditioned input (Wigstrom and Gustafsson 1985; Gustafsson and Wigstrom 1986). If this is so, then generation of LTP should take place even in the absence of tetanization, provided that single afferent volleys are presented in the test input concurrently with postsynaptic depolarization. This indeed was the case (Kelso *et al.* 1986; Sastry *et al.* 1986; Gustafsson *et al.* 1987). For example, Gustafsson *et al.* (1987) made intracellular recordings of EPSPs resulting from low-frequency, single-volley stimulation in the stratum radiatum or oriens, while pairing the volley with brief depolarizing current pulses injected into the CA1 pyramidal cells through the recording electrode. The EPSP became potentiated when the volley inducing the EPSP occurred together with the depolarizing current or preceded it by less than 0.1 s. Potentiation outlasted the conditioning period by at least 1 h. No LTP was detected when the depolarizing current preceded the EPSP.

Depolarization, like any other experimental manipulation, may have secondary actions. For example, it may induce release from adjacent afferent terminals. However, an additional piece of evidence in favour of its direct role in initiating LTP emerged from experiments in which the intracellular injection of hyperpolarizing current in the postsynaptic cell, during application of the conditioning tetanus, reversibly blocked LTP induction (Malinow and Miller 1986).

● At the beginning of this chapter LTP was defined as a *macroscopic* phenomenon, based on initial physiological observations. We can now redefine LTP on the *microscopic* level as follows: LTP is a long-lasting increase in the efficacy of an excitatory synapse that occurs when the synapse is activated while at the same time the postsynaptic membrane is sufficiently depolarized. Note that although it is depolarization, not spike activity, that is necessary in the postsynaptic cell, the requirement for concurrent pre- and postsynaptic events does conform to a broad interpretation of a Hebbian synapse.

What cellular and molecular mechanisms underlie this synaptic process? And do they serve a role in learning? We look at the mechanisms first and the behaviour later.

Cellular and molecular correlates: suggested roles for Ca^{2+}, *N*-methyl-D-aspartate receptors, second-messenger cascades, increased transmitter release, and synaptic morphology

In addition to endurance, rapid onset, input specificity, co-operativity, and associativity, models of LTP must take into account several observations made at the cellular and molecular level.

Ca^{2+} is required for LTP

This could be shown by manipulation of both extracellular and intracellular Ca^{2+}. Reduction of extracellular Ca^{2+} blocked the development of LTP in hippocampal slices (Dunwiddie *et al.* 1978; Wigstrom *et al.* 1979), even under conditions where synaptic transmission, facilitation, and PTP were still maintained (Dunwiddie and Lynch 1979). The effect was reversible and LTP was reinstated by tetanization once normal extracellular Ca^{2+} concentration was re-established. Actually, in other experiments, transiently increasing the extracellular Ca^{2+} was sufficient to induce an LTP-like phenomenon (Turner *et al.* 1982; Higashima and Yamamoto 1985). As to the intracellular role of Ca^{2+}, injection of the chelator EGTA into pyramidal cells blocked LTP; this effect was attributed to a reduction of free Ca^{2+} levels (Lynch *et al.* 1983).

Also relevant are the observations that an increased incorporation of radioactive Ca^{2+} into hippocampal slices correlated with LTP (Baimbridge and Miller 1981) and an increase in the number of Ca^{2+}-containing electron-dense deposits was evident, especially in dendrites (Kuhnt *et al.* 1985).

Receptors for *N*-methyl-D-aspartate (NMDA) are involved in the induction of LTP in some synapses

Receptors for glutamate and other, related, putative excitatory amino-acid transmitters in mammalian brain are classified into several types according to their responsiveness to agonists (Foster and Fagg 1984). One of these receptor types, potently activated by *N*-methyl-D-aspartate (NMDA), is involved in the induction of LTP in the hippocampus and the cortex.

The NMDA receptor complex bears multiple binding sites and is associated with an ion channel (Fig. 6.3). The same type of channel molecule, but in different conductance states, might be shared by other glutamate receptors (Cull-Candy and Usowicz 1987; Jahr and Stevens 1987). In addition to a site for glutamate, NMDA, and other glutamate agonists, the

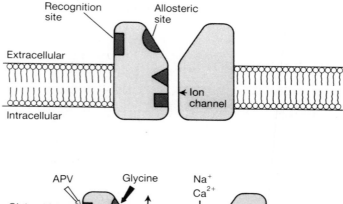

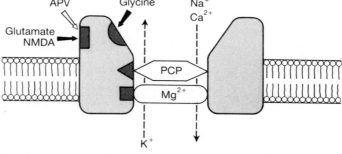

Fig. 6.3. The NMDA receptor–channel complex is a molecular switching device activated only when its agonist (probably glutamate) binds while the membrane is depolarized, for example during induction of LTP in hippocampal area CA1. The complex can, therefore, associate two neuronal events. This feature makes it an attractive candidate for a molecular component of learning systems. A schematic model of the receptor is depicted here. Under normal physiological conditions, the channel is blocked by Mg^{2+}. Sufficient depolarization removes the Mg^{2+} blockade and allows Ca^{2+} influx, as well as monovalent ion fluxes. The psychoactive drug phenylcyclidine (PCP) blocks the channel if the transmitter site is occupied by an agonist. NMDA, *N*-methyl-D-aspartate, an agonist; APV, D-2-amino-5-phosphonovalerate, an antagonist often used in studies of LTP and learning. Glycine binds to an allosteric site, possibly to facilitate agonist-induced channel opening. (Adapted from Foster and Fagg 1987.)

NMDA receptor also carries an allosteric site for glycine and sites for phenylcyclidine (PCP) and Mg^{2+} (Ascher and Nowak 1987). The open channel permits Ca^{2+} influx, as well as fluxes of monovalent cations (Dingledine 1983; MacDermott *et al.* 1986). The channel is blocked by physiological concentrations of extracellular Mg^{2+} (Mayer *et al.* 1984; Nowak *et al.* 1984). The NMDA receptor complexes are not, therefore, expected to participate in normal excitatory transmission. This is mediated in the hippocampus via amino–acid receptor types that respond to kainate and quisqualate and activate channels for monovalent cations (Collingridge *et al.* 1983). The NMDA-activated channels come into play only when Mg^{2+} is removed in a voltage–dependent manner. Indeed, the selective NMDA antagonist, DL-2-amino-5-phosphonovalerate (APV), prevented LTP in region CA1 of the hippo-campus, without significantly affecting unpotentiated synaptic responses (Collingridge *et al.* 1983; Harris *et al.* 1984; Wigstrom and Gustafsson 1984; Coan and Collingridge 1987; Huang *et al.* 1987). Similar results were obtained in the dentate gyrus (Morris *et al.* 1986; Errington *et al.* 1987) and in the visual cortex (Artola and Singer 1987). In contrast, tetanization in the presence of the quisqualate/kainate receptor antagonists 6,7–dinitro–quinoxaline–2–3–dione (DNQX) or 6–cyano–7–nitroquinoxaline–2–3–dione (CNQX), did not prevent expression of LTP after washout of the drug (Muller *et al.* 1988*a*; Kauer *et al.* 1988*b*). This proves that the quisqualate/kainate receptors are not required for the induction of LTP.

NMDA receptors have recently become almost synonymous with LTP in the scientific literature. Two reservations regarding the relevance of these receptors to LTP are, therefore, pertinent here. First, prevention of LTP by NMDA antagonists is not a universal phenomenon. For example, APV did not prevent LTP in the mossy fibre input to hippo-campal field CA3 (Harris and Cotman 1986). Second, there might be brain pathways in which NMDA receptors play a substantial role in normal excitatory transmission (Thomson 1986). Recent binding studies with radiolabelled glutamatergic ligands revealed the existence of two classes of NMDA receptors in the mammalian brain (Monaghan *et al.* 1988). It is possible that these classes differ in their role in transmission and plasticity.

The protein kinase C (PKC) cascade plays a role in LTP

Application of phorbol esters, potent activators of PKC, to hippocampal slices resulted in lasting potentiation of synaptic transmission, interpreted as LTP, in area CA1 (Malenka *et al.* 1986). Another PKC activator, oleic acid, induced LTP-like phenomena on iontophoretic application into the dentate gyrus of anaesthetized rats (Linden *et al.* 1986). The potentiation induced by PKC activators might be short-lived compared to LTP induced by high-frequency stimulation, raising some doubts about the earlier proposed equivalence of the drug-induced and the stimulation-induced phenomena (Muller *et al.* 1988*b*). Moreover, even the duration of the potentiation observed in experiments with PKC activators may be an overestimate, because these lipid-soluble activators are difficult to wash out. Drugs which inhibit PKC can block LTP (Lovinger *et al.* 1987; Madison *et al.* 1988; Malinow *et al.* 1988; Reymann *et al.* 1988). However, these drugs also affect other protein kinases (see below). One of these drugs, H-7, which inhibits the activated kinase, blocked established LTP in a reversible manner, i.e. potentiation was restored after the drug was washed away. This suggested that the relevant kinase(s), e.g. PKC, is persistently activated in LTP and that this activated state is independent of the kinase catalytic activity (Malinow *et al.* 1988; compare to models of kinase activation presented in Chapter 8). Injection of PKC itself into CA1 pyramidal cells in slice was reported to induce an LTP-like process, but it was not clear whether the effects were due to a transient pulse of PKC activity, or to the persistence of the injected active enzyme in the cell (Hu *et al.* 1987).

Another line of evidence for the potential role of PKC in LTP comes from experiments by Akers *et al.* (1986). They correlated LTP, induced by tetanization in the perforant path of the anaesthetized rat, with the subcellular distribution of PKC in the hippocampal region which is innervated by this path. In preparations from animals sacrificed 1 h after induction of LTP, but not in those from animals sacrificed 1 min after tetanization, there was a twofold increase in the activity of PKC in membranes and a corresponding decrease in cytosol. The sum total of membrane and soluble PKC activity was not significantly altered. The results suggested strong attachment of the kinase to membranes following initiation of LTP, which might lead to kinase activation and/or facilitated access to substrates.

A predominant substrate for PKC in synaptic membranes is a protein of molecular weight 43–50 kD, which suffers an identity crisis and was given different names by different groups, including F1 (Akers and Routtenberg 1985), B-50 (Zwiers *et al.* 1982), and GAP-43 (Jacobson *et al.* 1986; Snipes *et al.* 1986). A selective increase in the *in vitro* phosphorylation of this protein (called here F1) was found at 5 min and at 1 h, but not at 1 min, after induction of LTP in the dentate gyrus of anaesthetized rats (Routtenberg and Lovinger 1985; Lovinger *et al.* 1986). In these experiments, the perforant path was potentiated in the intact animal and potentiation was monitored at 1–60 min after the high-frequency stimulation. After the appropriate interval, the animal was rapidly frozen in liquid nitrogen, followed by dissection of the dorsal portion of the dentate. A membrane fraction was prepared and subjected to phosphorylation with radiolabelled ATP. There was a correlation between the level of ^{32}P incorporation into F1 and the magnitude and persistence of potentiation previously recorded *in vivo*. No such correlation was observed in membranes prepared from animals that had received low-frequency stimulation during training and hence did not develop LTP.

Additional second-messenger cascades are involved

The involvement of Ca^{2+} in LTP raises the possibility that Ca^{2+}/calmodulin–dependent protein kinase (PKB) is activated in the process. H-7 and additional drugs which were used as PKC inhibitors to block LTP (see above), inhibit PKB as well, and hence their effect implicates also PKB in LTP (Malinow *et al.* 1988).

LTP was associated with altered lipid metabolism (Bar *et al.* 1984). One possibility is that glutamate receptors are coupled to phospholipase C (PL_C), which hydrolyses phospho-inositides to inositol phosphate and diacylglycerol, and the latter compounds activate intracellular Ca^{2+} mobilization and PKC respectively (Collingridge 1987; see also Fig. 3.2). An additional PKC-activating pathway possibly involves phospholipase A_2 (PL_{A_2}). The PL_{A_2} inhibitor, mepcarine, shortened the duration (but not the initial magnitude) of LTP in the dentate. Oleate, an activator of PKC, reversed the effect (Linden *et al.* 1987). Oleate can be generated *in vivo* by PL_{A_2}. The latter enzyme also releases arachidonic acid, metabolites of which may act as second messengers (Piomelli *et al.* 1987). In primary cultures of striatal neurons, glutamate and NMDA, but not kainate and quisqualate, activated the arachidonic acid cascade (Dumuis *et al.* 1988).

An increased release of putative transmitters is correlated with LTP

Some of the results mentioned so far on the role of Ca^{2+} in LTP (e.g. blockade by EGTA injection into pyramidal cells) favour a postsynaptic mechanism for LTP. The involvement of NMDA receptors corroborates a postsynaptic mechanism, although such receptors, with their coupled messenger cascades, might reside in presynaptic terminals as well. A presynaptic role for Ca^{2+} in increasing transmitter release should, however, also be considered. Skrede and Malthe-Sorenssen (1981) investigated the correlation between LTP and release. Hippocampal slices were first loaded with D-[^{3}H]aspartate. This compound was used as a marker for L-glutamate, and is also a marker for L-aspartate, itself a putative excitatory transmitter (Foster and Fagg 1984). Repetitive electrical stimulation of the Schaffer collaterals, under conditions that could result in LTP, led to an increased resting and stimulus-evoked efflux of radioactivity.

Bliss and his colleagues demonstrated that the increased release of putative excitatory neurotransmitters does indeed correlate with LTP *in vivo* (Dolphin *et al.* 1982; Bliss *et al.* 1986). They used a 'push–pull cannula' that enabled a combination of local microperfusion with electrophysiological recording. In brief, an outer, miniature 'pull' tube, with recording electrode(s) glued to its outer surface, was inserted into the appropriate region of the brain. A slightly longer inner 'push' tube was inserted into the 'pull' tube. The diameter of the tubes was designed to leave a narrow spacing between the outer wall of the inner tube and the inner wall of the outer tube. Appropriate solutions were pumped into the inner tube, perfused the area near its tip, and drawn out, via the space between the outer wall of the 'push' tube and the inner wall of the 'pull' tube, into a fraction collector for further analysis.

The dentate of anaesthetized rats was thus perfused with [^{3}H]glutamine, a precursor of glutamate, and [^{3}H]glutamate in the perfusate was subsequently determined. A prolonged increase was found in the amount of the putative excitatory transmitter following induction of LTP by tetanization of the perforant path. Dolphin *et al.* (1982) reasoned that since glutaminase, an enzyme converting glutamine to glutamate, is concentrated in presynaptic terminals, most of the glutamate thus detected was of a presynaptic origin. However, other sources, for example glia, were not excluded. Using the same technique and preparation, but relying on fluorometric detection of endogenous amino acids instead of metabolism of a radioactive precursor, Bliss *et al.* (1986) found that the release of glutamate and aspartate was significantly increased for 1–3 h after tetanization. No significant changes were detected in the release of glutamine, glycine, and γ-aminobutyric acid (GABA).

A similar phenomenon of increased transmitter release was found by Lynch *et al.* (1985), using a combination of *in vivo* and *in vitro* methods. Again LTP was induced by tetanizing the perforant path in anaesthetized rats. Forty-five minutes after training, the animals were sacrificed, and hippocampal slices prepared. These were preloaded with radiolabelled glutamate and aspartate. The slices were depolarized with K^+, and the release of the putative transmitters from the slices measured. The release of glutamate was significantly enhanced in slices from LTP-animals compared to those from non-LTP controls. The release of aspartate was also slightly enhanced.

Morphological alterations correlate with LTP

A number of investigators have reported changes in dendritic synapses in the hippocampal formation following input tetanization (Fifkova and Van Harreveld 1977; Lee *et al.* 1980; Fifkova and Anderson 1981; Desmond and Levy 1983; Chang and Greenough 1984; Andersen *et al.* 1987). Rapid onset and enduring increases were reported in the shape and number of both spine and shaft synapses. The subcellular loci and the nature of the morphological changes seemed to depend on the method of preparation of the tissue for ultrastructural analysis.

Molecular models for the initiation and maintenance of LTP

The information accumulated so far indicates that the mechanism of LTP is complex, and probably involves multiple processes operating on both the presynaptic and the postsynaptic sides of the potentiated synapse. As already noted, it is not at all clear that the physiological phenomena classified as LTP in different systems represent a unitary process. Since most of what we know about LTP emerges from studies of the hippocampal formation, the following model refers to this brain structure. The validity of this model, even if proven for the appropriate hippocampal pathways, must, therefore, be retested in every other system that displays LTP.

Initiation

The effects of depolarization and of NMDA-receptor ligands raise an interesting molecular model for the initiation of LTP in the hippocampal area CA1 and in the dentate gyrus (Collingridge 1985; Fig. 6.4). As noted above, LTP in the mossy fibre input to hippocampal area CA3 does not involve NMDA receptors. This model is of special interest since it predicts

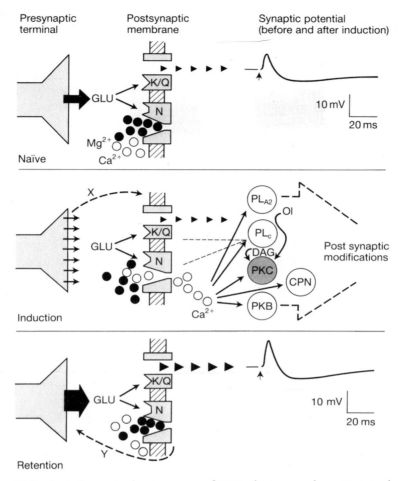

Fig. 6.4. A model for the induction and maintenance of LTP in hippocampal area CA1. In the postsynaptic membrane are two types of receptors for excitatory amino acids, the K/Q (kainate/quisqualate) and NMDA (N) receptors. When the Schaffer collaterals or commissural pathway are stimulated at low frequency, the transmitter (e.g. GLU, glutamate) binds to both types of receptors, but the N channel is blocked by Mg^{2+}. During tetanus, the massive depolarization removes the Mg^{2+} blockade and Ca^{2+} enters the cell, initiating, possibly together with second-messenger systems coupled to the K/Q and N receptors, cascades of molecular events that sustain LTP. These events in the postsynaptic terminal subsequently lead to an enhanced response to the transmitter. Activation of the NMDA-receptor complex is necessary but not sufficient for induction of a robust LTP; other factors (X), released by presynaptic activity, are necessary. Maintenance of LTP may depend also on changes that take place in the presynaptic terminal, and lead to a lasting increase in transmitter release. There might be factors that feed information back from the postsynaptic to the presynaptic cell, to initiate or sustain the postulated modifications in the latter. Second-messenger cascades depicted in the postsynaptic cell probably operate also in the presynaptic terminal, and are not marked on the diagram for the sake of simplicity. For example, the PKC cascade may play a prominent role in regulating transmitter release from the presynaptic terminal. CPN, calpain, a Ca^{2+}-activated protease; DAG, diacylglycerol; Ol, oleic acid; PKB, Ca^{2+} calmodulin-dependent protein kinase; PKC, Ca^{2+} phospholipid-dependent protein kinase; PL_{A2}, phospholipase A_2; PL_C, phospholipase C; X, unknown factor(s) released by the presynaptic terminal; Y, unknown factor(s) released by the postsynaptic terminal; arrowheads, net influx of monovalent cations. (Modified from Collingridge 1987; Collingridge and Bliss 1987.)

macroscopic properties of LTP induction. According to the model, the neuronal membrane has at least two types of receptor for excitatory amino acids, the kainate/quisqualate (K/Q) and the NMDA receptors. The endogenous transmitter, e.g. L-glutamate, interacts simultaneously with both receptors. Normal transmission is not mediated via the NMDA receptor, since its associated ion channel is blocked by physiological levels of extracellular Mg^{2+}. A brief, above-threshold depolarization removes the voltage-dependent Mg^{2+} blockade, leading to Ca^{2+} influx. This generates further depolarization, further removal of the Mg^{2+} block, and even greater Ca^{2+} influx, until the channel is blocked again following the regenerative repolarization of the membrane. The activated receptor itself is probably coupled to a second-messenger system. The latter, in concert with Ca^{2+}, initiates a cascade of molecular processes that sustain LTP.

Is activation of the NMDA-receptor complex sufficient for LTP induction? Direct application of NMDA to CA1 in a hippocampal slice, unaccompanied by afferent tetanization, produced only a short-lived (10–30 min) potentiation (Kauer *et al.* 1988*a*). Removal of Mg^{2+} from the medium, which was expected to open the NMDA channel, resulted in LTP in a hippocampal slice (Coan and Collingridge 1985; Huang *et al.* 1987), but the effect seemed to take place only in the presence of spontaneous afferent bursts (Neuman *et al.* 1987). Taken together with the APV-blocking experiments described above, these results indicate that activation of the NMDA-receptor complex is necessary but not sufficient for induction of a stable LTP. Additional factors, related to presynaptic stimulation, are required.

Even though activation of the NMDA-receptor complex may not be sufficient for triggering a robust LTP, the properties of this molecular complex can explain macroscopic characteristics of LTP induction. *Specificity* results from the requirement for transmitter occupancy of the NMDA receptor. *Co-operativity* and *associativity* result from the requirement for local depolarization to remove the Mg^{2+} blockade of the NMDA channel. Whether the kinetics of activation of the NMDA-receptor complex, combined with the kinetics of transmitter release and diffusion, also explains order dependency revealed in some experiments on associativity, is not yet clear.

Maintenance

Less is known about the maintenance of LTP than about its initiation. The data support multiple maintenance mechanisms that operate *in vivo* in a concerted manner in both the postsynaptic and the presynaptic terminals. Again, as is the case with initiation mechanisms, pharmacological activation of only one mechanism is not expected to result in a stable LTP (e.g. see earlier discussion on the proposed role of PKC).

On the *postsynaptic* side, Ca^{2+} influx and activation of Ca^{2+}-dependent protein kinase cascades and phospholipases could lead to multiple long-term changes in the cell. These include enhancement of the synaptic responses mediated via the quisqualate/kainate receptors (Kauer *et al.* 1988*b*; Muller *et al.* 1988*a*). The enhancement is selective; measurement of the EPSP in the presence of low Mg^{2+} or depolarization, to unmask the NMDA channel, revealed potentiation of only the non-NMDA component of the excitatory response. Note that this result cannot be explained by increased transmitter release, since enhanced release unaccompanied by postsynaptic receptor modification would be expected to result in potentiation of both components of the excitatory response. Nevertheless, the results do not exclude the possibility that transmitter release is enhanced in addition to the postsynaptic modification (and see below).

Another postsynaptic, Ca^{2+}-stimulated long-term change has been proposed, which involves activation of a Ca^{2+}-dependent protease, calpain. According to one facet of this model (Lynch and Baudry 1984), calpain exposes 'mute' glutamate receptors on the membrane, thus increasing the glutamate sensitivity. The assumption that LTP is accompanied by an increase in the number of glutamate receptors, based on results obtained from binding of radiolabelled glutamate (Baudry *et al.* 1980; Lynch *et al.* 1982), was, however, questioned (Sastry and Goh 1984; Lynch *et al.* 1985). Increased binding, if occurring at all, might reflect an alteration in the number of transport sites rather than conventional receptors (Pin *et al.* 1984). Nevertheless, other facets of the postulated Ca^{2+}-proteinase cascade(s) could be considered. These include the reorganization of the cytoskeleton which may affect dendritic morphology and synaptic efficacy (Lynch and Baudry 1984; Seubert *et al.* 1988; Siman and Noszek 1988), and activation of protein kinases, such as PKC (Melloni *et al.* 1985).

So much for the postsynaptic side. What about the *presynaptic* terminal? Assuming that LTP is initiated postsynaptically, and that enhanced transmitter release does contribute to its endurance, a mechanism should exist to relay information from the postsynaptic to the presynaptic side. One possibility is that the excitatory transmitter itself binds to presynaptic autoreceptors. APV depresses the evoked release of preloaded radiolabelled glutamate from hippocampal slices (but not *in vivo*), raising the possibility that NMDA receptors exist on the presynaptic membrane (Errington *et al.* 1987). Alternatively, factors diffusing back from the post- to the presynaptic side may do the job. These include proteins, secreted extracellularly during LTP (Duffy *et al.* 1981; Bliss *et al.* 1987); nitric oxide, or 'endothelium–derived relaxing factor' whose production may be triggered by activation of the NMDA receptor (Garthwaite *et al.* 1988); and lipids, which function as second messengers in the postsynaptic side (Dumuis *et al.* 1988). The intercellular messages might activate mechanisms in the presynaptic cell that enhance release. These could again be phosphorylation cascades, like the Ca^{2+}-phospholipids and Ca^{2+}-calmodulin cascades. PKC, for example, is well known to regulate transmitter release (Kaczmarek 1987). Its role in regulating enhanced release in LTP is, however, still unproven; increased synaptic efficacy, interpreted as LTP and induced by perfusion with a PKC activator, was not associated with a sustained increase in transmitter release (Aniksztejn *et al.* 1987). Finally, protein F1 may be involved in the mobilization of Ca^{2+} for release, or in presynaptic membrane recycling (De Graan *et al.* 1986; Scharma *et al.* 1986).

Additional factors contribute to the induction, maintenance, and modulation of LTP. These include the inhibitory transmitter GABA, which attenuates LTP (Wigstrom and Gustafsson 1983; Artola and Singer 1987), and neuromodulators such as norepinephrine, which facilitate LTP (Bliss *et al.* 1983; Hopkins and Johnston 1984; Robinson and Racine 1985; Gray and Johnston 1987). Especially intriguing is the possible role of peptides, such as analogues of the mast-cell degranulation peptide of bee venom (Cherubini *et al.* 1987). They might act as LTP-inducing cotransmitters, which are released only above a certain stimulus intensity threshold. The receptors for such peptides, and other cell-surface receptors that play a role in the induction and maintenance of LTP, may be unveiled by the usage of monoclonal antibodies that specifically block potentiation (Stanton *et al.* 1987).

The possible involvement of LTP in learning

Although LTP is a favourable candidate for a cellular mnemonic device in the mammalian brain, the data on its involvement in behavioural plasticity are still limited and mostly indirect. Four experimental approaches have been used to relate LTP to learning: (1) correlating the ability of an individual animal to sustain LTP with its ability to learn; (2) testing the effect of LTP on subsequent learning; (3) testing the effect of learning on subsequent LTP; (4) testing the effect of NMDA, or NMDA antagonists known to block LTP, on learning.

Correlation between the ability to sustain LTP and memory

The hippocampus is implicated in several kinds of memory, including spatial memory (see Chapters 11 and 14). Barnes (1979) subjected rats to a spatial memory task, which required the animal to locate an opening of a dark tunnel in a circular maze, using extramaze cues. The animals were chronically implanted with stimulating electrodes in their perforant path and recording electrodes in the dentate gyrus. A correlation was found in these animals between the ability to retain experimentally induced LTP and memory. In addition, the older rats displayed poorer retention in both LTP and behaviour. This hinted at some mechanism shared by behavioural memory and synaptic 'memory' measured by LTP, and the deterioration of this mechanism with age.

The effect of LTP on subsequent learning

In the above study by Barnes (1979), LTP impaired subsequent behaviour in old but not in young rats. The behaviour in question was spontaneous alternation in a maze: when a normal rat is placed in a T-maze, the probability is high that its second choice will differ from its first one. This spontaneous alternation was normal in old rats before LTP induction but disappeared after LTP induction. The interpretation was that spontaneous alternation involves synaptic modification in the hippocampus, that old rats have a limited pool of plastic synapses, and that LTP saturates this pool. In subsequent studies on the correlation between LTP and spatial memory in multiple-choice mazes, the induction of LTP in the perforant path impeded the acquisition of novel information in a working memory situation, and disrupted recently acquired information. LTP had no effect on well-established memory. In addition, LTP had no effect on the mere ability to use spatial cues for retrieval of food in a maze (McNaughton *et al.* 1986; these types of memory tests are further elaborated in Chapter 11). The above results, combined together, hinted at a competition for shared pathways between potentiation and learning; massive potentiation might 'mask' the acquisition of new data by brain system(s) that use the potentiated path.

A contrasting effect of LTP on subsequent learning was, however, reported in another, simpler learning paradigm. This was the classical conditioning of the eyelid and nictitating membrane (NM) reflex of the rabbit. (Again, the behavioural test is described in detail in Chapter 11.) The hippocampus was implicated in some aspects of experience-dependent modification of this reflex. Berger (1984) induced LTP in the perforant path of rabbits and later conditioned a subset of the animals to respond with the NM reflex to one, but not to another, tone. Animals in which LTP was induced learned faster than controls given only low-frequency perforant path stimulation. General enhancement of excitability by LTP was excluded, since the task was based on successful discrimination and not merely on enhancement of the response to the tone.

The effect of experience and learning on LTP

Skelton *et al.* (1987) monitored LTP in freely moving rats, trained to retrieve food in an operant conditioning situation. Recording of the response evoked in the dentate by perforant path stimulation was made every day after training. Enhanced population spike amplitudes were progressively detected in trained but not in free feeding animals. This differential effect persisted for at least 10 days after cessation of training.

In a study intended to test the relevance of LTP to a more 'diffused' experience, rats were implanted with a test electrode in the perforant path and a recording electrode in the dentate, and transferred from a dull rearing cage to a room with a variety of boxes and ramps ('complex environment'; Sharp *et al.* 1985). LTP-like changes were detected during the first days in the novel environment, which then subsided, but reappeared on transfer to another complex environment.

The modification by behavioural experience of some biochemical parameters implicated in LTP, was also considered as corroborating the suggestion that LTP-like mechanisms are involved in learning. This type of evidence is very indirect but worth mentioning. Classical conditioning of eyelid closure resulted in an increased number of [^{3}H]glutamate-binding sites in hippocampal synaptic membranes (Mamounas *et al.* 1984). In yet another study, Laroche *et al.* (1987) detected an increase in the K^+-evoked release of [^{3}H]glutamate from hippocampal slices prepared from rats which were classically conditioned to associate a tone with an electric shock.

Blockade of learning by an NMDA antagonist

Chronic intraventricular administration of APV impaired spatial learning in rats (Morris *et al.* 1986). The animals were placed in a large pool of opaque water and trained to locate an underwater platform. In contrast, the NMDA-receptor blocker did not affect performance in a visual discrimination task, in which the animals were trained to prefer a stable discriminable platform, placed visibly above the water, to a non-stable platform.

Ligands for the NMDA receptor affect experience-dependent brain development

Evidence for the involvement of the NMDA-receptor complex in brain development stems from studies of the visual system. In the primary visual cortex of a variety of mammals with overlapping binocular visual fields, or in the visual tectum of lower vertebrates, there are alternating columns of cells that respond only to input from one or the other eye. These ocular dominance columns are shaped during a critical developmental period, which can be disrupted by sensory deprivation or anatomical and pharmacological manipulations (Wiesel 1982; Singer 1987). During this critical period, the majority of neurons in the appropriate cortical layer lose their responsiveness to binocular stimulation. Monocular deprivation during the critical period abolishes the ability of most cells to respond to the deprived eye, and expands the columns receiving input from the normal eye. Wolf Singer and his colleagues (Kleinschmidt *et al.* 1987) found that local infusion of APV reduced this ocular dominance shift. In accordance with a postulated involvement of NMDA receptors in developmental plasticity, APV blocked the visual response of cortical neurons in the young cat more efficiently than in the mature cat (Tsumoto *et al.* 1987). The interpretation was that NMDA receptors are particularly important in the immature visual cortex, whereas non-NMDA excitatory amino-acid receptors play a greater role in mediating visual responses in the mature brain.

Similar conclusions on the role of NMDA receptors in developmental plasticity emerged from the studies of Martha Constantine-Paton and her colleagues on the development of retinotectal projections in the frog (Cline *et al.* 1987). In tadpoles with a third eye implanted adjacent to a normal eye, retinal ganglion cells from the normal and supernumerary eyes project to the same optic tectum and their terminals segregate into ocular dominance stripes. Chronic application of APV to the optic tectum resulted in a reversible desegregation of the eye-specific stripes, without blocking electrical activity or terminal morphology. In contrast, chronic application of NMDA produced stripes with sharper borders.

Taken together, the effect of NMDA-receptor blockade clearly indicates that this molecular complex participates in different aspects of the fine tuning of neuronal systems, ranging from experience-dependent brain maturation to learning. Interestingly, changes in an additional molecular component linked to LTP, were shown to correlate with development. This molecule is the PKC substrate F1, which has been implicated in axonal growth in normal development and in regeneration (Jacobson *et al.* 1986; Kalil and Skene 1986). Multiple mechanisms of long-term cellular change may, therefore, be shared by LTP and developmental neuronal plasticity.

The fact that drugs which block LTP disrupt some learning situations, does not imply that LTP itself is involved. Or is it? Although we have seen that tetanization of excitatory hippocampal pathways and behavioural plasticity do interact, models that implicate massive LTP in learning are problematic. It is difficult to envisage how massive tetanization, which rapidly saturates LTP in a major neuronal pathway, could encode specific and detailed information, unless the system has an extremely limited information capacity, which is of course not true for the brain. The cases in which massive LTP impeded subsequent learning are in line with this notion. In cases where facilitatory effects were detected, LTP was probably more relevant to general facilitation and attention than to learning and memory of specific information. In any case, further assessment of the role of LTP in the above cases requires information about the neuronal representations of the behavioural parameters, i.e. what role do the implicated potentiated pathways play in subserving the behavioural response?

But should we even seek a role in learning for massive LTP? The microscopic components, unveiled by the experimentally induced macroscopic phenomenon, are much more likely to play an interesting role in real life. (The same consideration applies to LTD; see, for example, Stanton *et al.* 1988; White *et al.* 1988.) The data of White *et al.* (1988), mentioned earlier, showing LTP in restricted, local dendritic domain, emphasize this point. LTP is initiated and sustained by multiple processes, operating in multiple subcellular loci. In the context of learning, some, or all, of these processes may operate in the relevant circuits. Some processes, i.e. activation of second-messenger and phosphorylation cascades, are, not surprisingly, similar to processes encountered in learning in invertebrates. A new type of molecular component does emerge here: the NMDA-receptor complex. It is a specialized switching device, activated only when an event (transmitter) occurs in a specific context (postsynaptic depolarization). It would be of great interest to search for additional molecular complexes with similar functional properties, since they are fit to play a role in learning.

7   *A different approach: neurogenetics*

Why *Drosophila?*

Although the studies of *Aplysia*, *Hermissenda*, or LTP in hippocampal cells all address different organisms and behaviours, they do share a common methodology, namely cellular analysis. A different, powerful, and complementary approach can also be used to unveil molecular mechanisms relevant to learning and memory. This approach is neurogenetics.

The rationale for the neurogenetic approach is straightforward, and was already hinted at in Chapter 1. Genes encode the macromolecules that make possible the acquisition, retention, and retrieval of neuronal information. Suppose some of these macromolecules play a much more significant role in learning and memory than in other physiological processes. In such a case, mutations in the appropriate genes should result in a relatively specific disruption of behavioural plasticity. Genetic, biochemical, anatomical, and electro-physiological comparison of normal and mutant organisms, differing in one gene only, may then reveal the identity of the defective gene product and its physiological role. This may eventually lead to the identification of identical or homologous mechanisms in other organisms as well.

The organism most suited for single-gene neurogenetic dissection is the fruit fly, *Drosophila melanogaster.* It has been the pet organism of geneticists since the turn of the century, was somehow overshadowed by micro-organisms in the 1960s and the 1970s, but has regained the favour of molecular biologists in recent years. Modern neurogenetic dissection of *Drosophila* behaviour was pioneered by Seymour Benzer (1967, 1973). The application of this approach to learning became possible only later, following the first demonstration that fruit flies can indeed learn (Quinn *et al.* 1974). Flies have long been considered by folk psychology as a token of stupidity. It became apparent, however, that similarly to many other organisms, flies too behave quite intelligently if only presented with the appropriate task.

What can *Drosophila* learn?

The first reliable learning test for *Drosophila* was an olfactory associative conditioning task which combined elements of both classical and instrumental conditioning. During training in this task, a group of flies was attracted by light into a test-tube containing one of two odorants, X or Y. Whenever entering the tube with odorant X, but not the tube with odorant Y, the flies received an electric shock. Thus, in this case, X was a shock-associated odorant, and Y a control odorant. During testing, the fraction of the flies avoiding odorant X (this time presented without a shock) was compared to that avoiding the control odorant Y. A reciprocal experiment, in which X was the control and Y the shock-associated odorant, was then conducted on a different group of flies from the same stock, to control for odour-bias and sensitization (Quinn *et al.* 1974). During testing, about two-thirds of the flies avoided the shock-associated odorant, whereas only one-third avoided the control odorant. The 'learning index', defined as the fraction of flies avoiding the shock-associated odorant minus the fraction avoiding the control odorant, was thus about 0.40. This test was used for the identification of almost all the learning mutants identified so far (see below).

A similar, more robust version of this test was later developed, which followed a strict classical conditioning procedure. Here, during training the flies were not given the choice of phototacting into the shock-tube or avoiding it (hence eliminating the instrumental component). Instead they were sequestered in the electrified odorant-containing chamber, and subjected to a controlled series of electric shocks in the presence of controlled air currents carrying the appropriate odorant (Tully and Quinn 1985; Fig. 7.1). During testing in this procedure, wild-type flies avoided almost completely the shock-associated odorant, but not the control odorant, yielding learning indices larger than 0.90. In the above tests, as in all the tests mentioned below, some strains of wild-type flies performed much better than others.

Drosophila were also shown to be capable of habituation, sensitization, classical conditioning, and operant conditioning in a variety of other paradigms, based on olfactory, visual, and mechanosensory cues (Dudai 1977; Menne and Spatz 1977; Aceves-Pina and Quinn 1979; Siegel and Hall 1979; Booker and Quinn 1981; Duerr and Quinn 1982; Gailey *et al.* 1982; Tempel *et al.* 1983; Kyriacou and Hall 1984; Heisenberg *et al.* 1985; Mariath 1985; Corfas and Dudai 1989).

What are learning mutants, and how are they isolated?

First, one must define a learning mutant. Many genetic aberrations are expected to interfere with learning and memory. However, most of these mutations might also be expected to interfere with development, metabolism, and sensory or motor capabilities. This makes it very difficult to prove that the genetic lesion indeed interferes with molecular systems necessary for learning *per se*. One must set, therefore, operational criteria for a learning mutant. The most straightforward criterion is the following: the mutation should affect acquisition and/or retention and/or retrieval, but the effect on these processes should be reasonably dissociated, under the appropriate experimental conditions, from the effects on other functions of the organism (Dudai 1988*a*).

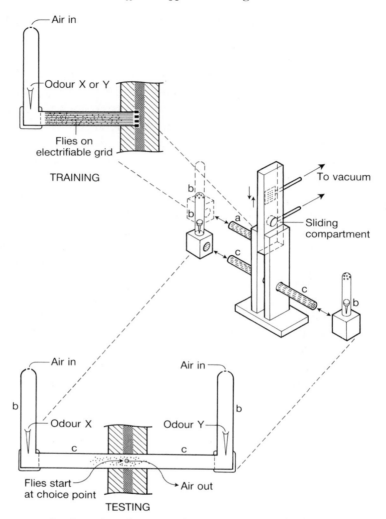

Fig. 7.1. An apparatus for classical conditioning of *Drosophila*. About 150 flies are sequestered in a training compartment (a) and trained by sequentially exposing them to two odorants, X or Y. The odorants are contained in small cups housed in tubes (b) and delivered in air current, drawn by vacuum. In training, the flies receive shock pulses during presentation of one but not the other odorant. In testing, the flies are transported via a sliding centre compartment to a T-maze choice point, between converging currents of X and Y, which flow through the arms of the maze (c). A learning index is calculated as the fraction of flies avoiding the shock-associated odour minus the fraction avoiding the control odour, averaged for two groups of flies, one trained to avoid X, the other to avoid Y. (After Tully and Quinn 1985.)

How could one isolate mutations that correspond to this definition? Two main methods are possible. In the first, mutants previously isolated by a variety of morphological, developmental, biochemical, or physiological criteria are submitted to learning tests. The ones that fail in the tests are considered putative learning mutants. Since these mutants have salient physiological abnormalities, the specificity of their deficits in learning is, a priori, questionable.

A second mutant isolation method is more straightforward. Here one treats flies with a mutagen, which penetrates germ cells and alters the DNA in the chromosomes. The progeny are screened for defects in learning. The mutagen is often a chemical, for example ethylmethane sulfonate (EMS), an alkylating agent that induces chemical changes in the DNA. Other classes of mutagens are X-rays, that break the DNA, or naturally occurring transposable DNA elements, such as the ones called p-elements, that 'jump' into genes and disrupt their structure and function. The physical nature of the defect in DNA, resulting from the mutagenesis, depends on the mutagen used. For example, under the appropriate conditions EMS induces mainly point mutations, i.e. replacement of one base in the DNA with another one, while X-rays induce mainly small chromosomal deletions. The frequency of mutations per chromosome is determined by the intensity of exposure to the mutagen.

Mutagens induce mutations practically at random in all four pairs of *Drosophila* chromosomes. Genetic breeding schemes are, however, available that preserve in the progeny of each mutagenized fly one mutated chromosome, while diluting the other mutated chromosomes with wild-type chromosomes. This generates lines of genetically identical individuals, each line carrying a different mutated chromosome. These lines of putative mutants are then submitted to a behavioural screen that determines their learning ability. If the flies fail to learn, they are further subjected to a battery of sensory and motor tests, to determine whether the poor learning might operationally be deemed specific.

Most learning mutants isolated so far were selected by quantifying the performance of the progeny of EMS-treated flies in the olfactory negative-reinforcement test described earlier. The mutagenesis and breeding scheme used in these studies selected specifically for mutations on the X (sex) chromosome. This was done because genetic manipulations of the X chromosome are easier than those of the other chromosomes. The first mutant so isolated was named *dunce* (*dnc*) (Dudai *et al.* 1976). Additional mutants, isolated in a similar manner, were *amnesiac* (*amn*) (Quinn *et al.* 1979) and mutants named after vegetables, such as *cabbage* (*cab*), *rutabaga* (*rut*), and *turnip* (*tur*) (Aceves-Pina *et al.* 1983). In addition, several mutations in the X or other chromosomes, which were originally identified because of morphological, neurological, or biochemical defects, were subsequently reported to interfere with learning in a relatively specific manner (Tempel *et al.* 1984; Heisenberg *et al.* 1985; Cowan and Siegel 1986).

The behaviour of learning mutants: general defects in acquisition and memory

Although learning mutants have lesions in different genes, and are not utterly identical in their behaviour, they do share a number of behavioural abnormalities.

Learning mutations affect performance in different learning tasks

The mutants *dnc*, *amn*, *cab*, *rut*, and *tur* are defective not only in their ability to associate odorants and electric shock, but also in their ability to learn other tasks. These include habituation and sensitization of a proboscis-extension feeding reflex (Duerr and Quinn 1982), learning of a preference for a sugar-associated odorant (Tempel *et al.* 1983), modification of leg position in response to an electric shock (Booker and Quinn 1981), and modification of whole-body orientation in response to an aversive heat source (Mariath 1985). These mutations thus affect processes that are common to different learning systems.

Learning mutations affect learning in the context of natural behaviours

Perhaps the most intriguing paradigms used in the study of *Drosophila* learning take advantage of behavioural modifications which normally occur in courtship. Courtship in *Drosophila* involves a complex series of behavioural acts, in which a variety of sensory signals are exchanged between the partners to prepare them for copulation. The major steps in courtship are innately programmed, but some responses are modifiable by experience. The experience may act as either a positive or negative reinforcement. An example of a positive reinforcement is the priming effect of the male's love song on the receptivity of females. An example of a negative reinforcement is the depression of the male's courtship following rejection by a fertilized female. The rejected male avoids any female, including receptive virgins, for some hours afterwards. Maturing males also learn to avoid the futile courtship of immature males (Hall 1986).

Several learning mutants of *Drosophila* do not behave normally in the above situations. For example, mutant males return quickly to court virgins, in spite of a recent disappointing encounter with an unco-operative fertilized female. These abnormalities were interpreted as learning disabilities (Siegel and Hall 1979; Gailey *et al.* 1982; Kyriacou and Hall 1984). The apparent importance of experience-dependent modification in courtship has led to the suggestion that learning had evolved in the fruit fly to contribute to efficient reproduction (Hall 1986). Indeed, the learning ability of various *Drosophila* wild-type strains is positively correlated with reproductive fitness (Hewitt *et al.* 1983).

Learning mutants isolated to date are affected primarily in early phases of memory formation

The most extensive studies on the effect of mutations on memory have utilized olfactory associative learning tests. In these tests, the mutations *amn*, *cab*, *dnc*, *rut*, and *tur* all appear to affect either acquisition, or memory, or both (Dudai 1979, 1983; Tempel *et al.* 1983; Tully and Quinn 1985; Dudai *et al.* 1987). Examples of the abnormalities in acquisition and memory are presented in Fig. 7.2. Similar results were obtained in non-associative learning.

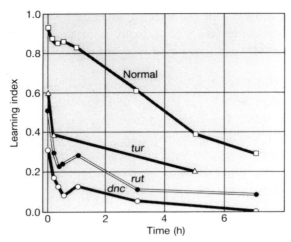

Fig. 7.2. Memory of normal and mutant *Drosophila*, trained in the classical conditioning task described in Fig. 7.1. Note the initial rapid memory decay in the mutants. (After Tully 1987.)

For example, in sensitization of a proboscis-extension feeding reflex, the memory of *dnc* and *rut* vanished at an abnormally rapid rate (Duerr and Quinn 1982). Habituation of a cleaning reflex, performed by the foreleg in response to mechanical stimulation of a thoracic bristle, was also abnormally short-lived in *rut* (Corfas and Dudai 1989).

Similarly to the situation in other organisms, in *Drosophila*, too, memory is multiphasic (memory phases in general are discussed at length in Chapter 8). In *Drosophila*, the multiplicity of memory phases can be inferred already from the kinetics of memory decay. Such analysis shows an early rapid decay phase of several minutes, followed by a slow decay phase, which may last for many hours (Dudai 1983; Tully and Quinn 1985). However, multiple memory phases are better demonstrated by the use of anaesthesia, especially cold anaesthesia (Quinn and Dudai 1976; Dudai 1977). If flies are placed on ice they pass out, but recover almost immediately after being returned to room temperature. Since this treatment does not have lingering deleterious effects on the flies' behaviour, it is preferred to other anaesthetic methods, such as exposure to high concentrations of N_2 or CO_2. When normal flies were cooled during the first minutes after training, their subsequent memory became feeble. In contrast, cooling the flies tens of minutes to a few hours after training did not significantly affect subsequent memory (Quinn and Dudai 1976; Dudai 1977; Tempel *et al.* 1983; Dudai *et al.* 1987). The anaesthesia-sensitive memory phase was operationally termed 'short-term memory', and the anaesthesia-resistant phase was termed 'long-term memory'.

The memory of the above-mentioned mutants decayed abnormally rapidly immediately after training, but only slowly afterwards, at an essentially normal rate (Fig. 7.2; Dudai 1983; Tully and Quinn 1985; Dudai *et al.* 1988). This indicated that the mutations affect an early rather than a late memory phase. A clearer indication for this dichotomy was found by testing the effect of post-training cold anaesthesia on the memory of the mutant *rut*. That part of memory which could still be detected in *rut* a few hours after training in the olfactory classical conditioning task was normally immune to cold anaesthesia, implying that the *rut* gene product is not crucial for maintaining memory in a long-term store (Dudai *et al.* 1988).

Results obtained in different types of learning and memory tasks indicate, nevertheless, that the effects of the mutations on acquisition and retention depend on the conditioning paradigm used (Dudai 1988*a*). Data obtained from visual conditioning illustrate this point. Training was performed in this case by exposing the flies to a colour in the presence of intense mechanical vibration, which served as a negative reinforcer. In contrast to olfactory conditioning, where the memory of *amn*, *dnc*, *rut*, and *tur* was abnormally weak, here the mutants displayed defective acquisition, but memory was normal when tested several hours after training (Folkers 1982).

The biochemical defects in *dunce, rutabaga, turnip,* and some other learning mutants: lesions in second–messenger cascades

Information is currently available on molecular defects in several learning mutants. In all these cases, the mutations affect second-messenger cascades.

dnc codes for an isozyme of cAMP phosphodiesterase

Several alleles of *dnc* are known, and all of them reduce or abolish the activity of an isozyme of cAMP phosphodiesterase (Byers *et al.* 1981; Davis and Kiger 1981; Davis and Kauvar 1984). The dnc^+ gene (i.e. the normal *dnc* gene) was cloned and the identity of the product as cAMP-phosphodiesterase was verified (Chen *et al.* 1986, 1987; Davis and Davidson 1986). The gene is very complex. It encodes at least six overlapping mRNA molecules, which are developmentally regulated. This raises the possibility that only one, or part, of the *dnc* mRNAs is/are important in learning, whereas other mRNAs encoded by the same gene play a role in other physiological processes, for example in development. Moreover, the DNA coding for these mRNAs spans a very large segment of the X chromosome, larger than 100 kb, and at least two other genes reside in a large intron that separates the *dnc* exons (Chen *et al.* 1987).

rut affects a subpopulation, or a functional state, of adenylate cyclase

The mutation *rut* decreases the level of adenylate cyclase in *Drosophila* by about 25 per cent (Dudai *et al.* 1983, 1984; Livingstone *et al.* 1984). The mutation abolishes the stimulation of adenylate cyclase by Ca^{2+}/calmodulin (Dudai and Zvi 1984; Livingstone *et al.* 1984; Livingstone 1985; Yovell *et al.* 1988). Additional properties, including responsiveness to Mg^{2+}

and to the activator forskolin, are also affected (Dudai 1985; Dudai *et al.* 1985). The biochemical data suggest that the defect is intimately associated with the catalytic subunit of the enzyme, and especially with the regulation of the catalytic activity by intracellular ligands.

tur affects a common denominator of several second-messenger cascades

The *tur* mutation drastically reduces the activity of membrane-associated protein kinase C (PKC), and of two membrane-associated phosphoproteins which are phosphorylated in a Ca^{2+}- and phospholipid-dependent manner (Smith *et al.* 1986). The *tur* mutation also reduces the activation of adenylate cyclase by guanyl nucleotides (Smith *et al.* 1986). In addition, the mutation reduces the level of high-affinity receptors for aminergic neurotransmitters (Aceves-Pina *et al.* 1983). The affinity of protein kinase A (PKA) for cAMP is lower than normal (Buxbaum and Dudai 1988). Most of these defects might be explained by assuming that the *tur* mutation damages a protein that regulates the activity of several second-messenger systems, for example a guanyl nucleotide-binding protein. It is not yet clear whether all the biochemical and behavioural defects in the mutant map to a single chromosomal locus.

Lesions in aminergic transmission are correlated with defects in learning

The enzyme dopa decarboxylase is required in *Drosophila* for synthesis of the aminergic neurotransmitters and neuromodulators serotonin and dopamine. It has been reported that mutations in *Ddc*, the locus that codes for this enzyme, affect acquisition in an olfactory conditioning task (Tempel *et al.* 1984). It is not clear, however, how general the effect on learning is (Tully 1987). Another mutation, *ebony*, has abnormally high dopamine and was reported to reduce performance in an olfactory conditioning paradigm (Dudai 1977). Again, it is not yet clear how general is the effect of this mutation on learning.

Why are the mutants stupid?

Standing alone, the information obtained from studying *Drosophila* mutants does not provide a concrete mechanistic model for learning and short-term memory. Such a model requires cellular analysis of a neuronal system that subserves learning. This is not yet feasible in *Drosophila*, because its nervous system is complex and very compact. The main heuristic conclusion that can be drawn is that perturbations in the cAMP cascade, and possibly in other second-messenger cascades, disrupt learning and short-term memory. Learning is disrupted when the level of the second messenger is chronically decreased (e.g. *Ddc*, *rut*) or increased (*dnc*). Metabolic cascades usually operate in optimal, not in peak performance; this means that there is an optimal balance in the activities of various components of these cascades, and perturbances, even if they increase the metabolic activity of one component, may end up by decreasing the overall output of the system. Quantitative kinetic modelling of the operation of the cAMP cascade indeed shows that abnormally increasing the level of cAMP results in lower efficiency in the activation of PKA (Buxbaum and Dudai 1989).

It is plausible to assume that the biochemical processes that are defective in learning mutants are obligatory for the ongoing, real-time molecular operation of learning systems. An alternative explanation should, however, be considered. It is possible that the molecular lesions cause developmental defects, resulting in neuroanatomical aberrations, e.g. altered neuronal wiring, and that these in turn cause learning abnormalities. Determination of the time window during which the normal products of 'learning genes' must function, is thus crucial for understanding how the mutations affect learning.

It was indeed reported that in both *dnc* and *rut*, brain structures called 'mushroom bodies', which are considered as 'high-level' sensory integration centres, develop abnormally. Neurological mutants with abnormal mushroom bodies have defective olfactory learning in spite of being able to respond to olfactory cues (Heisenberg *et al.* 1985). In normal flies, there is a transient increase in the number of neuronal fibres in the mushroom bodies during the first weeks of adult life. The absolute number of fibres during this period depends on the previous experience of the fly; flies reared in social isolation, or deprived of olfactory experience by masking of their antennae, have fewer fibres in their mushroom bodies (Technau 1984). In *dnc* flies, the number of fibres was initially higher than in the wild type, but it decreased steeply during the first weeks of life until it reached the level characteristic of socially deprived normal flies. In *rut* flies, the number of fibres was low at eclosion and did not significantly change during the first weeks of life. In both mutants, isolation had no effect on fibre number (Balling *et al.* 1987).

It cannot yet be concluded whether the abnormal development of mushroom bodies is at all related to learning, and if so, whether it is the cause of the learning defect, or its consequence. An argument in favour of the latter possibility is the observation that *dnc* and *rut* disrupt learning in different sensory modalities and in paradigms which can normally be performed by a headless fly. The mutants thus affect basic behavioural plasticity processes independent of the presence of the mushroom bodies.

Several observations favour the hypothesis that the defective biochemical processes in *dnc* and *rut* (and possibly also in other conditioning mutants) play a role in the ongoing molecular operation of learning systems. Under certain conditions, initial learning scores of the mutants approach normal values, indicating a defect in memory rather than in acquisition. The neuronal wiring required for acquisition is, therefore, likely to be intact. In addition, the short-term treatment of normal flies with drugs that alter cAMP metabolism, such as caffeine, disrupts learning (Folkers and Spatz 1984; Dudai *et al.* 1986). It is unlikely that such treatment induces morphological changes in the nervous system; these results actually corroborate the assumption that normal ambient levels of cAMP are required for normal learning.

This is not to say that second-messenger cascades do not play a prominent role in development; they surely do (reviewed in Dudai 1987). Mutations that affect these cascades may, therefore, affect development, too. Indeed, some alleles of *dnc* drastically reduce female fertility, and *rut* partially suppresses this effect (Livingstone *et al.* 1984). The crucial point is that the effects on development, when they do exist, can be dissociated from the effects on learning. This is the case for *dnc*, where suppressor mutations, including *rut*, suppress female sterility but do not restore normal learning (Byers 1980).

The notion that the ongoing activity of second-messenger cascades is critical in learning is in accordance with the cellular and molecular models presented in the previous chapters. The *Drosophila* data do not suffice to refute any of these models. They fit best into the *Aplysia* data, implicating the cAMP cascade in acquisition and retention. They also corroborate the general conclusion from *Aplysia*, that basic molecular and cellular building blocks are shared by non-associative and associative learning. A piece of information especially relevant in this context is provided by *rut*. This mutant implicates the Ca^{2+}/ calmodulin stimulation of adenylate cyclase in learning. This is exactly the property of adenylate cyclase which was deemed crucial for the molecular convergence of the CS and US in *Aplysia* (see Chapter 4). Note, however, that *rut* is defective in non-associative learning as well. The *Aplysia* model does not predict a role for Ca^{2+}/calmodulin stimulation of adenylate cyclase in habituation and sensitization. Other defects in adenylate cyclase, such as reduced non-Ca^{2+} stimulated activity, or defects in Mg^{2+} and forskolin binding sites, may contribute to the defective non-associative learning in *rut*.

Conclusions from the neurogenetic dissection of learning

● Taken together, the data obtained from the neurogenetic dissection of *Drosophila* learning indicate that components of second-messenger cascades play relatively specific roles in elementary learning and short-term memory (Fig. 7.3). These results corroborate the data obtained from the cellular analysis of learning in molluscs, and in some respects also the data from mammalian LTP. The behaviour of the mutants, excluding learning and memory, is reasonably normal. This suggests that the components of second-messenger cascades that are affected in the mutants do not play critical roles in 'house keeping' mediation of information, but rather in *modulation*.

It should come as no surprise to find that lesions in components of ubiquitous cellular systems, such as the cAMP and the Ca^{2+}-phospholipid cascades, exert relatively specific effects on behavioural plasticity. These cascades are heterogeneous, containing subtypes of enzymes, receptors, regulators, and substrates (Dudai 1987; Hunter 1987; Nishizuka 1988). Delicate microdissection by the use of mutations is bound to unravel such intricate complexity, and to assign specific physiological roles to specific components of the cascades.

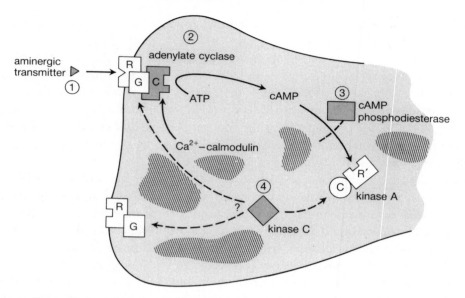

Fig. 7.3. A scheme of a hypothetical synapse, depicting the established and presumptive molecular lesions in the mutants *Ddc* (1), *rut* (2), *dnc* (3), and *tur* (4). C, a catalytic subunit; G, a GTP-binding protein; R, a receptor for neurotransmitter or neuromodulator; R′, the regulatory subunit of PKA.

The *rut* and *tur* mutations affect cross-talk between second-messenger cascades. This suggests that such a cross-talk is especially critical in acquisition and short-term memory. It is also of interest to note that although the mutations affect learning in different paradigms and sensory modalities, the exact effect on acquisition and/or short-term memory depends on the paradigm and modality. This indicates that elementary building blocks are shared by different learning systems, but the exact role of these building blocks differs with the system. Furthermore, the fact that none of the available mutations completely abolishes learning, suggests that there is no exclusive mechanism of learning and short-term memory, but rather that several acquisition and retention processes always operate in parallel.

Finally, one interesting advantage of the neurogenetic analysis is its ability to address the phylogenetic basis of learning. Mutations can pinpoint developmental and functional links between biological processes that at first do not seem related. For example, certain mutations that affect biological clocks and disrupt circadian rhythms also disrupt some learning tasks (Hall 1986). Is this because biological clocks play a role in learning or because the two processes share molecular building blocks, which fulfil different cellular roles in each case? Also, genetic analysis may tell us why learning had developed during evolution (see also Chapter 1). As has already been mentioned, in *Drosophila* the quest for more efficient mating might have driven the genus to develop its intelligence.

8   *On the life span of molecules and memories*

Where is molecular memory retained?

All the studies mentioned so far have implicated second-messenger cascades and post-translational modification of proteins in learning. Some studies went further and suggested how cellular memory is read out, but none of these studies identified the exact locus where molecular memory is retained. The following are selected possibilities. Transmitter molecules may linger after release and continuously activate postsynaptic receptors. Receptors may shift into a stable conformational state with higher affinity for the transmitter. Enzymes which synthesize second messengers in response to receptor activation, and protein kinases stimulated by these second messengers, may switch into a persistently active state. The concentration of the second messenger may decay only slowly into its pre-stimulation level. And, finally, proteins phosphorylated by the kinase may remain phosphorylated, and hence activated, long after the trans-synaptic stimulus that had triggered their modification vanished.

In those few cases where the site of molecular retention was experimentally investigated, the answers were equivocal. Take *Aplysia* (see Chapter 4): the phosphorylated, closed S channel, central to memory readout, appeared to be rapidly dephosphorylated and re-opened. It was therefore reasoned that molecular memory is retained upstream of the channel along the cAMP cascade. Persistent activation of adenylate cyclase, the enzyme synthesizing cAMP, was proposed as a retention mechanism. However, this was not supported by subsequent *in vitro* analysis of the time course of cyclase activation by transmitter transients. Furthermore, no change was detected in the activity of cAMP-phosphodiesterase, the enzyme which degrades cAMP. All this must still be reconciled with the observation that cAMP levels remain elevated for many minutes after serotonin application. In the meantime, we return to square one: how does the cell remember to rephosphorylate persistently the S channel? A possible mechanism, which has recently gained favour with experimenters and theoreticians alike, is persistent activation of protein kinase.

Protein kinases as short–term memory devices

Protein kinases are activated by second messengers such as cAMP, Ca^{2+}, and lipids (see Fig. 3.2). Any activation of the kinase that outlasts a transient increase in the level of the appropriate second messenger can subserve memory (Schwartz and Greenberg 1987). Two types of mechanisms could be considered for this purpose. (a) A second messenger-induced modification of the kinase increases its sensitivity to the second messenger, so that low concentrations of the latter, previously ineffective, now suffice to cause dramatic activation of the enzyme. (b) The active kinase becomes independent of its second-messenger activator. Such kinase molecules are dubbed 'autonomous'.

Increasing the sensitivity of a protein kinase for its second-messenger activators

This type of mechanism has been especially considered for the cAMP-dependent protein kinase (PKA). As was noted earlier (see Chapter 4), the PKA complex is composed of two types of subunits—a catalytic subunit, C, and a regulatory subunit, R (Fig. 8.1(a)). There is heterogeneity in R, and possibly also in C (Nestler and Greengard 1984; Dudai 1987), but this heterogeneity is neglected here for the sake of simplicity. R inhibits C; cAMP binds to R, dissociates it from C, and hence activates the kinase. R is intramolecularly phosphorylated by C, and this lowers its affinity for C.

The inhibition of C by R implies that any long-term change in the availability of R results in long-term modification of the kinase activity. Thus, a decrease in the ratio of R to C releases more free, active C in the presence of low concentrations of cAMP. Two types of processes were suggested for long-term alterations in the availability of R (Fig. 8.1(a)). One possibility is that the phosphorylated, dissociated R is dephosphorylated only slowly, and therefore does not readily reassociate with C (Buxbaum and Dudai 1989). Another possibility is that the ratio of total R to C decreases, for example due to proteolysis activated by the cAMP cascade. A change in the R/C ratio was indeed detected in sensory neurons subserving the GSW reflex in *Aplysia* 24 hours after sensitization (Fig. 8.1(b),(c); Greenberg *et al.* 1987).

Persistent activation of a kinase may also be due to a subcellular translocation that exposes the enzyme to activators and co-factors. An example of the role of translocation in the generation of a non-autonomous (or partially autonomous), highly sensitive kinase, is provided by the Ca^{2+} phospholipid-dependent kinase system (PKC). Following prolonged exposure to the hormone angiotensin II, adrenal glomerulosa cells became more responsive to subsequent exposure to the same hormone, or to a drug that increases intracellular Ca^{2+} (Barrett *et al.* 1986). This increased sensitivity was attributed to translocation of PKC from the cytosol to the membrane, resulting in sustained association with membrane phospholipids, or covalent modification of the kinase in the membrane. Note that PKC translocation and activation were also correlated with LTP (see Chapter 6).

Generation of autonomous protein kinases

An extensively studied example of kinase autonomy is the activation of type II $Ca^{2+}/$ calmodulin-dependent protein kinase (PKB) by autophosphorylation (Saitoh and Schwartz

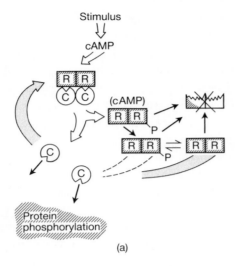

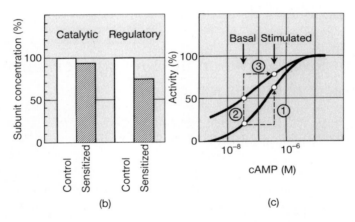

Fig. 8.1. Protein kinases may store memory by remaining active after the level of their second-messenger activators subsides. Possible mechanisms for persistent activation of cAMP-dependent protein kinase (PKA) are shown here. (a): The enzyme is composed of a dimer of regulatory subunits (R) associated with two monomers of the catalytic subunit (C). R inhibits C and cAMP relieves that inhibition by dissociating R from C. Therefore, any change in the effective ratio R/C modifies the kinase activity. Such change may result, for example, from phosphorylation of R which lowers its affinity for C, or from proteolysis of R. (b): Twenty-four hours following sensitization training of the GSW reflex in *Aplysia*, the ratio R/C is changed. (Adapted from Greenberg *et al.* 1987) (c): Although the change in R in (b) is relatively small, it can lead to a dramatic increase in the availability of free C and hence in kinase activity. This graph depicts the dependence of free C (assumed to be proportional to kinase activity) on cAMP, as a function of the R/C ratio. The lower curve is for a normalized ratio of 1, and the upper curve for a ratio of 0.75. Following short-term sensitization, which does not change the ratio R/C, stimulation by cAMP results in the above-basal activation by pathway 1. A reduction of 25 per cent in R (as seen after long-term training in (b)) clearly leads to enhanced activity (pathway 2). The system can still show further response to new pulses of the sensitizing transmitter (pathway 3). (After Greenberg *et al.* 1987.)

1985; Lai *et al.* 1986; Lou *et al.* 1986; Miller and Kennedy 1986; Schworer *et al.* 1986; Hashimoto *et al.* 1987; Miller *et al.* 1988). The activity of the autophosphorylated enzyme becomes partially, or even completely, independent of Ca^{2+}; the requirement for the cation and the sensitivity to it are restored following dephosphorylation. Autophosphorylation of PKB was specifically correlated with two types of experience-dependent modifications: one is visual adaptation in *Drosophila* (Willmund *et al.* 1986); the other is kindling, a lasting predisposition to epileptiform convulsions observed in the vertebrate brain following repeated focal electrical stimulation (Goldenring *et al.* 1986). (It should also be noted that persistent activation of PKB, by an unknown mechanism, probably not involving autophosphorylation, was suggested to occur in LTP; see Malinow *et al.* 1988 and Chapter 6.)

Other mechanisms that may produce an autonomous kinase are subcellular translocations and proteolytic cleavage (see above). Proteolysis, in contrast to autophosphorylation and translocation, is irreversible. It was reported to generate an autonomous form of PKC (Kishimoto *et al.* 1983; Melloni *et al.* 1986). Interestingly, the protease was Ca^{2+} dependent, so that the second messenger which activated the kinase also activated its modification. Again, Ca^{2+}-dependent proteolysis and activation of PKC was proposed to occur in LTP (Malinow *et al.* 1988).

Molecular turnover: a limit on the retention of cellular changes by post-translational modifications

Memories based solely on post-translational modification of proteins, such as the ones described so far, are expected to deteriorate quite rapidly. This is due to two processes. First, covalent modifications are often readily reversible. Second, even an irreversibly modified protein molecule has a limited life span, ranging from minutes to weeks. Yet we know that memories may last for months, years, even a lifetime. Such memories cannot rely on the stability of individual copies of covalently modified proteins. They must be immune to molecular turnover.

Two major types of mechanism that might grant the cell immunity to molecular turnover, are described below (Fig. 8.2). These are activation of *regenerative molecular cascades*, and *alteration of gene expression*. Both mechanisms may be turned on by the same second-messenger systems that operate in short-term memory.

A hypothetical mechanism of immunity to molecular turnover: regenerative molecular cascades

In regenerative molecular cascades, demodified molecules are remodified, and newly synthesized copies are post-translationally modified. A model suggested by Crick (1984*a*)

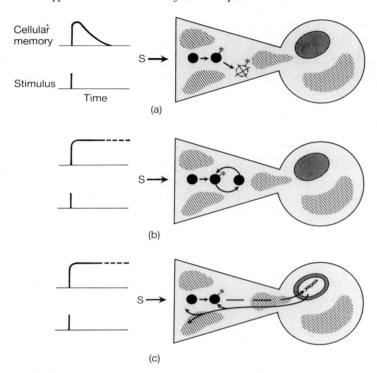

Fig. 8.2. Three types of models for elementary mechanisms of cellular memory, with short (a) and long (b, c) duration. (a): Post-translational modification of protein(s) (denoted by asterisk) alter neuronal function. Molecular turnover, however, erases the memory. The duration of memory does not, therefore, outlast, even under optimal conditions, the life span of the modified protein molecule. This type of mechanism was illustrated in earlier chapters, e.g. in *Aplysia*. (b): A regenerative molecular cycle, which replenishes copies of the modified protein, can grant immunity to molecular turnover. Hypothetical examples are provided in the text. (c): Another class of models proposes that gene expression is modulated to generate new proteins, which alter neuronal activity. S, stimulus. (For further discussion, see text.)

illustrates such a cascade. Suppose a protein, which we take the liberty of naming dihydromemorase, is essential for controlling synaptic efficacy. Dihydomemorase is a dimer, composed initially of two identical unmodified monomers. Each monomer can be replaced independently of the other by the cellular mechanisms responsible for molecular turnover. In addition, each monomer can be post-translationally modified in the dimer, for example by phosphorylation. (Other covalent modifications, e.g. methylation, would do as well, but we choose phosphorylation because of its established role in short-term memory.) We shall denote the unmodified monomer as (−), and the modified monomer as (+). Suppose that when the two monomers are phosphorylated [(+,+)] dihydromemorase is active, whereas when the monomers are non-modified, the protein is non-active.

Now come two additional, central assumptions. (1) There is a hypothetical dihydro-memorase kinase, which phosphorylates one monomer only if the other monomer is modified. Thus, it turns $(+,-)$ into $(+,+)$ but does not act on $(-,-)$. (2) The transient stimulus which modifies synaptic efficacy in learning causes phosphorylation of both monomers, so that $(-,-)$ turns into $(+,+)$. It is easy to see that this system is theoretically capable of maintaining the modified state of dihydromemorase indefinitely. This is because whenever molecular turnover turns $(+,+)$ into $(-,+)$, the kinase restores it into the $(+,+)$ form. Memory persists, in spite of the continuous turnover of monomers.

A conceptually related mechanism was suggested by Lisman (1985). In this model, the postulated system is a bistable molecular switch constructed of a protein kinase, kinase-1, and a protein phosphatase. Kinase-1 is crucial for controlling neuronal function. It exists in either an active or an inactive state. The transition from the inactive to the active state is catalysed by another protein kinase, kinase-2, or by intermolecular phosphorylation catalysed by another, active molecule of kinase-1. The phosphatase is capable of dephosphorylating kinase-1, rendering it inactive again.

Initially, kinase-1 is inactive, i.e. the switch is turned 'off'. In learning, a trans-synaptic stimulus triggers activation of kinase-2. The latter phosphorylates kinase-1, which in turn phosphorylates other molecules of kinase-1. Given the appropriate initial level of phosphorylation by kinase-2, and the appropriate ratio of phosphatase to kinase-1 activity, the switch will turn 'on', and will remain so indefinitely, in spite of the molecular turnover. Extinction would require blockade of kinase-1 (for example, by an endogenous inhibitor), or an increase in the ratio of phosphatase to kinase activity. Lisman (1985) simulated his model on paper, and demonstrated that under the appropriate conditions, a pulse activating kinase-2 can indeed set the system into a regenerative cycle, placing the switch in a stable 'on' position.

Central to the proposed bistable molecular switch is the ability of the kinase to undergo intermolecular autophosphorylation. As yet there is no experimental evidence that such an autocatalytic process plays a role in memory, or even in neuronal function in general. In contrast, the autophosphorylation discussed earlier in the context of short-term memory models, and for which there is experimental evidence, was intramolecular, and hence its effect is still limited by the life span of the protein.

Another possible mechanism of immunity to molecular turnover: alteration in gene expression

A different mechanism for generating enduring memory, in spite of the limited life span of proteins, might be based on alteration in gene expression, resulting in a long-term or permanent modification in protein synthesis and hence in neuronal properties. Ample evidence indicates that the expression of genes in the mature organism is influenced by the environment. It is, therefore, tempting to consider a role in memory for alteration in gene expression.

Before proceeding to discuss experimental evidence for the macromolecular synthesis hypothesis, we should clarify what role the newly synthesized proteins could have. We assume that these proteins modify properties of neuronal systems, by stabilizing and extending mechanisms used in short-term memory, or by initiating new types of cellular change. Note that the neuronal loci subserving short- and long-term memory are expected to overlap, at least at first, but need not be identical. An example from *Aplysia* illustrates this: short-term habituation of the GSW reflex was portrayed as homosynaptic depression (see p. 60), but long-term habituation involves additional central neurons and heterosynaptic depression (Montarolo *et al.* 1988). However, regardless of which cellular loci are modified in short- and long-term memory, one point is clear: macromolecules subserve changes in the representational properties of neuronal circuits, but do not directly encode a specific representation. The latter is encoded in the connectivity and molar activity of the circuit. In other words, similar molecular changes may subserve different modifications in different internal representations, depending on the *cellular context* in which these molecular changes take place.

At this point, we should be alerted to a radically different conceptual framework that could entertain the macromolecular synthesis hypothesis. The idea is that neuronal memory, similar to genetic memory, is a macromolecular code. In other words, here macromolecules are specific internal representations. This idea emerged against the background of much excitement about breaking the genetic code; it gained much publicity and evoked much debate in the 1960s and 1970s, but has since declined and is currently mainly of historical interest. The 'macromolecular code', or the 'macromolecular representation' hypothesis, led to an intensive search for macromolecules that encode specific memories; it culminated in claims that such memories can indeed be transferred by brain extracts and peptides from one individual to another, be it in worms, fish, or rodents (Ungar 1970). These claims were not substantiated by data from other groups (Byrne *et al.* 1966). It is not unlikely that successful 'memory transfers' were due to the transfer of hormones, controlling general arousal and fear, from trained to naïve individuals.

The macromolecular representation hypothesis made some early and partial contributions to the popularity of other types of experiments, which tested the correlation of alterations in nucleic acid with neural activity and behavioural experience (Hyden and Egyhazi 1962; Hyden and Lange 1965; Chapouthier 1983), and the effect of macromolecular synthesis inhibitors on learning and memory (see below). The experiments on the correlation of the chemistry of nucleic acid with behaviour resulted in reports of qualitative experience-dependent alterations in the composition of RNA in the brain, but the interpretation was not obvious. The experiments on the effect of macromolecular synthesis inhibition, further discussed below, did propose a role for RNA and protein synthesis in memory. However, the interpretation of the data did not require the assumption that specific memories are encoded in macromolecules. As a matter of fact, most researchers engaged in the latter types of experiments did not operate at all within the conceptual framework of 'macromolecular representation'. Rather, the conceptual framework that guided them was the old-time analogy, which keeps recurring in our discussion, between learning and growth.

Evidence from systemic pharmacology

The first indication that RNA and protein synthesis is needed for long-term memory came from experiments in which drugs were administered, either systemically or intracerebrally, in peri-training periods (Dingman and Sporn 1961; Flexner *et al.* 1963; Barondes and Jarvik 1964; Agranoff *et al.* 1965; Barondes and Cohen 1966; later studies are reviewed in Rainbow 1979; Davis and Squire 1984). The results of these experiments showed that the inhibitors block long-term memory if administered during training or immediately afterwards. A very effective inhibition of synthesis (> 90–95 per cent) was required to obtain the effect. The toxins were progressively less effective if given at progressively longer intervals after training. Acquisition and short-term memory were usually not disrupted.

The interpretation of systemic or even organ-directed pharmacological experiments is complicated. The metabolic specificity of the administered toxin, and the relevance of its phenotypic effect to its suspected primary molecular receptor, are questionable. For example, several RNA and protein synthesis inhibitors used in memory studies exert general toxic effects, induce abnormal electrical activity in the brain, and interfere with aminergic transmitter systems (Cohen *et al.* 1966; Flexner and Goodman 1975; Wetzel *et al.* 1976). However, taken together, the results obtained with different drugs, conditions, and species did form an internally consistent body of evidence that implicated protein synthesis in long- but not in short-term memory (Davis and Squire 1984). But, even so, the identity of the cellular targets of the drugs remained unclear. The effects on memory, and hence presumably on neurons, could have been secondary to the action of the toxins on non-neuronal targets.

● The afore-mentioned problems have severely complicated the interpretation of the data and, following an initial surge of observations, somewhat diminished the enthusiasm about the systemic pharmacological approach. Renewed interest emerged only in recent years following two developments: (1) the findings that the environment can modulate gene expression in the adult nervous system; and (2) the development of methodologies that made it possible to reinvestigate the effect of inhibitors of RNA and protein synthesis on identified neurons in circuits that subserve memories. Let us first address the latter development.

Evidence from cellular studies

Following the identification of molecular and cellular correlates of acquisition and short-term memory in molluscs, several research groups have turned their attention to the mechanisms of long-term memory in these animals. The findings confirmed and extended the earlier systemic pharmacological studies on the role of RNA and protein synthesis in long-term memory. The most extensive data stem from the investigation of relatively long-term sensitization of defensive reflexes in *Aplysia* (see Chapter 4).

When *Aplysia* is sensitized by a few shock trains delivered every day for four successive days, its GSW or tail withdrawal reflexes remain sensitized for at least 1 week (Frost *et al.* 1985*a*). Cellular analysis reveals that the same sensory–to–motor neuron synapses which are modified in short-term sensitization are also facilitated in long-term sensitization (Frost *et al.* 1985*a*; Scholz and Byrne 1987). Moreover, similarly to the short-term modification, relatively long-term sensitization is associated with increased excitability (Dale *et al.* 1987) and a reduction in K^+ currents, including the S current, in sensory neurons (Scholz and Byrne 1988). cAMP can initiate relatively long-term sensitization (Schacher *et al.* 1988; Scholz and Byrne 1988). However, the level of cAMP, which appears elevated in the short term, is not elevated in long-term sensitization (Bernier *et al.* 1982).

The GSW reflex system has a powerful experimental advantage in that the monosynaptic sensory-motor component of the reflex can be reconstituted in cell culture (see Chapter 4). This advantage was recruited for the analysis of the role of macromolecular synthesis in long-term sensitization (Montarolo *et al.* 1986; Dale *et al.* 1987; Schacher *et al.* 1988). In a representative set of experiments, a brief application of serotonin resulted in short-term facilitation in the synapse between the sensory and the motor cell, lasting for a few minutes. Several successive applications of the putative neurotransmitter within 1.5 hours, resulted in facilitation for more than 24 hours. The latter modification was regarded in these studies as long term.

Two inhibitors of protein synthesis, anisomycin and emetine, and two inhibitors of RNA synthesis, actinomycin D and α-amanitin, were tested for their effect on both short- and long-term sensitization. These drugs effectively inhibited protein synthesis and RNA synthesis, respectively, in *Aplysia* ganglia, without significantly affecting the electrical properties and resting potential. The inhibitors blocked facilitation 24 hours after the long-term training, but had no effect on short-term facilitation. Furthermore, there was a critical time window for the inhibitors' action; these drugs were effective only if applied during the application of serotonin in training, but not if applied before or after training (Fig. 8.3).

These results demonstrated that inhibitors of macromolecular synthesis block the consolidation of cellular memory into a relatively long-lasting form, by acting on the neurons that retain the memory. Such a conclusion concerning the cellular locus of macro-molecular synthesis inhibitors could not have been inferred from the earlier, systemic pharmacological investigations in other organisms.

The dilemma of gene-expression models of memory

Before proceeding to enquire which genes might be activated in consolidation, it is advisable to note some difficulties raised by gene-expression memory models. If the cell's genome is altered in memory, then either all the synapses of the neuron are altered or, alternatively, some mechanism exists that enduringly restricts the effect of the new gene product(s) to selected synapses. Both assumptions are problematic. Altering all the synapses

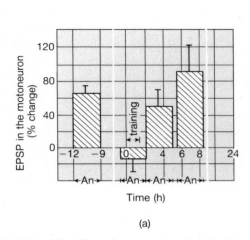

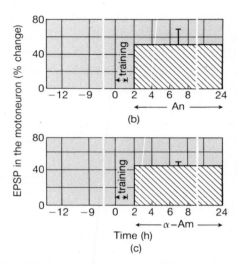

(a)

(b)

(c)

Fig. 8.3. The effect of a protein-synthesis inhibitor on relatively long-term facilitation in the reconstituted monosynaptic component of the GSW reflex in *Aplysia*. For details of the reflex and its components, see Chapter 4. The sensory-motor connection was facilitated by repetitive applications of serotonin during the interval marked 'training' in the diagram. Facilitation was measured 24 hours after the beginning of training, and is expressed as the change in the amplitude of evoked EPSP in the motoneuron. Hatched areas, the period during which the drugs were applied. Application of anisomycin (An) during training, but not before or after training, prevented the facilitation at 24 hours (a), but had no effect on facilitation immediately after training (not shown). Inhibition of protein synthesis by An or RNA synthesis by α-amanitin (α-Am) for the entire period following training did not block facilitation at 24 hours (b, c), if the drugs were not present during the critical time window of training. (After Montarolo *et al.* 1986.)

following experience-dependent modification of only part of them, seems to be a wasteful way of managing storage capacity. On the face of it the second possibility, namely selective deposition of gene products in individual synapses, is more appealing, and is substantiated by pieces of circumstantial evidence. Polyribosomes in neurons are selectively positioned beneath synaptic junctions, hence providing individual synapses with a protein synthesis machinery (Steward and Levy 1982). Moreover, dendrites contain a cytoskeletal system for the selective transport of RNA to synaptic sites (Davis *et al.* 1987). However, if novel RNA, or ribosomes, or proteins, are funnelled into only selected synapses, then what marks these synapses in the first place? And if such a primary long-term change occurs in the synapse, why is gene regulation needed at all?

One possibility is that initial events in the synapse create a local nucleation centre for assembly of an active protein synthesis machinery and/or for accumulation of new RNA or protein molecules. Furthermore, structural alterations may then occur retrogradely from the synapse towards the cell body, directing the transport of newly synthesized mRNA or proteins only to the modified synapse. Finally, the possibility that synapses have autonomous DNA, for example in mitochondria, should not be neglected.

Which genes are presumably activated in memory consolidation?

Attempts are already under way to identify proteins whose synthesis is altered in consolidation (Castellucci *et al.* 1988). This is not easy to do, because the relevant proteins may exist in minute amounts, whereas changes detected in more abundant proteins may reflect only secondary processes. In the meantime, suggestions have been made regarding the nature of the postulated critical gene products.

The effect of the macromolecular synthesis inhibitors is not due to depletion of constitutively expressed proteins

It could be argued that inhibitors of RNA and protein synthesis prevent consolidation simply by blocking the replenishment of constitutively expressed neuronal proteins. This is unlikely. The critical time window is narrow. Levels of proteins with a half-life of days or weeks would hardly decrease by inhibition of protein synthesis for such a short period. Squire and Barondes (1976) have demonstrated that the amnesic effect of a protein synthesis inhibitor is not even due to depletion of a hypothetical constitutive protein with a postulated short half-life of 10 minutes. They carefully manipulated the schedule of drug administration in mice, to create two experimental situations. In one situation, inhibition of more than 90 per cent of brain protein synthesis was established only for a few minutes at the time of training. In this situation, depletion of the hypothetical protein at the time of training was expected to be small. In the other situation, inhibition had been established long before training and was already declining, so that the actual inhibition of protein synthesis was below 90 per cent at the time of training, but depletion of the hypothetical, rapidly turned-over protein was expected to be considerable.

The experimental results indicated that the magnitude of protein synthesis inhibition at the time of training, and not the level of the hypothetical protein, was critical for blockade of consolidation. The conclusion was, therefore, that the inhibitors block the synthesis of protein(s) induced by training, rather than the replacement of constitutive protein(s).

Consolidation may involve the expression of early genes that regulate late genes

A narrow time window and a requirement for protein synthesis are suggestive of developmental processes. It is generally assumed that in such processes a temporal and spatial programme of gene expression demands the precise interaction of regulatory and structural genes. In a few systems, it has indeed been demonstrated that early genes are induced, which regulate the expression of late effector genes. The most studied examples concern the life cycle of DNA tumour viruses, such as SV40, Herpes, and adenoviruses (Botchan *et al.* 1986).

In these cases, detailed interactions between sequentially expressed genes have been analysed. For example, in adenovirus a group of proteins coded by the gene E1A must be expressed in the first stages of infection to stimulate transcription of later genes (Berk 1986).

The prevalent view is that interactions between early and late genes similar to those revealed in viruses, also take place in normal development of animal cells. Much work in this field was performed on the response of mammalian fibroblast cell lines to stimulation by growth factors (Cochran *et al.* 1983; Rozengurt 1986; Lau and Nathans 1987). Evidence was found here for the temporal expression of genes, but direct interactions between early and late genes have not yet been demonstrated.

In several additional systems, a critical period was found during which protein synthesis is required for normal development. One example is the induction of insect growth and moulting by the hormone ecdysone (Richards and Ashbruner 1984). Here gene induction was followed by observing chromosomal puffing. Some genes puff later than others. Administration of ecdysone concurrently with a protein-synthesis inhibitor prevented the induction of the late genes. If the inhibitor was administered a few hours after ecdysone, the late genes were induced normally. A somewhat different example is the development of lordosis, a cyclical, hormonal-induced reflex in mammals. Lordosis is a characteristic sexual receptivity posture of the female. Its emergence in the menstrual cycle involves induction by oestradiol of specific proteins in the brain, and can be effectively blocked in rats by protein-synthesis inhibitors, provided that they are administered during a time window of a few hours following exposure to the hormone (Parsons *et al.* 1982).

The proven and hypothetical sequential activation of genes in various types of developmental processes has led Goelet *et al.* (1986) to suggest that during consolidation, early genes are induced that regulate the expression of late effector genes. Nuclear proto-oncogenes, such as *c-myc* and *c-fos*, were proposed as prototypes of such nuclear switches. These genes encode nuclear proteins that affect transcription or mRNA processing. Their activity is short lived, yet they presumably induce long-lasting cellular changes (Weinberg 1985; Verma 1986; Halazonetis *et al.* 1988; Sassone-Corsi *et al.* 1988). Moreover, these genes are known to be induced by ion fluxes, transmitters, hormones, and second-messenger systems, the same molecular stimuli that trigger and establish short-term memory (Table 8.1).

Although this idea is intellectually appealing, it must be considered with the appropriate caution. First, as yet there is no experimental evidence that nuclear switches in general, and proto-oncogenes in particular, do play a role in memory consolidation. Most of the data on the environmental regulation of these genes stem from experiments on *c-fos*; but the extreme sensitivity of *c-fos* expression to many different treatments leaves the physiological meaning of these data obscure.

In addition, the effect of protein synthesis inhibition on the induction of proto-oncogenes is rather complex. In some cases the inhibition led to superinduction rather than suppression, and in other cases there was no effect at all (Barka *et al.* 1986; Greenberg *et al.* 1986*a*). Such complexity, of course, does not exclude the possibility that interference, in one or the opposite direction, with the induction of such genes does disrupt memory. The effect of protein-synthesis inhibitors on the molecular process of learning and memory may itself be heterogeneous. This is hinted by results from experiments with *Hermissenda*. I_A and $I_{K(Ca)}$ in the B photoreceptors are depressed in conditioned phototaxis (see p. 83). These currents can also be depressed by a substantial increase in intracellular Ca^{2+}. The recovery from Ca^{2+}-mediated depression was delayed by anisomycin and cycloheximide (Alkon *et al.* 1987*a*). In other words, inhibition of protein synthesis prolonged biophysical changes similar to those correlated with short-term memory in this system. The authors took this to suggest that in memory consolidation in *Hermissenda*, turning off the synthesis of some proteins prolongs the effects of acquisition so that subsequent consolidating mechanisms, probably involving ultimately the synthesis of other proteins, can proceed. This hypothetical prolongation of acquisition may be due to a reduction in the level of proteases or phosphatases, which normally degrade or reopen, respectively, the closed channels. It is still possible that the above observations reflect a non-specific reduction in proteolysis or dephosphorylation, rather than a mechanism relevant to memory.

The nuclear genes presumed to be induced during consolidation must not, of course, be identified proto-oncogenes. They may be any type of transcription-regulating proteins whose function is modulated by second-messenger systems. Such proteins were discovered (Imagawa *et al.* 1987; Lee *et al.* 1987; Montminy and Bilezikjian 1987; Yamamoto *et al.* 1988), and the DNA elements that presumably bind them were identified in the promoter region of several neuropeptide genes (Comb *et al.* 1987) and the tyrosine hydroxylase gene (Lewis *et al.* 1987).

• Taken together with the observations summarized in Table 8.1, these data make a strong case for the general ability of second-messenger systems to regulate both post-translational modifications of proteins and gene expression. It is, therefore, highly likely that both facets of the action of second-messenger systems play a role in memory (Berridge 1986; Goelet *et al.* 1986; Comb *et al.* 1987; Dudai 1987).

Actually, in phylogeny, cAMP regulation of gene expression probably preceded cAMP regulation of cellular activity via post-translational modification. In prokaryotes, the sole established function of cAMP is the induction and repression of genes (Bostford 1981). This is done by binding of the cyclic nucleotide to a receptor protein, termed 'catabolite repressor protein' (CRP) or 'cAMP-binding protein' (CAP). The binding is followed by interaction of CAP with DNA sequences that regulate transcription or adjacent promoters. There is much homology between CAP and the regulatory subunit of PKA in eukaryotes. This has led to the hypothesis that the role of cAMP in the regulation of transcription was conserved in eukaryotes, and that its role in post-translational modification was added only later in evolution to enable prompt physiological responses to extracellular stimuli (Nagamine and Reich 1985). The details of this hypothesis might need a revision, because current data implicate the catalytic subunit of PKA in induction of mammalian gene expression

Table 8.1. Selected examples of modulation of gene expression by extracellular stimuli

Gene product	Tissue	Stimulus	Effect	Reference
A. Receptors and channels				
Nicotinic acetylcholine receptor (α–subunit)	Skeletal muscle	Denervation	+	Merlie et al. (1984)
				Klarsfeld and Changeux (1985)
	Cultured myotubes	Tetrodotoxin	+	Klarsfeld and Changeux (1985)
Na$^+$ channel	Skeletal muscle	cAMP, forskolin	+	Harris et al. (1988)
		Denervation	+	Harris et al. (1988)
				Cooperman et al. (1987)
B. Enzymes				
Tyrosine hydroxylase	Adrenal medulla, locus coeruleus	Reserpine	+	Mallet et al. (1983)
	Sympathetic ganglion	Reserpine	+	Black et al. (1985)
Ornithine decarboxylase	Phaeochromocytoma	NGF	+	Greenberg et al. (1985)
C. Peptides and hormones				
Enkephalins	Adrenal cells	Reflex stimulation	+	Kanamatsu et al. (1986)
		cAMP	+	Quach et al. (1984)
		Nicotine, forskolin, Ca^{2+} ionophore	+	Kley et al. (1987)
			+	Kley et al. (1987)
Substance P	Basal ganglia	Dopaminergic denervation	+	Young et al. (1986)
	Basal ganglia	Dopaminergic denervation	−	Young et al. (1986)
		Haloperidol	−	Bannon et al. (1986)
Somatostatin	S. cervical ganglion in culture	Veratridine	−	Roach et al. (1987)
	Brain cell culture	Forskolin	+	Montminy et al. (1986)
TRH	Hypothalamus	Chemical thyroidectomy	+	Koller et al. (1987)
		Triiodothyronine (T$_3$)	−	Koller et al. (1987)
TSH	Pituitary	Chemical thyroidectomy	+	Koller et al. (1987)

D. *Proto-oncogenes*

Gene	Tissue/cells	Stimulus		Reference
c-*fos*	Brain	Seizures	+	Dragunow and Robertson (1987)
				Morgan *et al.* (1987)
	Spinal cord neurons	Cutaneous stimulation	+	Hunt *et al.* (1987)
	Phaeochromocytoma	K$^+$ depolarization	+	Greenberg *et al.* (1986*b*)
				Morgan and Curran (1986)
		Na$^+$ and Ca^{2+} channels agonists	+	Morgan and Curran (1986)
		NGF	+	Kruijer *et al.* (1985)
	Salivary gland	β-adrenergic stimulation	+	Barka *et al.* (1986)
	Thyroid cell culture	TSH, cAMP	+	Tramontano *et al.* (1986)
c-*myc*	Thyroid cell culture	TSH, cAMP	+	Tramontano *et al.* (1986)
	Phaeochromocytoma	NGF	+	Greenberg *et al.* (1985, 1986*a*)

In most cases, the results are based on determination of the level of the appropriate mRNA. The question of enhanced transcription vs. increased stabilization of mRNA is neglected for simplicity. In a few cases, the results are based on immunocytochemical detection of the relevant gene product. In the experiments on enkephalins, the data on the peptide and its precursors were combined in the table. Reflex splanchnic nerve stimulation of the adrenal medulla (Kanamatsu *et al.* 1986) was by insulin hypoglycaemia. Dopaminergic denervation (Young *et al.* 1986) was done by treatment with the toxin 6-hydroxydopamine. cAMP was the 8-bromo or the bituryl derivatives. Tetrodotoxin blocks voltage-gated Na$^+$ channels, reserpine depletes catecholamines, forskolin activates adenyl cyclase, veratridine is a voltage-gated Na$^+$ channel agonist, and haloperidol is a dopamine antagonist. TRH, TSH-releasing-hormone; TSH, thyroid-stimulating hormone; NGF, nerve growth factor; +, increase; −, decrease in gene expression.

(Riabowol *et al.* 1988), whereas the role of the regulatory subunit in this context remains unclear. Nevertheless, the general notion remains valid; namely, that the original function of the cAMP cascade in gene expression was conserved in phylogeny while a new function, in post-translational modification, was added. Other second-messenger systems might have followed a similar evolutionary route.

Long-term memory may require the synthesis of novel isoforms of the same proteins that participate in short-term memory

If early genes induce late genes that sustain memory, what are the products of the late genes, and how do they function in long-term retention? Likely candidates are new isoforms of the enzymes, receptors, or ion channels that participate in short-term memory, but with altered affinity for their ligands. One model proposed that a new regulatory subunit of PKA, R, with higher affinity for cAMP, is induced during consolidation in *Aplysia* (Kandel and Schwartz 1982). Such a novel variant of R would endow the kinase with higher sensitivity towards cAMP, resulting in continuous phosphorylation of substrate proteins, such as K^+ channel(s), even in the presence of resting levels of cAMP. In a conceptually related model, Greenberg *et al.* (1987) suggested that a protease might be induced that degrades R, causing a decrease in the R/C ratio (see p. 120). Alternatively, a repressor of the gene that codes for R might be induced.

   Other candidates are genes whose products control the efficacy of transmission, for example by altering the balance between co-released transmitters, or the number and/or sensitivity of receptors for neurotransmitters (Changeux 1986; Black *et al.* 1987; Changeux *et al.* 1987).

Long-term memory and morphological plasticity

In considering candidates for proteins whose synthesis is required to sustain long-term memory, one should take into account the ample evidence that associates structural changes in neurons with the effect of training. Two examples are the changes in synaptic active zones and in the number of synaptic varicosities detected in habituated and sensitized *Aplysia* (Bailey and Chen 1983, 1988*a,c*), and the changes in the shape and number of synapses observed in LTP (see p. 100). Additional examples will be provided in later chapters. Morphological alterations can be classified into two categories: (1) changes that are confined to existing connections and leave the connectivity of the network anatomically fixed (yet functionally altered); and (2) changes that lead to the growth of new anatomical connections (or possibly also to the elimination of existing ones).

   Both processes take place in response to environmental change and injury (Cotman 1983; Cotman and Nieto-Sampedro 1984; Greenough 1984). It becomes more and more apparent that the nervous system, including the adult mammalian CNS, is endowed with a remarkable capacity to modify its anatomical circuitry. An especially intriguing example is provided by results that have recently·been obtained with a novel imaging methodology. Sophisticated light-microscopy techniques were combined with low-light level video recording, digital image processing, and intracellular injection of non-toxic vital dyes, to allow repeated observations of identified pre- and postsynaptic elements *in situ*, in the nervous system of a live animal. This allowed an assessment of the stability over time of the synaptic elements. The findings demonstrated that in both sympathetic and para-sympathetic ganglia of rat, nerve endings are subject to considerable remodelling over time (Purves and Voyvodic 1987).

It should be noted that the relevance of anatomical remodelling in a neuronal circuit to specific alterations in the representations encoded by this circuit is not a trivial issue (see also Chapter 9). It is, however, reasonable to assume that even relatively minor anatomical alterations in circuits might modify representations, and hence that the structural changes observed in neuronal systems after learning are indeed relevant to memory.

Both intracellular and extracellular macromolecules can contribute to morphological plasticity. Major candidates for intracellular macromolecules controlling morphology are cytoskeletal elements (Lynch and Baudry 1984). Major candidates for extracellular proteins that play a role in tissue remodelling are cellular adhesion molecules (CAMs) (Edelman 1984) and growth factors (Henderson 1987).

There is no shortage of additional candidates. Brief mention is made here of only three examples, in which some experimental evidence links the proteins in question to behavioural plasticity. The first is a set of proteins termed S-100. Those are small, acidic, Ca^{2+}-binding polypeptides which are abundant in brain. Some isoforms of S-100 are confined to glia whereas others are located in neurons. These proteins are also secreted into the extracellular space. Forms of secreted S-100 or related polypeptides have been shown to promote neurite extension (Kligman and Marshak 1985). One of the intracellular roles proposed for S-100 is the remodelling of cell structure by altering the assembly of micro-tubular matrix. An increase in the content of S-100 has been reported to correlate with training, and antibodies to S-100 were reported to inhibit learning (Donato 1986).

Other candidates which could be mentioned in this context are ependymins. These glycoproteins are released into the extracellular fluid and tend to polymerize at low Ca^{2+} and form a fibrous insoluble matrix. An increased synthesis of ependymins was reported to correlate with training and learning in goldfish and rodents (Shashoua 1985). Ependymins were postulated to precipitate on axonal or dendritic loci at which Ca^{2+} is depleted due to enhanced electrical activity, and to mark physical targets for synapse formation (ibid.).

Finally, additional macromolecular candidates for controlling neuroanatomical plasticity are proteoglycans, e.g., similar to the product(s) of the *Drosophila* gene *period*. Mutations in this gene affect biological clocks and impair experience-dependent modification of courtship (see Chapter 7). The gene *per* was reported to affect intercellular communication via gap junctions (Bargiello *et al.* 1987).

Dendritic spines as potential sites of memory-correlated morphological change

A strategic locus of morphological alterations that might store memory is the dendritic spine. Spines are postsynaptic specializations found on many types of neurons in the brain (Fig. 8.4). Morphologically, they can be described as being constructed of two major parts; a head, or bulb, measuring 0.5–1.0 μm in width, and a stalk, measuring 0.5–1.5 μm in length, which connects the head to the dendritic shaft. Spines have a characteristic intracellular organization, which differs from that of the shaft. The main elements inside a spine are: membranous saccules and tubules, which together form a structure called the spine apparatus; a network of filamentous material which probably includes contractile elements; and a postsynaptic density, to which the filamentous material attaches. Often spines are classified according to their overall shape, i.e. stubby, thin, or mushroom-shaped spines; but

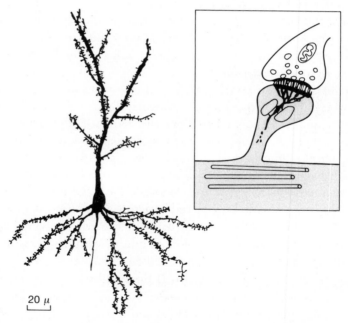

Fig. 8.4. Dendritic spines are postulated strategic loci for morphological changes that subserve long-term memory. The diagram shows a camera lucida drawing of a pyramidal neuron from the rat cortex, with a typically large number of spines. (From Peters and Kaiserman-Abramof 1970.) *Inset:* Schematic representation of a dendritic spine (shaded) with the characteristic filaments and sacs comprising the spine apparatus, and the postsynaptic density. The elongated tubes in the dendritic shaft are microtubules, which participate in the formation of spines. The postsynaptic terminal synapsing onto the spine is also shown. Spines usually bear one excitatory synapse; changes in their morphology may attenuate information transfer from the synapse to the shaft and soma. (After Westrum *et al.* 1980.)

there is a continuous spectrum of head and stalk shapes and their combinations (Wilson *et al.* 1983).

In the mammalian cerebral cortex, as well as in other brain parts, the majority of excitatory afferent synapses are made on to spines. In most cases, the spine receives one synaptic contact. In some cases there is also a second synapse, which might be inhibitory. There is no evidence for spines with only inhibitory synapses. The reason for the evolution of spines has been a subject of discussion for many years now. Some authors suggested that spines serve a structural role in optimally accommodating dense synaptic contacts (Swindale 1981). Other authors, however, have pointed out that the spine neck might have an important functional role in regulating synaptic efficacy. This is because the geometry of the neck affects the electrical resistance between the head and the shaft. In other words, the magnitude of somatic depolarization due to activation of the excitatory synapse on the spine head is expected to be a sensitive function of the neck length and diameter. Stretching or narrowing the neck should reduce, and shortening or widening the neck should increase, the somatic depolarization (Rall 1978; Koch and Poggio 1983). The effect of changes in individual spines is small, but the cumulative effect of changes in adjacent spines is expected to become significant. The spine shape may, therefore, be a 'cellular memory'.

Data implicating changes in spine shape in LTP were mentioned in Chapter 6. Interestingly, there was even a report of a rapid shortening of the spine stems in the honeybee's brain following one-trial learning (Brandon and Coss 1982). In some other cases where spine morphology was correlated with behavioural experience, changes were found in the head volume, and in the size of the postsynaptic density, with or without alteration in stalk length and diameter (Fifkova 1985). The modification of cellular matrix proteins was suggested to contribute to the morphological transformation of spines. In addition, non-consolidated, rapid, and reversible twitching of contractile spine proteins was postulated to subserve very short-term memory or transient working memory (Crick 1982). There is, however, no evidence for this sort of brain twitching.

Concluding remarks on molecular and cellular processes in learning: elementary mechanisms in search of content

From the discussion of molecular and cellular mechanisms of learning and memory, in this and the previous chapters, two major conclusions emerge. First, there is no single mechanism of acquisition and/or storage. Several cellular mechanisms operate concurrently and integratively. The classification of memory into only two discrete phases, short and long term, is naïve, and represents an artificial categorization imposed on a natural continuous process. In all probability, formation of a stable memory is analogous to a developmental process, in which multiple extracellular signals initiate cascades of events in the membrane, the cytosol, and the nucleus, which gradually modify cellular properties on a time-scale of seconds to years. One should expect, therefore, to identify a continuous spectrum of interlocked molecular mechanisms of memory, with different endurances. Sophosticated behavioural studies (Baddeley 1976, 1986; Klatzky 1980), clinical data on amnesia (see Chapter 15), and systemic pharmacology (McGaugh 1966; Gibbs and Ng 1977; Martinez *et al.* 1981; Frieder and Allweis 1982; Mizumori *et al.* 1987), all provide strong support for the multistep, multichannel nature of memory in different species.

The second conclusion is that neurons use second-messenger cascades essentially as many other cells do, to translate extracellular signals into cellular change. There are, of course, extremely critical neuronal specializations, both molecular (e.g. an extensive repertoire of receptors and ion channels) and structural (e.g. neurites, synapses). But, on top of it all, and as already noted in Chapter 3, there is an impressive universality in the mechanisms utilized by cells in molecular communication, and in the ways cells exploit both genetic and epigenetic instructions.

It is not surprising, therefore, that major aspects of the cellular and molecular studies of learning and memory have merged, in recent years, with the study of cellular communication and differentiation in general. This is both gratifying and disillusioning. The gratification lies in the fascinating discoveries of intricate and sophisticated molecular devices; in the beautiful unity of cellular processes in ontogeny and phylogeny; and in the ability to apply a plethora of advanced methods and concepts, developed in other branches of biology, to memory research. The 'disillusion' relates to the notion, had it ever been seriously considered, that one could understand neuronal memory solely by adhering to molecular analysis. Internal representations are ultimately stored in neuronal circuits, not in molecules. In other words, the individuality and *content* of memories (i.e. their semantics) are not reducible to molecular cascades. We must, therefore, shift now into higher levels of neuronal organization.

PART III

Windows to the Architecture of Memory Systems

9   *On the complexity of internal representations*

Readdressing the differences between simple and complex internal representations

At this stage in our discussion we should readdress the question of internal representations. In Chapter 1 we stated that ultimate comprehension of learning requires an understanding of how the relevant information is represented in the nervous system. In the case of very simple learning systems, this goal is conceptually approachable. Consider again the GSW reflex in *Aplysia*: the neuronal system that subserves it is innately wired to encode a representation of cutaneous tactile stimuli and a behavioural response to these stimuli. Note that the representation is here regarded as the neuronally encoded information on both the sensory stimulus and the motor response that this sensory stimulus deserves. On the face of it, the representation is trivial: we know that pressure applied to cutaneous mechano-receptors is encoded in spike trains, which ultimately elicit EPSPs in motoneurons, which in turn fire to activate the appropriate muscles. It is this knowledge of the basic scheme of the circuit that enabled us to correlate changes in excitability of circuit elements with the modified behaviour. In the simplified model of the monosynaptic component of the circuit, memory was an altered correspondence between spike trains in the central mechanosensory neurons and EPSPs in the follower motoneurons.

In reality, even here the situation is far from being trivial. In intact *Aplysia*, the representation of stimulus and response is encoded in the coherent activity of scores, if not hundreds, of neurons in the central nervous system (see Chapter 4). This emphasizes the point that the information on activation of the relevant mechanoreceptors is integrated into other information that *Aplysia* gets from its senses about the world, and it is the sum-total of all this integrated dynamic information that guides the behavioural response. This is clearly illustrated in sensitization and classical conditioning, where other pieces of information about the world modify the GSW reflex. The sensory-to-motor neuron path can be considered as a separate system only for the sake of reductionist analysis. At present, a global description of the neuronal encoding of the representation which guides the reflexive response *in situ* is not yet feasible, even though the system appears at first to be so simple.

Disregarding the *in situ* complexity, three properties of the GSW reflex and its modification by experience deserve attention. (1) The core of the neuronal system that subserves the reflex can be traced to *identified neurons.* (2) The system is *committed* and *dedicated.* The neurons are innately programmed to deal with a particular and limited type of internal representation. (3) Learning is a *modification of the probability* of a reflexive response. Memories in which what is stored is an altered probability of a reflex, are often termed, appropriately, *reflexive memories* (Fig. 9.1).

Only simple cases of habituation and sensitization, and rather simple cases of associative conditioning, are reflexive. Most memories are not. There is no accepted term

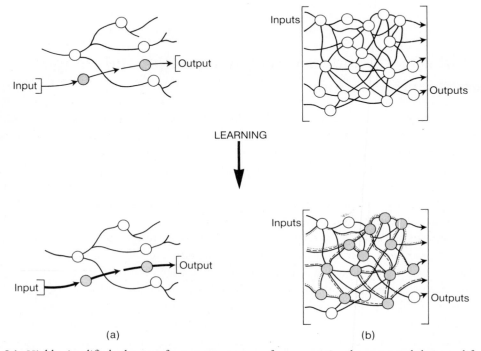

(a)   (b)

Fig. 9.1. Highly simplified schemes of two extreme types of representational systems and their modification by experience. (a): A most simple reflexive system. It is localized, and innately committed to encoding a primitive representation of a certain type of *stimulus* and the appropriate *response* to this stimulus. Learning alters the probability of the reflexive response (heavier connection after learning implies facilitation). (b): A small portion of a distributed neuronal system. Such a system can encode innately programmed or learned representations, or both. Ideally, a system that learns is initially uncommitted to specific representations, although it may generate pre-representations by endogenous spontaneous activity. After learning, various combinations of neurons are capable of being stabilized in different connectivity patterns (exemplified by the dotted and broken lines, respectively), which encode internal representations in an ensemble code. Any neuron may participate in multiple representations, and the information is distributed among many neurons. Due to the rich connectivity, activation of part of the neurons may yield a meaningful output. Analysis of internal representations and learning in reflexive systems is conceptually and sometimes practically feasible; in contrast, it is not yet clear how distributed representations are encoded in the brain. (For further discussion, see text.)

encompassing all types of non-reflexive learning and memory. The situations and neuronal systems involved are very heterogeneous and justifiably defy a unifying terminology. Sometimes non-reflexive memory systems, or subsets of them, are referred to as 'representational'. This is regarded here as inappropriate, since all memories, by definition, are representational. Categorizations commonly used in describing non-reflexive learning are based on behaviours and the interpretation of these behaviours by the observer, rather than on neuronal encoding principles and learning algorithms. Examples will be given in Chapter 15. In the meantime, for lack of a better alternative we shall occasionally use the loose term 'complex memory systems' to denote those learning systems which involve internal representations far more complex than those encoded in sensory-to-motor connections. The representations involved will be termed 'complex internal representations'. This terminology simply reflects our ignorance about the involved representations. The main issue is that the picture as promoted here is not of a sharp dichotomy between reflexive and non-reflexive learning, but rather of a gradual continuum between simple and complex. This is the situation in the real world: a gradient of complexity exists in memory systems in individuals and between species. We must know a lot more before attempting further to classify complex learning systems on the basis of neurobiological criteria. Hopefully, information provided in the coming chapters will be relevant in that respect.

The cardinal question: how are complex representations encoded?

The study of complex memory systems is thus inseparable from the study of internal representations, and hence from the study of processes by which events and their relationships are encoded in the brain and used to guide behaviour. This is the crucial problem of cognitive sciences, and the cardinal question shared by the neurosciences, psychology, and philosophy. Voluminous literature exists, speculations and models are continuously generated, defended, and devastated, but relevant experimental data are still much needed. Only a few issues are noted below. For recent comprehensive discussions of facts and hypotheses concerning this problem, see Ballard (1986), Churchland (1986), Rumelhart and McClelland (1986a), Skarda and Freeman (1987), and the section on Further reading.

Some principles of central neuronal processing: parallel as well as serial processing, and sensory spaces mapped in the brain

Several principles of neuronal coding operations do emerge from the experimental study of central processing in sensory systems, such as somatosensory (Nelson *et al.* 1981; Welker and Van der Loos 1986; Pons *et al.* 1987), electroreceptive (Heiligenberg and Bastian 1984), visual

(Hubel and Livingstone 1987; DeYoe and Van Essen 1988; Zeki and Shipp 1988), and auditory systems (Konishi 1986, 1989). Some of these principles are:

1. Generally speaking, central sensory systems transform the neural representation of a stimulus from the form generated by the peripheral sensory apparatus to a form which subserves higher-level brain functions, such as perception and guidance of action.

2. Different qualities of sensory stimuli, e.g. form, colour, and motion in the visual system, may be kept separate or partially separate and processed in parallel in the brain. Parallel processing may involve different types of computations performed concurrently in the same anatomical locus, and/or similar types of computations performed concurrently in different anatomical loci.

3. In addition to parallel processing of stimulus qualities, there is serial processing. This may be hierarchical, so that the analysis becomes more global at certain stages of processing. For example, neurons in the mammalian inferior temporal cortex synthesize information funnelled by stations along the cortico-cortical pathways from the striate cortex, and process both spatial and spectral information throughout a much larger portion of the visual field than neurons in earlier stations. Globality of analysis revealed in recordings from single neurons may become quite striking; for example, in the temporal cortex of both primate and non-primate species neurons were identified that responded preferentially to different features of faces (Gross *et al.* 1972; Kendrick and Baldwin 1987; Perrett *et al.* 1987; see Chapter 14).

4. Serial pathways running in parallel communicate reciprocally with each other at both low- and high-levels of analysis. These pathways also communicate reciprocally with subcortical structures. The coherence of percepts is expected to depend on such multistage integration.

5. Sensory spaces are mapped in the brain. This means that attributes of the sensory space are represented so that there is a systematic variation in their value across at least one dimension of the neural structure (e.g. Fig. 9.2). Neurons in maps use a place code; mapping may involve transformation from a spike code (see the definitions on page 42, and the example in Fig. 9.3).

Multiple maps for each modality exist in different brain regions. The maps may differ in the representation of different stimulus attributes and in the type of computations performed on these attributes. Mapping probably facilitates the architecture of neuronal connectivity and the strategy and speed of computations. It does not seem indispensable for computation, because many functional neuronal modules in the brain do not show a systematic variation in their selectivity for a sensory quality (Knudsen *et al.* 1987). It is not safe, however, to conclude that a map does not exist, as the neurons may still display a systematic variation in their tuning for a yet unidentified attribute.

6. Central maps are conventionally classified into two types. In a *topographical map* (also called a *projectional map*), central neurons are arranged so that their spatial relationship

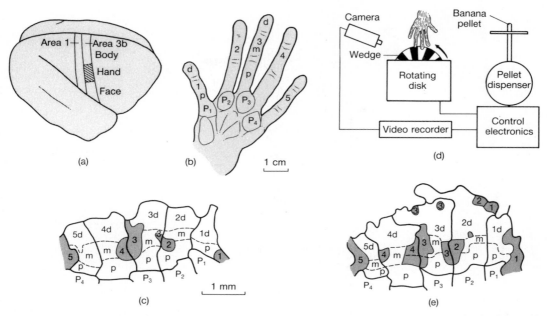

Fig. 9.2. A central topographical map of sensory space and its modification by experience. (a): A simplified lateral view of the right neocortex of the owl monkey. Areas 1–3b in the primary somatosensory cortex contain a highly ordered somatotopic representation of the body surfaces and some deep body structures. The location of the hand representation is marked by hatching. (b) The hand surface of a normal adult owl monkey. Numbers 1–5 denote the digits (1 is the thumb); d, m, and p, distal, middle, and proximal phalanges; P_{1-4}, the palmar pads at the base of the digits. (c) A map of the representation of hand surfaces indicated in (b), in area 3b of the somatosensory cortex. The map has been rotated 90° counterclockwise with respect to (a). Grey areas, dorsal (hairy) skin on each digit. (d) The behavioural apparatus used for studying the effect of differential stimulation of restricted skin surfaces of the hand on the representation of these surfaces in area 3b. The monkey was trained to maintain contact with a rotating disk in order to get a reward. Only the distal aspect of the distal segment of digits 2, 3, and occasionally 4, contacted the disk. (e) The cortical hand representation of the same monkey as in (c), following about 20 weeks of daily training (1.5 hours per day) in the apparatus depicted in (d). Note the remodelling of the map and the marked expansion of the representations of the distal aspects of digits 2, 3, and, to a lesser degree, 4. (Adapted from Jenkins and Merzenich 1987.)

conserves the spatial relationship of the sensory epithelium. For example, in the primate primary somatosensory cortex, the surface of a hand is represented in a systematic two-dimensional neuronal topography in the appropriate cortical field (Fig. 9.2(a)–(c)). The pertinence of topographical mapping to the representational code is not obvious. Synchronous input from epithelial loci which are adjacent to each other plays a role in the parcellation of topographical maps (Clark *et al.* 1988). This suggests that the topography of the map is relevant to the representation of correlated events. However, the map may result in the first place from ontogenetic constraints that do not bear directly on the neuronal coding (Konishi 1986; Rakic 1988).

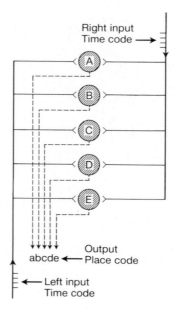

Fig. 9.3. A schematic model of a computational, centrally synthesized map of acoustic representations, which converts spike time code into place code. The circuit uses delay lines and coincidence detection for measuring and encoding interaural time differences. Neurons A–E are arranged in an array and fire maximally only when signals from the left and from the right arrive simultaneously. Temporal information about the acoustic signal is encoded by spike timing. The axonal path increases in opposite directions for the two sources, thus creating a left–right asymmetry in transmission delays. When binaural disparities in the acoustic signals exactly compensate for this asymmetry, the neurons fire maximally. The output of the neurons does not use spike timing to encode time, but rather the position of the neuron in the array signals the interaural time difference for which the neuron responds maximally. (From Konishi 1986.)

In contrast, in a *computational map* (also called a centrally synthesized map), the selectivity of neurons for attributes of the sensory stimulus is not related to the topography of the sensory epithelium, but rather is synthesized by central neuronal circuits (Konishi 1986, 1989; Knudsen *et al.* 1987). The central system which measures and encodes interaural time difference in the barn owl is but one example of a computational map. Here spike time code is transformed into place code by a brain-stem circuit that uses neural delay lines and coincidence detectors (Fig. 9.3; Sullivan and Konishi 1986).

Maps recombine to produce higher-order maps (e.g. Sparks and Nelson 1987; Wagner *et al.* 1987). Topographical and computational maps of different modalities interact, and may also be aligned, as is the case for visual and auditory maps in the barn owl, or visual, auditory, and somatosensory maps in the mammalian superior colliculus (ibid.). This indicates that the brain uses general strategies for processing place codes regardless of the modality and of whether the maps were topographical or computational initially (Konishi 1986).

Experience-dependent modification of central maps in the adult brain

• Substantial experimental evidence for experience-dependent modification of central maps stems from studies of topographical somatosensory and motor maps in the mammalian brain (Wall *et al.* 1986; Jenkins and Merzenich 1987; Merzenich 1987; Merzenich *et al.* 1987; Calford and Tweedale 1988; Clark *et al.* 1988; Pons *et al.* 1988; Sanes *et al.* 1988). Most of these studies used peripheral anatomical lesions and manipulations to modify the central maps, but the effect of behavioural experience was also investigated. An intriguing example is provided by studies by Michael Merzenich and his colleagues on the reorganization by experience of the primary cortical somatosensory representation of the hand in the adult owl monkey (Jenkins and Merzenich 1987; Merzenich 1987). The monkeys were trained to contact a rotating aluminium disk with 0.35-mm wedges (Fig. 9.2(d)). The rotating disk was mounted close to the cage so that the animal could contact it only with the distal tips of two or three fingers. The animal was rewarded with a tiny banana pellet if it maintained continuous digital contact with the disk for 10–15 s at a time. Accumulated experience resulted in a marked reorganization of the somatosensory map of the hand and expansion of the fields corresponding only to the stimulated distal aspects of the appropriate digits (Fig. 9.2(c),(e)).

In spite of the afore-mentioned generalizations concerning central neuronal processing, our understanding of this issue is still very rudimentary. Promising progress has been made in recent years, for example in studies of the auditory and somatosensory systems mentioned above, and in combining empirical and theoretical approaches in the study of the visual system (Koch *et al.* 1986; Ullman 1986; DeYoe and Van Essen 1988; Zeki and Shipp 1988). It is likely that the mechanisms revealed in the central processing of sensory information also apply to manipulation of internal representations in general. As long as these mechanisms and their function remain cryptic, our understanding of complex memories is bound to remain incomplete, even if neuronal substrates and cellular correlates of these memories are identified.

Some phenomenological properties of complex memory systems

In the mean time, some assumptions may be made regarding the nature of neuronal systems that store and use complex internal representations. Several phenomenological properties of complex memories are of help here. These interrelated properties emerge from experimentation on perception and memory, as well as from everyday experience.

Completion: Pattern recognition as well as complex recall is often evoked by partial and different stimuli. For example, memory of a familiar face can be revived by very few facial features. Similarly, only a few notes may bring to memory Beethoven's Fifth or 'The Emperor'.

Associativity: Completion may occur because parts of the stimulus elicit by association other characteristic features of the sensory pattern. In addition, often a stimulus or a memory evokes a chain of memories, whose interrelationship is not always apparent.

Noise tolerance: A familiar pattern, such as the shape of an everyday object, may be recognized even if the clues are embedded in a noisy background.

Graceful degradation: Related to the above is the observation that the gradual elimination of sensory data, or the gradual accumulation of errors in these data, is not fatal to recognition, up to a certain point. The term 'graceful degradation' is also used to describe the relative immunity of the output of complex memory systems to partial physical damage to the system (and see, in this context, Chapter 10).

Endurance: It is a common experience that some memories last for decades. Moreover, memories that seemed to vanish may come back vividly after many years.

Fast retrieval: Vivid memories are often retrieved within fractions of a second.

Quick alternation: Memories may alternate very quickly in our mind.

Very high capacity: There is no evidence that in a normal, healthy person, long-term memory can be saturated.

A note on distributed memories and ensemble encoding

The phenomenological properties of complex memories suggest that they are not encoded in a few, dedicated and exclusive neurons and connections, similar to those encoding the elementary components of simple reflexive memories, but are rather distributed in systems of many neurons and connections (see Fig. 9.1). Such memories are termed, therefore, *distributed memories.* The term 'distributed memory systems' bears the connotation of *ensemble coding.* In an ensemble code (also called population code) an event is represented over many distributed units by the collective pattern of activity of these units, so that the role of each unit is minute and any unit may participate in the representation of many events. Sets of coactive neurons that may store distributed memories are termed *neuronal assemblies* (Hebb 1949; Palm 1982; Rummelhart and McClelland 1986*a*).

Three points deserve attention here. First, some properties of complex memory systems, such as completion and noise tolerance, can be shown theoretically to emerge in hierarchical networks that store complex concepts in single units, i.e. in so-called "gnostic" or "grandmother cells" (Baum *et al.* 1988). Therefore, the phenomenology of complex memory systems cannot *per se* be used as a decisive argument in favour of ensemble coding. Second, the term 'distributed systems' became very popular in neurobiology, and is already used even in the description of relatively simple reflexes (Camhi 1988; Frost *et al.* 1988). However, in many neuronal systems currently dubbed 'distributed', the term is used merely because the system is physically distributed and processes information in parallel, and not because there is experimental evidence for ensemble coding (for examples of exceptions, see the ensemble coding of saccadic eye movements by neurons in the superior colliculus, in Lee *et al.* 1988, and the ensemble coding of arm movement by neurons in motor cortex, in Georgopoulos *et al.* 1988). And third, the term 'assembly' is introduced here more as a concept than as a concrete physical entity with well-defined parametric characteristics. For example, one may wonder what is the time-window during which interconnected neurons must fire in order to be considered as members of the same assembly? Time windows in the range of milliseconds to seconds are regarded as appropriate by various authors. The term 'assembly' as used in this book refers to a variety of multi-neuron systems of different sizes, the output of which may be based on different codes and their combinations, ranging from ensemble coding to hierarchical abstraction in 'gnostic' units.

Various mathematical models were suggested for describing the dynamic organization, representational properties, and modification by inputs of assemblies of neuron-like units. A rich selection of examples is provided in Rumelhart and McClelland (1986*a*) and Anderson and Rosenfeld (1988). This formal, model-building approach to neuron-like networks and learning far exceeds the scope of the present discussion. The following notes are merely a brief sketch of the conceptual framework of this flourishing branch of research on pattern recognition and learning. For our purpose, suffice it to say that an important group of models exploits analogies with large systems of interacting degrees of freedom, as studied in statistical mechanics. In many models, the system is portrayed in layers of units, consisting of an input ('primary sensory') layer, intermediate ('central', 'hidden') layer(s), and an output layer. The excitability of each neuron-like unit, or the strength of its connections, 'synapsing' on other units, is represented by a binary or continuous variable. The assembly is thus formally represented as a matrix of real numbers. The collective computational behaviour of the network emerges from the rich interconnections between the individual units.

In the process of learning, the network matches a given set of values presented as input, to a desired state that serves the role of output. The matching of output to input is done by algorithms based on modification rules that adjust the weight of the connections in the matrix. Variants of Hebb's rule (p. 70) might, for example, be considered as local modification rules. Under the appropriate conditions, the neuron-like network self-organizes in a stable connectivity pattern, corresponding to the desired input–output relationship. The system can accommodate many stable connectivity patterns corresponding to different representations, and any given unit can be shared by different representations. If later presented with partial or noisy information in the input layer, the system may relax into the appropriate stable state yielding the corresponding output, and is hence capable of 'recognition'. It can be demonstrated that such systems display properties analogous to the phenomenological properties of biological memory systems, such as associativity, completion, etc. (Tank and Hopfield 1987). Moreover, given the appropriate learning algorithm and input, the hidden neuron-like components of such a type of network display properties simulating those of brain neurons that perform analogous tasks (Zipser and Andersen 1988). Simulated networks are even capable of learning rudimentary linguistic tasks (Rumelhart and McClelland 1986*b*), although the relevance of this to real human language is strongly disputed (Prince and Pinker 1988).

Learning as a selective, Darwinistic mechanism

The relevance of current mathematical models of parallel distributed processing (also called 'connectionist models') to the biology of memory is still questionable. These are not genuine neuronal models, only simulations of the behaviour of abstract, interconnected units. Connectionist models sacrifice fidelity for simplicity. For example, some models assume symmetrical coupling between units, full connectivity between units, and/or binary cell variables. Even more important, the current models do not take into account the intricate integrative, input and output properties of real neurons (see Chapter 3). Therefore, even when the models succeed in simulating a modifiable neuronal behaviour, no claims are made that the brain actually uses the same learning algorithm as the model (Zipser and Andersen 1988). Also important is the fact that connectionist model networks must be informed a priori about what is and what is not a desired output; how such instruction could be achieved in the brain is an intriguing problem (Skoyles 1988). Modelling is, however, relevant to three neurobiological notions that merit special attention in the present context.

1. Models show that stable representations can indeed, in principle, be acquired and stored in non-dedicated ensembles of computational units that have primitive neuron-like properties. There is no need to adhere to a switchboard picture of learning systems, in which the connections are dedicated lines linking exclusive units with unique physical addresses, similar to those considered in simple reflexive memories. In the modelled assemblies, as in the postulated neuronal ones, perception and memory are content-addressable and reside in the coherent activity of many interconnected neurons (John 1972; Abeles 1982; Rumelhart and McClelland 1986*a*; von der Malsburg 1987; Anderson and Rosenfeld 1988).

2. Models show that networks can, in principle, self-organize stably on the basis of local rules and do not have to be driven, or supervised. by global programs. This implies that massive alterations in representational properties in learning can be based on local modifications and need not invoke global changes.

3. Robust memories can be, in principle, established by a statistical process in systems with many degrees of freedom. This could be taken as being in line with a view which portrays learning of complex representations as the selection of alternatives from a repertoire of spontaneously generated pre-representations in neuronal networks (Young 1979; Changeux *et al.* 1984; Heidmann *et al.* 1984; Changeux 1985; Edelman 1987). Suppose an organism attends to a sensory stimulus. The sensory receptors activate a set of neurons in the central nervous system. This activated neuronal set encodes a 'primary central percept' of the stimulus. If in learning the primary percept is directly stabilized into an enduring representation, i.e. the primary percept and the stored representation are identical, then the mechanism is said to be instructive, because the external milieu directly instructed the nervous system to form the internal representation. In contrast, if the primary percept initiates a process by which a stable representation is selected among spontaneously generated configurations of neuronal activity, then the mechanism is said to be selective. In the latter case, the external milieu established internal order indirectly (Changeux *et al.* 1984; Changeux 1985).

It is worthwhile to reconsider here briefly the notion of representation in the light of models of learning by selection. Intuitively, we regard internal representations as stylized reflections of the external world. However, in the above neuronal models, representations are internal states which are actually generated by the nervous system regardless of whether or not the external world is there at all. The representations could be regarded as 'anticipated' internal versions of the external world. Some authors would claim that because these internal states are produced endogenously and do not directly mirror the external world, they are not 'representations' in the classic sense of this term (see, for example, the discussion in Skarda and Freeman 1987). However, this is not the stand taken here. In the present context, any pattern of neuronal activity that provides the nervous system with structured information about the world, is a bona fide representation (Chapter 1).

Learning by selection is, of course, deeply anchored in Darwinian philosophy. Here, again, emerges the similarity between learning and development, which was noted in previous chapters. In development, too, mature connections in the peripheral and central nervous system are in some cases selected among a larger number of wiring alternatives (Changeux and Danchin 1976; Mariani 1983; Heathcote and Sargent 1985). A principle similar to that operating in species phylogeny is thus suggested to operate in neuronal ontogeny. In species phylogeny, genetic memory is stabilized by the selection of spontaneously generated genetic pre-representations. This genetic memory generates, in ontogeny, a core of dedicated neuronal systems, as well as the neuronal substrates and mechanisms for the generation of pre-representations, which are in turn selected by learning. Learning by selection may operate on a local level, for example via Hebbian synapses, and on a global level, for example via a bias towards co-operative electrical activity within and between assemblies. As we shall see in Chapters 12 and 13, there are critical developmental periods for the stabilization of some internal representations and hence, probably, for stabilization of some neuronal assemblies.

The need for novel techniques for monitoring neuronal assemblies in action

● What cellular and organizational mechanisms operate in assemblies to generate plasticity, stabilization, selection, gating, etc.? As far as local mechanisms are concerned, there is no need to exclude cellular and molecular mechanisms similar to those that endow dedicated reflexive systems with plasticity. Ion channels and second-messenger cascades also operate in neuronal systems which subserve complex representations. On the other hand, it is likely that novel molecular mechanisms were superimposed on more primitive ones in response to novel evolutionary demands. For example, whereas in reflex modification there is an advantage to relatively enduring traces, in some instances involving the use of temporary complex representations, such as in working memory, there is a need for a reversible transiency of traces. Special molecular and cellular mechanisms may be required for implementing this transiency, but they were not yet identified (spine twitching, mentioned in Chapter 8, is but one hypothetical candidate).

Regardless of whether or not novel molecular mechanisms evolved to cope with memory in complex representations, one point is clear: major efforts in the search for mechanisms of experience-dependent modification in complex representations should be directed at the level of intra- and interassembly interactions. A major obstacle is the lack of satisfactory research technologies. Neurophysiology has for years directed attention to single cells and connections, because they succumb to the classical electrode. In the study of complex representations there is much need for methods for recording coactivity of many neurons. There were some limited attempts to record from several neurons simultaneously by using microelectrodes, and the results have suggested that the activity of single neurons in the brain should be assessed in the context of the activity of other neurons (Abeles 1982; Bach and Kruger 1986, Llinas 1988). Several research groups are currently engaged in improving the technique and theory of multi-electrode recording in the brain (Kruger 1983; Abeles and Gerstein 1988). In addition, in recent years novel optical imaging methods have been developed that permit monitoring the activity of many neurons in the brain by the use of voltage-sensitive dyes, or by relying on intrinsic optical signals which exist in brain tissue and correlate with electrical activity (Grinvald 1985; Blasdel and Salama 1986; Grinvald *et al.* 1986). These, together with additional types of sophisticated techniques for monitoring regional and global neuronal activity, are discussed in Chapter 10. Improved versions of such techniques, when combined with the appropriate behavioural experiments, may cast light on assemblies and on the correlation of their activity with behaviour.

A complementary experimental approach is to search for anatomical loci in the vertebrate brain, in which electrical, metabolic, and/or structural changes occur which are correlated with complex learning. The search for anatomical engrams should tell us where the relevant assemblies are located (or, more accurately, distributed). It should also assist us in understanding how assemblies in various brain regions interact to generate and modify representations, and to guide behaviour.

10 *In search of the topography of engrams*

Early attempts to localize functions in the brain

As noted in Chapter 1, the search for anatomical sites of memory has been with us since ancient times. Its scientific era can be traced back almost 200 years, to organology and its descendant, phrenology. Organology is the hypothesis that different mental faculties are localized in specific organs in the brain. Phrenology further suggests that the size of each cortical organ is reflected in the overlying skull. The practical conclusion of phrenology was, therefore, that mental faculties could be read from the contours of the skull (Boring 1950; Davies 1971; Brazier 1988). The merit of organology does not, of course, rest on the validity of phrenology. It is the latter doctrine that had, unfortunately, gained much popular attraction and, rightly so, scientific disgrace. The great proponent of organology and the founding father of phrenology was the German-born anatomist Franz Gall (1758–1828). It is due to phrenology that Gall was destined to acquire a bad reputation in scientific circles. He was, nevertheless, a gifted and respectable neuroanatomist, and versions of his organology, considering the mind as functionally modular, are well echoed in the modern cognitive sciences (Marshall 1980; Fodor 1983; Gardner 1983). Memory *per se* was not regarded by Gall and his followers as a unitary faculty residing in a specific organ in the brain. In contrast, perception of causality and imitation, so intimately associated with learning, were considered as independent intellectual faculties localized in the frontal lobe.

Although phrenology was wrong, it did boost interest in the localization of function in the brain. The experimental evidence was derived from neurosurgical manipulations in animals, and clinical observations of human patients. The conclusions reached during the nineteenth century ranged from the assignment of general functions to global brain parts, such as perception and memory to the cerebral lobes (Pierre Flourens, 1820s), to the assignment of specific faculties to specific brain areas, such as speech to the third frontal convolution of the left cerebral hemisphere (Paul Broca, 1860s; for further historical accounts see Boring 1950; Brazier 1988). Systematic attempts to localize memory in the brain benefited tremendously from the development of robust tests for animal learning (see Chapters 1 and 2). The pioneers in combining neuroanatomical manipulations with experimental animal psychology were two Americans, Shepherd Franz (1874–1933), and his student and colleague, Karl Lashley (1890–1958).

154

Lashley, and lesions as experimental tools

Franz found that in cats and monkeys trained to perform simple motor tasks or operate puzzle boxes, removal of the frontal lobes interfered with recent but not old memory. The lost habits could still be relearned. He also reported that destroying the frontal lobe in one hemisphere was not sufficient to erase habits. Franz also investigated the effects of brain lesions on cognitive functions in human patients. In general, his conclusion was that the exact localization of function in the brain is impossible (Franz 1912).

Franz's line of research was followed and much extended by Lashley (1929, 1950). Lashley conditioned rats and monkeys to perform various tasks, damaged parts of their brains either before or after training, and measured the effect of the lesions on acqustion and retention. The majority of his studies were concerned with the effects of cortical lesions on the performance of rats in two types of tests. One test was brightness discrimination, in which the animal was trained to prefer an illuminated to a dark alley in a training box, in order to avoid an electric shock and obtain food. The other test was orientation in a maze. Several types of food-baited mazes, differing in their complexity, were used; the most commonly used one is depicted in Fig. 10.1.

In brief, Lashley found that it was the amount of cortical tissue damaged, rather than its localization, that was significant for the performance of animals in the learning tests. This was especially valid for the maze experiments, from which most of Lashley's conclusions were drawn. The findings led him to propose two principles. (1) Cortical areas are

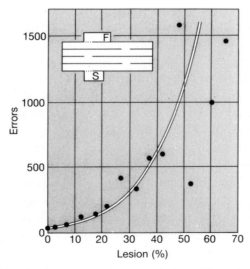

Fig. 10.1. A maze with eight culs-de-sac employed extensively by Lashley in his famous engram experiments (*inset*), and the performance of rats in this maze as a function of the percentage of destruction of cortical area. S, start; F, food. (Adapted from Lashley 1929, 1950.)

essentially *equipotential* for learning, i.e. they can substitute for each other as far as learning is concerned. (In the brightness discrimination test there was, however, differentiation of cortical function, since the occipital cortex was important for retention.) (2) The reduction in learning is roughly proportional to the amount of tissue destroyed. This is known as the *mass action* principle. Lashley also observed that the more complex the behavioural task, the more effective is removal of a given chunk of cortex (Lashley 1929).

Lashley concluded, therefore, that memory traces, or 'engrams', cannot be localized to specific groups of cells and/or pathways. In summarizing more than 30 years of experimentation, he even went further to remark that: 'This series of experiments has yielded a good bit of information about what and where the memory trace is not. It has discovered nothing directly of the real nature of the engram. I sometimes feel, in reviewing the evidence on the localization of the memory trace, that the necessary conclusion is that learning just is not possible. It is difficult to conceive of a mechanism which can satisfy the conditions set for it. Nevertheless, in spite of such evidence against it, learning does sometimes occur' (Lashley 1950).

Lashley's conclusions about the engram were already being questioned at the time of their publication (Hunter 1930). The main objections were, first, that the behavioural tasks were not well defined in terms of the sensory inputs required for successful performance: the animal could succeed in these tasks by using very different sensory channels and behavioural strategies. For example, a rat can learn a maze by using visual, olfactory, auditory, and cutaneous stimuli. Therefore, removal of cortical regions important for one modality would not prevent learning by using other modalities. The second objection was that the cortical lesions were not delicate enough to differentiate the contribution of distinct functional divisions.

Disappointed with the evasive engram, Lashley came to favour field theories of learning, which regarded sensations and reminiscences as patterns of excitations being reduplicated throughout the cortical surface, 'much as the surface of a liquid develops an interference pattern of spreading waves when it is disturbed at several points' (Lashley 1950). This view was compatible with Gestalt theories. Gestalt was a school of holistic psychology maintaining that perceptual qualities should not be dissected and reduced into perceptual atoms, but rather considered as a whole (Koffka 1935).

Gestalt field theories proposed that mental representations are formed and retrieved by continuous current flows on the supposedly homogeneous polarized layer of cortex. Lashley, in the later part of his career, put this proposal to experimental test. He placed strips of gold foil or gold pins on the visual cortex of the rhesus monkey. The metal was expected to short-circuit the cortex or generate local depolarization foci on it. Had the aforementioned version of the Gestalt field theory been correct, the treatment should have disrupted visual perception. This did not happen (Lashley *et al.* 1951). The conclusion was, therefore, that the flow of electric currents in the large field of cortical surfaces is not critical for cerebral integration. Similar conclusions, following similar experimental arguments, were reached at about the same time by additional research groups, including those of Roger Sperry and Karl Pribram. Together, the data demonstrated that Gestalt field theories, when applied to brain physiology, were naïve.

- Lashley's seminal contribution to the study of learning was not in his concrete conclusions, but rather in his methodological and conceptual framework. On the methodological side, he turned much attention to the potential inherent in the combination of neuroanatomical lesions and animal psychology. He was not the first to use this inter-disciplinary combination, but his systematic research programme, as well as his controversial results, stimulated considerable use of these experimental tools. As we shall see later, and in contradiction to Lashley's pessimistic conclusions, lesions can indeed be instrumental in delineating parts of engrams. Lashley also established standards in experimental neuropsychology, by incorporating into his protocols careful post-mortem verification of lesions, and statistical analysis of data. On the conceptual side, Lashley argued that dedicated effector–affector connections could not be considered as elementary building blocks of high brain function. In contrast, he promoted theoretical notions that have survived and been revived, including distributed, multiple internal representations, and emergent network properties (Lashley 1949, 1950; see Chapter 9). Part of these notions were extended, modified, and propagated by one of Lashley's students, whose name we already know: Donald Hebb.

Penfield, and electrical stimulation of the conscious brain

A different approach to the localization of the engram has been practised, starting in the mid-1930s, by an eminent Canadian neurosurgeon, Wilder Penfield, and his colleagues (Penfield and Perot 1963). Penfield's work should be considered against two bodies of earlier observations. One was the studies pioneered in 1870 by Gustav Fritsch and Edouard Hitzig in Berlin, on the behavioural effects of electrical stimulation of specific cortical areas in the dog (Brazier 1988). The second and more pertinent body of observations was that visual and/or auditory hallucinations could occur in epileptics in the course of seizures. This was clearly documented as early as the eleventh century, and might actually have been already reflected in some biblical accounts of the Prophets. Interest in this phenomenon arose again towards the end of the nineteenth century, following reports by the British neurologist John Hughlings Jackson (1888) of epileptics who had seizure-associated hallucinations which could be interpreted as vivid reoccurrences of past experiences.

Penfield and his team extended the study of hallucinations associated with spontaneous seizures, to what they called 'experimental responses' evoked by electrical stimulation of the cortex. Brain stimulation was performed in the course of surgery that was intended to alleviate severe seizures. The seizures were unresponsive to medication and the goal of the operation was to remove the epileptic focus. To delineate this focus, the surface of the cortex, or some inner layers, were stimulated with a small electrode. In the case of cortical stimulation, the area of each contact was approximately 1.5 mm². Since the brain apparently contains no pain receptors, surgery could be performed under local anaesthesia, with the patient reporting his or her sensations. For example, since electrical stimulation of speech areas may cause a transient loss of speech, this method was useful in ensuring that no parts of cortical speech areas were damaged during the operation.

Penfield and his associates operated on more than 1100 patients. In about 8 per cent of these patients, electrical stimulation of the brain evoked various types of hallucinations, including reports that were interpreted as visual or auditory memories, or both (Fig. 10.2). In general, the response was obtained when stimulation was at the temporal lobe. What follows is but one example, cited from Penfield and Perot (1963), which details the reactions, during surgery, of R.B., a 21-year-old man. Stimulation was performed in two loci, marked 5 and 7, in the left temporal lobe:

5. Patient did not reply.
5. Repeated. 'Something.'
5. Patient did not reply.
5. Repeated. 'Something.'
5. Repeated again. 'People's voices talking.' When asked, he said he could not tell what they are saying. They seemed to be far away.
5. Stimulation without warning. He said, 'Now I hear them.' Then he added, 'A little like a dream.'
7. 'Like footsteps walking—on the radio.'
7. Repeated. 'Like company in the room.'
7. Repeated. He explained 'It was like being in a dance hall, like standing in the doorway—in a gymnasium—like at the Kenwood Highschool.' He added, 'If I wanted to go there it would be similar to what I heard just now.'

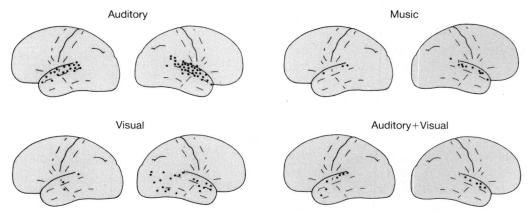

Fig. 10.2. A schematic map of cortical locations where electrical stimulation elicited auditory, visual, and combined auditory-visual responses in a selected group of Penfield's patients. Also shown are loci of stimulation that elicited well-defined musical experience. (Adapted from Penfield and Perot 1963.)

7. Repeated. Patient said, 'Yes, yes, yes.' After withdrawal of the stimulus, he said it was 'like a lady was talking to a child. It seemed like it was in a room, but it seemed as though it was by the ocean—at the seashore.'

7. Repeated. 'I tried to think.' When asked whether he saw something or heard something, he said, 'I saw and heard. It seemed familiar, as though I had been there.'

5. Repeated (20 minutes after last stimulation at 5). 'People's voices.' When asked, he said, 'Relatives, my mother.' When asked if it was over, he said, 'I do not know.' He explained it seemed like a dream.

5. Repeated. Patient said, 'I am trying.' After withdrawal of the electrode he said, 'It seemed as if my niece and nephew were visiting at my home. It happened like that many times. They were getting ready to go home, putting their things on—their coats and hats.' When asked where, he said, 'In the dining room—the front room—they were moving about. There were three of them and my mother was talking to them. She was rushed—in a hurry. I could not see them clearly or hear them clearly.'

7. Repeated. 'Radio. I think it was Philadelphia; it was news.' He said it was at his home.

Generally speaking, the reaction of subjects to focal brain stimulation was unpredictable. Repeated stimulation at the same locus sometimes evoked responses in the same sensory modality but differing in content. In many cases, the experience had the subjective flavour of a dream. Was this a memory, or an hallucination? Penfield concluded that in stimulating the appropriate cortical loci, he randomly activated faithful reproductions of past experiences. Since the so-called memories were random, it was impossible to establish whether a specific engram was lost following surgery, and hence whether it was stored in the excised piece of brain tissue. However, since excision of the stimulated areas did not appear to affect the general memory capabilities of patients, Penfield reasoned that engrams are not encoded in the stimulated foci, but are rather somehow connected to these foci.

Later studies by other groups (Halgren *et al.* 1978; Gloor *et al.* 1982), indicated that mental events similar to those observed by Penfield could be obtained by direct stimulation of the hippocampal formation and amygdala, which are parts of the so-called 'limbic system'. Limbic structures have critical roles in learning (see Chapters 11, 14 and 15). The conclusion was that the vivid experiences of Penfield's patients may have resulted from activation of the limbic system by widespread after-discharges of temporal lobe stimulation. We will be in a better position to evaluate the role of temporal cortex and limbic structures in memory, after acquainting ourselves with the material covered in the coming chapters. In the mean time, the tentative conclusion is that, in spite of the fascinating data, brain stimulations à la Penfield could not determine the location of engrams.

Images of the brain in action

Novel, powerful imaging techniques now enable aspects of brain function to be monitored during the performance of cognitive tasks. These techniques can, therefore, provide information on the involvement of brain regions in situations which require learning and memory.

A most useful technique for ·imaging dynamic aspects of human brain function is positron emission tomography (PET) scanning (Brownell *et al.* 1982; Phelps and Mazziotta 1985). This technique uses isotopes that decay by the emission of positrons (positive electrons). The emitted positron collides with an electron to produce two gamma rays, which travel at 180° apart. Coincident detection of the two gamma rays permits the position of their source to be determined. In practice, a biologically relevant compound, labelled with the appropriate isotope, is administered to the subject; the emitted radioactivity is monitored by an external ring scanner and used to compute a map of the distribution of the compound in the tissue. The isotopes used have a short half-life, in the range of minutes, so that large doses of radioactivity can be administered within acceptable safety limits. This is done to permit detection of minute amounts of isotope. The short half-life necessitates access to an accelerator to produce the tracers on site. The technique is, hence, expensive and limited to relatively few clinical and research centres.

The compounds in which the radionuclide are incorporated, depend on the objective of the test. For example, $C^{15}O_2$ is used for measuring blood flow, ^{15}O for measuring cerebral oxygen consumption, ^{11}C-labelled amino acids for measuring protein synthesis, 2-deoxy-2-[^{18}F]fluoro-D-glucose for measuring glucose metabolism, and ^{18}F-labelled agonists and antagonists for mapping receptors for neurotransmitters.

PET scans can be used to monitor the effect of sensory and cognitive stimuli on the distribution of each of the above compounds in the brain. This is done by administering the sensory stimuli, or the psychological tests, during the scan, and comparing the maps obtained in the same individual in control and test situations. In most cases, 2-deoxy-2-[^{18}F]fluoro-D-glucose ([^{18}F]2DG) is used (Mazziotta *et al.* 1982, 1984; Phelps and Mazziotta 1985; Kushner *et al.* 1988). 2-Deoxy-D-glucose (2DG) is an analogue of glucose which is taken up by the cells and phosphorylated by the enzyme hexokinase, similarly to glucose, but cannot be further metabolized to fructose 6-phosphate. Therefore, deoxyglucose 6-phosphate is trapped in cells proportionally to their glucose consumption (Sokoloff *et al.* 1977). It is assumed that a high metabolic rate is indicative of increased neuronal activity. In the original 2DG method, the brain was sectioned and processed for autoradiography at the end of the experiment, and the map was constructed from the autoradiograms. This auto-radiographic mapping is extensively used in investigating metabolic aspects of brain function, including memory, in laboratory animals (John *et al.* 1986; see also Chapters 12 and 14). The introduction of the positron-emitting isotope ^{18}F enabled the use of 2DG in PET scans, and yielded maps of glucose metabolic rates in the conscious human brain. These maps are reconstructed from images of planar sections at various brain levels. A schematic, highly simplified map is depicted in Fig. 10.3.

● The major conclusions of [^{18}F]2DG PET scan studies are as follows. (1) Mental tasks correlate with changes in glucose metabolic rate in many brain regions. The identity of the

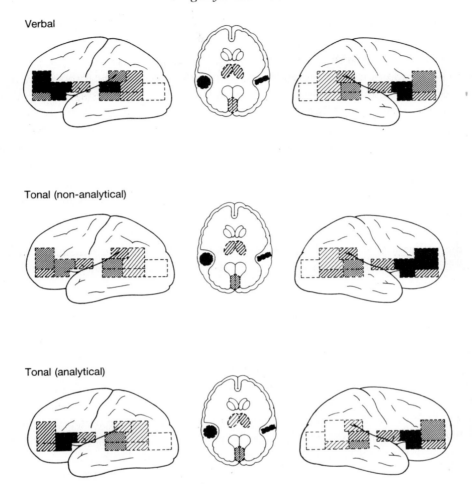

Verbal

Tonal (non-analytical)

Tonal (analytical)

Fig. 10.3. A schematic landscape of glucose metabolism in the brain of individuals engaged in two types of mental tasks. The diagrams are simplified versions of [¹⁸F]2DG distribution maps, constructed from [¹⁸F]2DG PET scans. The analysis was confined to brain regions outlined by broken lines. The shading of each field depicts the glucose metabolic rate, relative to hemispheric mean glucose metabolic rate determined for each individual in control, non-stimulated conditions. Blank circumscribed fields indicate rates below the mean; the darker the shades, the greater is glucose consumption. Shown are lateral views of the left (*left*) and right (*right*) cortical hemispheres, constructed from planar scans, and a planar cross-section (*middle*). The only subcortical region marked is the thalamus (middle of cross-section). The statistical significance of the activations compared to control, and of the asymmetries between hemispheres, were omitted here for simplicity. The verbal task was listening to a Sherlock Holmes story. The tonal task involved presentation of tone sequence pairs, each containing three to five tones of different frequencies, but of equal intensity, duration, and interval. Following a brief pause, the subject had to identify whether the pairs were different. Some of the subjects were musically naïve and considered to perform the task in a non-analytical strategy. Other subjects were musically sophisticated, and considered to process the musical material using an analytical strategy. The metabolic landscape is asymmetrical and different in each task. (Modified from Mazziotta *et al.* 1982.)

labelled regions may depend on the nature of the task. (2) Metabolic changes take place in primary and secondary cortical sensory regions involved in processing the relevant sensory modality. For example, the striate, inferotemporal, and posterio–parietal cortex are active in visual tasks. In addition, changes are detected in other cortical and subcortical regions. (Additional information on primary and secondary sensory processing in the brain is provided in Chapter 14.) (3) The frontal cortex is involved in many different types of mental task. (4) The hippocampal and parahippocampal regions increase their glucose consumption in tasks that clearly involve memory. (5) There is metabolic asymmetry in the brain. Usually, tasks involving linguistic processing are correlated with a relative increase in glucose consumption in the left hemisphere, whereas visual tasks and some musical tasks preferentially activate the right hemisphere. (6) There are individual differences in the metabolic maps. The differences are not only in fine details, but also in gross features. A map obtained in a specific task may depend on the cognitive strategy used by the subject. In an illustrative study, the utilization of glucose in the brain of musically sophisticated individuals, who analyse scores while listening to music, differed from that of musically naïve individuals (Fig. 10.3).

Another imaging technique, specifically designed to measure regional cerebral blood flow, was also used in mapping human brain activity in mental tasks (Chien 1985). This technique, radioangiography, monitors the concentration of a gamma–emitting isotope of an inert gas, ^{133}Xe, in the blood. The gas is usually injected into the carotid artery, and its clearance from various cortical regions is followed by a multichannel detector. The rate of clearance is proportional to the rate of blood flow in the region. It is assumed that the higher the blood flow, the higher is oxygen consumption and hence metabolic activity. An illustrative study is presented in Fig. 10.4. It again shows multiple, asymmetric loci of cortical activity; the identity of these loci depended on the mental task.

Methods for real-time mapping of global aspects of brain electrical activity are also potentially useful in searching for engrams. A classic techique, used extensively in diagnosis and research for many years now, is electroencephalography (EEG). Here macroelectrodes are used to record summated electrical activity of large groups of neurons in the cortex. Recording can be done from the scalp (classical EEG), from the exposed cortical surface (electrocorticogram), or from structures inside the brain (depth electrogram). The classical EEG has the advantage of being completely non-invasive. Voluminous literature exists on the theoretical and clinical aspects of spontaneous and event-related EEG, but its discussion far exceeds the scope of this book (e.g. see Niedermeyer and Lopes da Silva 1982). Two comprehensive research programmes studying EEG correlates of learning are exemplified in John (1967) and Skarda and Freeman (1987). It is of interest to note, especially with respect to our discussion in the previous chapter, that results from the above research programmes, as well as from other similar studies, emphasize the role of endogenous molar activity of the brain in determining the outcome of training. Clearly, the brain is not sitting in quiescence, waiting for the world to tell it something; rather, there is a continuous, dynamic interaction

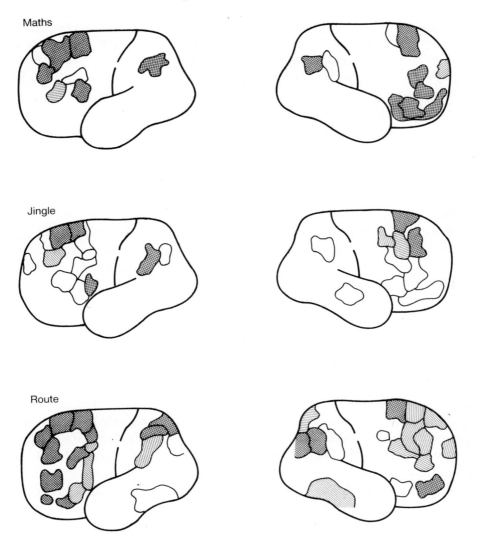

Maths

Jingle

Route

Fig. 10.4. Regional blood flow in human cortical areas, measured by the ^{133}Xe-clearance technique (see text) during the performance of three mental tasks. The first task (Maths) was to start by thinking of the number 50 and then continuously subtract 3 (i.e. 50, 47, 44, etc.). The second mental task (Jingle) was to think about a jingle and to jump every second word in a closed loop. The authors used the following highly meaningful phrase, which they advertised as the best known Danish jingle: 'Ogger gogger gummi klokker erle perle pif paf puf'. In the third mental task (Route) the subjects imagined that they walked out of their front door and then walked alternatively to the left and the right every time they reached a corner. The schematic maps of the left (*left*) and right (*right*) cortical hemispheres depict areas with increased blood flow in each task. The darker the patch, the larger is the increase in blood flow as compared to the flow in a rest period. Note the involvement of patches of the frontal cortex in all tasks, and of the inferotemporal cortex in the route visual imagery task, mainly in the right hemisphere. This region is discussed again in Chapter 14. (Adapted from Roland and Friberg 1985.)

between intrinsic neuronal activity and sensory information (see also a similar conclusion based on cellular electrophysiology in Llinas 1988). Another major conclusion, which is especially relevant to the present discussion, is that even relatively simple conditioning involves activity of neuronal systems in widespread anatomical loci (John 1967). In recent years, the resolution and sophistication of EEG have improved, and computerized, multiple-loci skull EEG now provides rather detailed topographical maps of potentials evoked in the human cortex by sensory and cognitive stimuli (Duffy 1987).

A conceptually related method is magnetoencephalography (MEG) (Sato and Smith 1985). Since active neurons produce electric currents, they also produce magnetic fields that are measured by MEG. The magnetic fields are very weak, and their measurement requires magnetically shielded cages and extremely sensitive detectors which operate at very low temperatures (about 4K). MEG, similarly to EEG, is non-invasive. Both methods provide complementary information on electrical activity in the brain. It can be shown that MEG detects some sources of electrical activity, for example neuronal populations in cortical sulci, with a better sensitivity and spatial resolution than EEG. In both methods spatial resolution is practically measured in centimetres. MEG, like EEG, can generate topographical maps of cortical potentials correlated with cognitive function.

Real-time optical imaging of neuronal activity, already mentioned in Chapters 4 and 9, should also be mentioned in the present context. Two variants of optical imaging are currently available. One monitors the electrical activity of multiple neuronal sites by using voltage-sensitive dyes and a computerized photodiode array photodetection system or video recording (Cohen *et al.* 1978; Grinvald 1985; Blasdel and Salama 1986). The other method uses similar detecting systems but is based on slow intrinsic changes in optical properties of active cortical regions (Grinvald *et al.* 1986). These signals originate from several sources. Some originate from changes in haemoglobin absorption due to local, activity–dependent capillary recruitment; other signals are due to light scattering resulting from activity-dependent alteration in ion movement, transmitter release, the volume of extracellular space, etc. Both optical imaging methods can currently map only relatively small areas of neural tissue. Most importantly, both methods require removal of the skull overlying the brain surface. Therefore, the methods can be used only in laboratory animals or, potentially, in brain surgery. These methods hold promise in monitoring and analysing the activity of neuronal ensembles in localized 'interesting' brain areas, and might be useful in studying distributed networks within engrams (see Chapter 9).

Current limitations of imaging techniques

Two main factors currently limit the exploitation of various types of imaging techniques in searching for engrams. These are, first, limited temporal and/or spatial resolution, and second, the complexity of information in maps of real-time electrical activity of multi-neuron systems, and the lack of satisfactory theoretical concepts and tools which could guide and further the fruitful analysis of this information.

1. Neurophysiological investigations of acquisition and retrieval necessitate temporal resolution in the millisecond-to-second range, and spatial resolution in the submillimetre range. Neither of the global imaging techniques provides such temporal and spatial resolution. In PET scans, temporal resolution is practically measured in minutes and spatial resolution in centimetres. In regional blood flow measurements with ^{133}Xe, the resolution is coarser, and deep brain structures are practically inaccessible. EEG and MEG provide temporal resolution in the millisecond range but spatial resolution is currently in the centimetre range; furthermore, these methods do not provide direct information on sources of electrical activity deep in the brain. Real-time optical imaging using voltage-sensitive dyes has temporal resolution in the submillisecond range and spatial resolution of about 50 μm; real-time optical imaging of intrinsic optical signals has a similar spatial resolution, and temporal resolution in the range of seconds. However, as noted above, these methods are not applicable to the neuropsychological investigation of human subjects, and can be used to analyse only limited patches of cortex on the brain surface.

2. Maps obtained by sophisticated EEG, MEG, or real-time optical recording contain an immense amount of complex data on brain electrical activity. Most of these data are not understood in terms of cellular, assembly, or global brain activity. There is much need for better theories and analytical tools for the analysis of multineuron systems; this is essential for a better understanding of the rich information that some novel imaging techniques already supply.

The conclusion is that at present, imaging techniques only provide us with global landscapes of brain activity, often averaged over periods and areas in which the brain does a lot of computations. However, because of the potential advantages of these techniques in brain research and in the clinic, significant progress is expected in the coming years.

One can infer from global imaging studies performed so far that mental tasks involve widespread activation of many brain regions. This reminds us of some of Lashley's conclusions. But because of the methodological limitations outlined above, we do not yet know which regions are activated within psychophysiologically relevant time windows, and which are only secondary to engrams. Therefore, in the present state of the art, the most powerful methods for searching for engrams still depend on cellular electrophysiology, combined with well-controlled behavioural tasks, neuropharmacology, and ablations. These experiments are, naturally, conducted on laboratory animals, not on humans. The data presented in the following chapters indeed emerge from combinations of classical neuroanatomical and electrophysiological techniques. In some of these studies, we are bound to re-encounter the same cortical and subcortical regions that appear active in PET scans of humans engaged in mental activity. But in these cases, we will be able to gain better insight into the function of the region in discrete behavioural operations. The studies most relevant to the operation of the human brain are, naturally, those conducted on primates. However, faithful to our 'bottom–up' exposition of the subject, we will start our search for engrams in the vertebrate brain with more primitive vertebrates and with simple learning tests.

11 *Fragments of engrams and of memory systems in the vertebrate brain*

The strategy for identifying engrams of relatively simple conditioning in the vertebrate brain

As a matter of fact, we have already searched for simple anatomical engrams, and found them—in molluscs. In both *Aplysia* and *Hermissenda*, identification of fragments of the neuronal trace was a prerequisite for cellular and molecular analysis of learning. Most of the emphasis in that research, and in our discussion of it, was, however, placed on cellular and molecular processes, rather than on the topography of the trace.

Analysis of relatively simple conditioning in vertebrates also succeeded, in some cases, in identifying fragments of engrams. These studies have again addressed learning on different levels; but because of the great complexity of the vertebrate brain, cellular and molecular analysis proved more difficult than in certain invertebrates' preparations. The studies on vertebrates have, therefore, contributed more information on the macro-organization of learning systems than on cellular and molecular mechanisms. In this chapter, we focus on the organization of engrams as revealed by these studies.

Often, the basic research strategy for identifying engrams of relatively simple conditioning in vertebrates is straightforward, and similar to the strategy used in studying the much simpler invertebrates.* First, neuronal systems essential for the relevant behaviour are identified. This is followed by attempts to identify loci of plasticity in the essential systems, as well as in brain regions known to connect with them. Especially useful methods in this respect are inference of function from dysfunction by using ablations, and correlation of electrophysiological activity with behaviour (see Chapter 2). Guided by Occam's razor, a parsimonious causal seriality in the memory trace is generally assumed. Hence, if neuronal activity evoked by the conditioned stimulus (CS) has a shorter onset latency in area X than in area Y, X is considered to activate Y. Moreover, if, following training, changes occur in the activity of X and Y, the changes in the former are considered to contribute to those in

---

* A search for engrams was also conducted in rather complex invertebrates which are not discussed here, among them the octopus (Young 1965) and the honeybee (Menzel 1983; Menzel and Mercer 1987).

the latter. Note that these assumptions are not necessarily valid. For example, X and Y may be located in parallel processing lines.

The answer to the question: 'Where in the brain should one start the search for engram candidates?' depends, of course, on the nature of the conditioned stimuli and responses. In some cases, a reasonable starting point is the input pathway for the CS, because it is the easiest to follow. We start our survey of engram fragments in the vertebrate brain with an example of a study that focused on the CS input pathway. Moreover, this study deals with a fundamental form of conditioned response common to many classical and operant conditioning situations. This is 'fear conditioning' (see Chapter 2).

Conditioned heart rate in the pigeon: elements of a 'fear conditioning' trace

Many paradigms have been developed over the years for studying fear conditioning. The parameter used to quantify the conditioned response (CR) varies. In humans, measurement of the galvanic skin response is common and used, for example, in lie detectors (Saxe *et al.* 1985). In other mammals, startle is a useful indicator (Brown *et al.* 1951; Tischler and Davis 1983). The use of tachycardia or bradycardia is also widespread, although care must be taken since they may be a function of motor activity in addition to fear (Black 1959). Indeed, the relevant system most extensively studied in terms of memory trace is conditioned heart rate in the pigeon (D. H. Cohen 1984).

In this system, light is used as the CS and footshock as the unconditioned stimulus (US). The standardized behavioural protocol, used by David Cohen and his colleagues in their comprehensive analysis of the reflex, is the following (Cohen and Goff 1978): birds are semirestrained and placed in a dark chamber. Whole-field illumination is presented for 6 s on a panel in front of the chamber. This is followed immediately on termination by a 0.5-s mild electric shock to the foot. The UR is an increase in heart rate from a basal rate of 150–190 beats per min to 300–450 beats per min. Intertrial interval is 2.5–5.0 min. Conditioning develops within the first ten trials and reaches a plateau after about 30 trials. The CR is a transient increase of 10–20 beats per min in basal heart rate, with an onset latency of < 1 s and a peak during the final 2 s of the CS.

Tachycardia is but one component of the response

Other components include bursts of high blood pressure and rapid respiration. The behavioural and neuronal analysis was, however, confined in this study to changes in heart rate. So even in studies of this 'diffuse' response, only one behavioural parameter was followed. In addition to the associative changes, non-associative changes in heart rate also took place in response to light during training. These non-associative modifications, which were a reaction to novelty or an 'orienting response' (see Chapter 2), rapidly habituated.

The neural input and output channels, and temporal characteristics of their response

Cohen and Pitts (1968) used ablations and drugs to demonstrate that the final common path for the CR is composed of both the parasympathetic (vagal) and sympathetic innervation of the heart. Cells of origin of the pre- and postganglionic sympathetic innervation, and of the preganglionic vagal cardiac innervation, were identified, and their response studied during conditioning (Gold and Cohen 1981, 1984; D. H. Cohen 1984). Both sympathetic and vagal neurons responded to light prior to training. Following training, sympathetic neurons displayed an increased phasic response to light, whereas vagal neurons displayed a decreased response to light. The picture that emerged was that the conditioned accelerated heart rate is largely driven by a short-latency (*c.* 0.1 s), short-duration (*c.* 0.3–0.4 s) burst of activity by the cardiac sympathetic neurons in response to light onset.

Multiple ascending CS pathways display functional plasticity

The organizational flowchart of the neuronal systems that mediate the CS and the CR in the heart rate conditioning of the pigeon, was worked out in considerable detail (for a simplified scheme, see Fig. 11.1). As was previously noted, analysis of plasticity in this system was until now focused on the CS pathways. At least two, and possibly three, visual pathways mediate the CS from the retina to cortical areas. The thalamofugal pathway leads to the visual Wulst, which is the analogue of the mammalian striate cortex, and the tectofugal and pretectofugal pathways lead to extrastriate regions (Fig. 11.1). Circumscribed lesions were used to demonstrate that each of these pathways is capable of transmitting effective CS information (D. H. Cohen 1984). The next stage in the analysis was to determine whether these pathways are modifiable, or are merely input lines. The retinal output, with an onset discharge latency of about 20 ms, showed no training-induced modification (Wild and Cohen 1985). In contrast, electrophysiological correlates of conditioning were found in the tectofugal and the thalamofugal pathways.

The tectofugal pathway: this consists of a massive retinal projection to the contralateral optic tectum, a tectal projection to the nucleus rotundus of the thalamus, and a rotundal projection to the ectostriatum (Fig. 11.1). The activity of neurons in the ectostriatum was studied first, the rationale being that if neuronal activity in this region is invariable, then the midbrain and the thalamic relays of the pathway would also be invariant with training (Wall *et al.* 1985).

The activity of neuronal units in the ectostriatum was, therefore, correlated with presentation of the CS. The peristimulus neuronal activity was quite heterogeneous and complex. However, there were clear, short-latency transient responses to light onset. This finding, combined with the previous observation that the CR is largely driven by phasic neuronal activity immediately following the onset of the CS, led to restriction of further

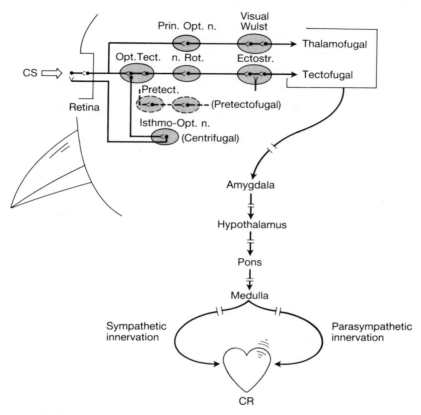

Fig. 11.1. A simplified flowchart of the neuronal system subserving the CS and the CR in the conditioning by light of heart rate in the pigeon. Analysis of plasticity was focused in this system on the CS pathways. Ectostr., ectostriatum; Isthmo-Opt., isthmo-optic nucleus; n. Rot., nucleus rotundus; Opt. Tect., optic tectum; Pretect., pretectal region; Prin. Opt. n., principal optic nucleus of the thalamus, which is considered equivalent to the mammalian dorsal lateral geniculate (LGN). (Modified from D. H. Cohen 1984.)

analysis to an on–light phasic response period of < 0.5 s. Ectostriatal cells were classified on the basis of their responses during this time window. About 80 per cent of the cells were responsive to the CS. About 20 per cent of these increased their discharge in response to light. They were termed 'type I' neurons. The rest, termed 'type II', decreased their discharge in response to light onset. The response latencies were *c*. 75 (type I) or 50–75 (type II) ms, and the burst duration was highly variable, with a mean of *c*. 150 (type I) or *c*. 230 (type II) ms.

The neurons that were unresponsive to light did not change their properties with training. In contrast, some type I neurons in conditioned birds displayed an enhanced discharge to light, whereas type I neurons in control birds displayed attenuation of response with repeated exposure to unpaired light and shock. The response of type II neurons remained constant over training. In conditioned birds these neurons displayed rapid attenuation during the initial training trials, but not afterwards. About 80 per cent of the ectostriatal neurons were also responsive to the US when tested at the end of training.

Since the above results indicated that the tectofugal pathway is modifiable by experience and is not merely an input line, Wall *et al.* (1985) went on to analyse the nucleus rotundus, the thalamic relay of this pathway (Fig. 11.1). Almost all the neurons in this nucleus were responsive to the CS. One-third of the neurons were classified as type I, because they showed an increased discharge in response to light. Their latency of response was *c.* 40 ms, and the mean burst duration *c.* 110 ms. The rest were classified as type II. They had an onset latency and burst duration essentially similar to those of type I neurons. Most type I neurons displayed enhancement of CS-evoked responses with training in conditioned birds, but attenuation in control birds. The CS-evoked response of type II neurons remained unchanged in conditioned birds, but the response was attenuated during initial non-paired presentations of CS and US in control birds. Similarly to the situation in the ectostriatum *c.* 75 per cent of rotundal cells were responsive to US at the end of training. The overall conclusion of these experiments was that experience-dependent modifications in neuronal activity are reflected already at the thalamic relay of the tectofugal path.

The thalamofugal pathway: in this pathway analysis was focused on the principal optic nucleus of the thalamus which, because of its equivalence to the mammalian dorsal lateral geniculate nucleus is abbreviated here as LGN (Gibbs *et al.* 1986). The approach was similar to that described above for the tectofugal system. Approximately 60 per cent of the tested LGN neurons showed light-evoked increase in discharge, at a mean onset latency and a mean duration of *c.* 30 ms. Most of the rest showed decreased discharge on light onset with an onset latency of 30–50 ms and duration of 25 to over 500 ms. Almost all the light-responsive cells also responded to the US. About half of these US-responsive cells displayed decreased discharge and the rest increased discharge to the shock.

A substantial part of the LGN neurons that responded to light with an increased discharge, displayed enhancement of the light-evoked response with training. Such conditioned enhancement, however, occurred only in the neurons that also decreased their discharge in response to shock. In control birds, there was either no change in the light-evoked response or attenuation of it. The response of LGN neurons whose discharge was decreased by light did not significantly change with conditioning, whereas in control birds it was attenuated. One additional point that deserves attention is the fact that the input mediating the decreased response of LGN neurons to shock probably originates from the locus coeruleus (Cohen *et al.* 1982), which is a source of aminergic innervation (Mason 1984). This hints at the role of biogenic amines in fear conditioning.

Tentative conclusions

● Only parts of the neuronal paths that subserve heart rate conditioning in the pigeon, and only limited aspects of neuronal responses in the various stations of these paths, have so far been analysed. Primary loci of change causally correlated with the CR, as well as loci of retention of memory, have not yet been established. Nevertheless, the findings have already resulted in some conclusions. The pathways that transmit information on the CS are not merely input lines; they undergo training-correlated functional modifications. The magnitude, but not the direction, of the CS-evoked neuronal responses is altered. In the CS pathways there are multiple anatomical loci in which plasticity is reflected. In each of these loci many neurons change their activity with training. One might expect to identify multiple US pathways converging on the CS pathways. Candidate engram elements are, thus, localizable, but the whole system is spread out. And finally, associative modifications are superimposed on non-associative modifications, which are encountered particularly during the initial phases of training. Note the similarity of some of these conclusions to those that emerged from our discussion of modifiable reflexes in molluscs (see Chapter 4).

Pathways mediating conditioned fear responses are activated in many 'specific' conditioning situations, but are often overlooked. We will sin similarly in the following discussion of conditioning of 'specific', overt motor reflexes in mammals. Extensively studied paradigms of such motor conditioning use twitching of eye and facial muscles as the CR. One example is eyeblink conditioning in the cat.

Conditioned eyeblink and nose twitch in the cat

Mammals blink their eyes when a potentially noxious stimulus is applied to the eye or to an adjacent cutaneous area. Such a stimulus can, therefore, be used as a US. It can be paired with a neutral, or essentially neutral CS, such as a moderate tone. With proper pairing, the CS will come to evoke blinking. This simple, robust, and easily quantifiable paradigm has long attracted the attention and energy of psychologists and physiologists alike (Martino 1939; Hilgard and Marquis 1940).

A version of an eyeblink paradigm, which is based on a short-latency response, has been employed by Charles Woody and his colleagues in studying anatomical engrams and cellular correlates of conditioning in the cat's brain (Woody 1982, 1984). In these experiments, the CS was typically a 60–70-dB, 1-ms auditory click. A louder click became an effective US for eyeblink and had to be avoided. The US was typically a glabella tap, produced by a solenoid-driven metal rod, that struck a screw previously implanted in the glabella. It elicited either an eyeblink or a more complex response consisting of an eyeblink plus nose twitch (Brons *et al.* 1982). The CR was measured by observation and by recording electromyograms (EMG) from the appropriate facial muscles, i.e. orbicularis oculi (eye) and levator oris (nose). Additional motor CRs, as well as visceral fear responses, were not systematically analysed.

In training, paired CS and US 20–400 ms apart were presented every 10 s. The CR was acquired pretty slowly, and four to six sessions of 150 pairings per day were required to reach a criterion of 80 per cent. Interestingly, the rate of acquisition of the CR could be accelerated by repetitive presentation of the US alone prior to the conditioning training (Matsumura and Woody 1981).

The US led to an UR in the eye and nose muscles mentioned above, with an onset latency of 9 ms (Brons *et al.* 1982). The CS led to a CR in these muscles with an onset latency of 20–40 ms (Woody and Brozek 1969; Woody and Engel 1972). This short latency suggested a relatively simple neuronal trace.

Information flow in the neuronal system

The CS gave rise to responses in the coronal pericruciate cortex with an onset latency of 13 ms. This is a sensory-motor cortex which subserves facial movements and will here be called, for simplicity, 'motor cortex' (Fig. 11.2). The onset latency of evoked responses in the facial nucleus in the brain-stem (Fig. 11.2), the final common efferent of the reflex, was 17 ms. The onset latency of evoked response in the midlateral and suprasylvian cortex was short enough to merit the assumption that it contributed to activation of the motor cortex.

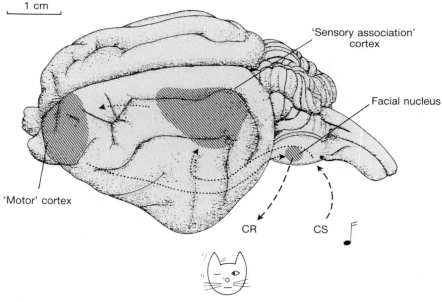

Fig. 11.2. A laterodorsal view of the cat brain, depicting three areas in which neuronal activity correlated with eye-blinking conditioning was identified by Woody and his colleagues. These areas are the coronal pericruciate cortex, dubbed here 'motor' cortex; the midlateral and suprasylvian cortex, dubbed here 'sensory association' cortex; and the facial nucleus. The flow of information suggested by these studies is schematically illustrated by dotted lines, in a highly simplified way: these lines do not represent actual anatomical paths. The view of the cat brain is modified from Tusa *et al.* (1978). A left lateral portion of the cerebellum was removed in the diagram. For the delineation of the 'motor' and 'sensory association' cortices, see Woody (1970) and Woody *et al.* (1976).

The midlateral and suprasylvian cortex is a polysensory area, but will here be called 'sensory association cortex' (Fig. 11.2). Stimulation of the motor cortex with sufficient current produced muscle activity with an onset latency of 7–8 ms. The shortest onset latency of the conditioned response was calculated, therefore, to consist of a latency of about 13 ms of motor cortical neurons fired by the click, about 4 ms of conduction from the motor cortex to the facial nucleus, and about 4 ms from the facial nucleus to the muscles (Woody 1970; Woody *et al.* 1970, 1976; Woody and Yarowsky 1972).

Instead of using a tone as a CS, one could pair a weak electrical stimulation to the motor cortex, that did not produce detectable muscle response in naïve animals, with a glabella tap. This training resulted in eyeblink in response to the weak electrical stimulus, but not to a similar stimulus applied at an adjacent cortical area. The onset latency of the CR was 12 ms or less. Since direct activation of CR from the motor cortex required only up to 8 ms (see above), this implied that under these training conditions, *c.* 4 ms or less were required to access learned information. Such a short time suggested that plastic changes involved in eyeblink conditioning could take place in, or near, the motor cortex (Woody and Yarowsky 1972).

Changes in excitability and activity are detected in the sensory association cortex following training

Unit activity and excitability were studied in the sensory association cortex (see above) of naïve, conditioned, and pseudoconditioned cats (Woody *et al.* 1976). The cells were heterogeneous in their response to auditory stimuli. In about one-quarter of the sampled cells there was no response, and in up to half, the response was a decreased rate of discharge. Only cells that displayed an increased rate of discharge in response to auditory stimuli were further analysed. In these cells, following conditioning, spontaneous and evoked unit activity increased, and the threshold for spike elicitation by a weak extracellularly injected current decreased. The effects were most pronounced in those units that specifically responded to the CS. Increased excitability and activity were also detected in pseudoconditioned animals, but the effects were smaller than in the conditioned animals.

Both associative and non-associative changes are detected in the motor cortex following training

Similarly to units in the sensory association cortex, units in the motor cortex were heterogeneous with respect to their response to the auditory stimuli, but here fewer were unresponsive and only very few displayed decreased activity (Woody 1970; Woody *et al.* 1970; Woody and Engel 1972). After conditioning, the CS evoked greater activity in units in the region subserving the CR. In addition, the threshold for eliciting a muscular response by electrical microstimulation of the cortical area, and the threshold for neuronal spike initiation by intracellular current injection, were lowered in conditioned animals. The effect was specific for cells projecting to the muscles performing a specific CR. Thus, when the US was the usual glabella tap, a lower threshold to microstimulation was predominant in cortical cells projecting to the orbicularis oculi (eye) muscle, whereas when the US was a puff of air directed at the nose, the effect was predominant in cells projecting to the levator oris (nose) muscle (Woody and Black-Cleworth 1973).

A closer look at the experience-dependent modifications in motor cortex excitability revealed their complexity (Brons and Woody 1980; Brons *et al.* 1982). The enhanced excitability of the CR-projective neurons following paired CS–US presentations persisted through behavioural extinction. This was attributed to latent memory unveiled only by saving, i.e. facilitation of acquisition during retraining. Unpaired presentation of the CS was associated with increased CS-evoked unit activity. Unpaired presentation of the US increased the excitability of CR-related neurons to intracellular stimulation, though the effect was transient. The US–CS pairing was not, therefore, obligatory for short-term changes, and had only long-term differential effects. But the effects of training were even more complex. Unpaired US caused long-term (*c.* 1 month) decreased excitability to extra-cellular stimulation, which again was not correlated with conditioning or extinction of the behavioural response. In addition, repeated presentations of the US alone produced long-term increase in the incidence of functional units projecting to the UR-related facial muscles. This result may be relevant to the behavioural observation, mentioned above, that repeated presentation of the US alone facilitated later conditioning.

All together, it could be concluded that associative changes in the motor cortex are superimposed on non-associative modifications. Cellular mechanisms involving second-messenger cascades were proposed to account for excitability changes in motor cortex neurons, but they are not the focus of our discussion here (Woody *et al.* 1978, 1984).

Associative and non-associative changes are also detected in motoneurons in the facial nucleus

Similarly to the picture obtained from the analysis of cortical neurons, complex effects of training were detected in the motoneurons of the facial nucleus in the brain-stem (Matsumura and Woody 1982). The excitability was measured by monitoring the threshold of the intracellularly injected current required to elicit a spike. When measured 1 week after training, excitability was increased in animals trained by US–CS pairing, as well as in animals trained by presentation of the US alone. However, 4 weeks after training, the effect was detected only in the paired animals. Again it appears that associative changes are super-imposed on non-associative changes.

Cortical lesions affect acquisition of the eyeblink response to auditory click

Reversible block of activity by application of 25% KCl to the rostral cortex reversibly suppressed the CR but not the UR (Woody and Brozek 1969). The resultant effects on other cortical or subcortical areas were not established. In other experiments, extensive bilateral lesions of the rostral sensory-motor cortex, but not of the caudal cortex, impaired acquisition of the short-latency eyeblink CR (Woody *et al.* 1974). The rate of spontaneous blinking, the threshold of blinking to glabella tap, and the latency of the unconditioned blink response to loud sounds, were not impaired. The threshold of unconditioned blink response to a loud sound was raised, although the cats were not deaf. The effects of the lesions on the UR to loud noise, and the fact that in most operated cats the lesions affected subcortical structures, complicated the interpretation of these experiments. A recent brief report states again that ablation of the motor cortex prevents conditioning in this paradigm (Woody *et al.* 1988). Taken together, the results of the lesion experiments are in line with the notion that the cortex contributes to conditioning of the short-latency eyeblink response to auditory click in the cat.

Stimulation of the hypothalamus alters the rate of acquisition and the CR onset latency

When a brief electrical stimulation of the lateral hypothalamus was paired with the CS–US presentation, the rate of conditioning increased dramatically (Kim *et al.* 1983). Instead of 500–1000 CS–US pairings, only about 20 were now needed to reach a criterion of 80 per cent CR to the tone. The amplitude of the UR and the CR were also enhanced. The onset latency of the CR depended on the interstimulus intervals used, and ranged from 20 to > 300 msec (Hirano *et al.* 1987).

Single-unit analysis in the motor cortex corroborated the behavioural findings (Woody *et al.* 1983). Fewer than 10 CS–US pairs combined with hypothalamic stimulation were sufficient to augment selectively the cortical unit response to the CS. The very rapid acquisition and increased intensity remind one of fear conditioning. In a brief report, Berthier *et al.* (1982) indeed showed that a similar hypothalamic stimulation was mildly aversive, and could serve as an effective negative reinforcer in the cat. This is in accordance with the results obtained in many other studies, where hypothalamic stimulation effectively modulated motivation and, at the appropriate loci, induced aversive responses (Valenstein 1973).

Tentative conclusions

• Similarly to the conditioned heart rate in the pigeon, here, too, only partial information is available on the trace. It is not clear whether the activity recorded in the cortical areas is causally related to the behavioural change (and see below), or is merely correlated with it. Neither a primary site of learning nor a permanent storage site has been established. Nevertheless here, too, several tentative conclusions emerge. Modification of the eyeblink and nose twitch reflex in the cat can be subserved by circuits of different complexity; hypothalamic input can markedly modulate a candidate trace and the behavioural response. A differentiation thus begins to emerge between subtypes of learning, and candidate circuits subserving them, in an apparently simple reflexive learning. Plastic changes are detected in multiple loci, in both sensory and motor parts of the candidate trace. There is a massive participation of neurons in each locus. Again, as in heart rate conditioning in the pigeon, associative changes are superimposed on non-associative changes.

Much attention has been devoted in recent years to a different type of conditioned eyeblink paradigm, which offers some experimental advantages. This paradigm is concerned with conditioning of eyelid reflexes in the rabbit.

Conditioning of the nictitating membrane reflex in the rabbit

The albino rabbit is especially suitable for studying eye reflexes. It is docile, has widely open eyes, and does not blink them too frequently. Two related reflexes have been used as the CR. One is the closure of the external eyelid (Schneiderman *et al.* 1962), the other the nictitating membrane extension (Gormezano *et al.* 1962). The nictating membrane (NM), or the third, internal eyelid, is a curved plate of cartilage covered with glandular epithelium, which is drawn from the inner canthus laterally across the cornea when a noxious stimulus is applied to the eye. Its extension is largely passive and results from contraction of the retractor bulbi muscle, innervated by the abducens and the accessory abducens nuclei in the brain-stem (Disterhoft *et al.* 1985). The NM extends only momentarily and rarely passes the midline of the pupil. The overt behavioural response can, therefore, be measured with greater sensitivity than with the external eyelid closure. However, measuring the overt NM response is still less sensitive than recording the EMG from the appropriate muscles.

Several conditioning procedures have been developed in this system (Gormezano *et al.* 1962; Schneiderman *et al.* 1962; Disterhoft *et al.* 1977; McCormick *et al.* 1982*c*; Yeo *et al.* 1984; Levinthal *et al.* 1985). They differ in the CS (tone or light), and in the training schedule (delayed or trace conditioning, differing in the stimulus and/or interstimulus intervals). The training schedule determines the behavioural response parameters and, as will be noted below, may also determine the identity of parts of the neuronal system which subserves learning.

The US is typically an air puff to the eye. It is easily controllable and permits artifact-free recording from neurons during its application. As we should already expect, the CR is not limited to the routinely measured response. In addition to eyelid closure and nictitating membrane extension, the air puff may evoke twitching of facial and neck muscles, and may transiently change heart and respiration rates (McCormick *et al.* 1982*c*).

Since classical conditioning of the NM and eyelid response (NM conditioning) is the most studied paradigm of reflexive memory in the mammalian brain, the methodology and the parametric characteristics deserve proper attention. To illustrate them, a procedure employed by Richard Thompson and his colleagues is described (McCormick *et al.* 1982*c*). This study was selected because both the NM and the eyelid response were determined concomitantly.

Rabbits were anaesthetized and a headstage wired for electrophysiological recordings connected to their skull. In addition, a small loop of silk thread was sutured into the nictitating membrane of one eye. After recovery, the animals were restrained and placed in a soundproof chamber. A headgear, containing the proper electronics and an airpuff-outlet nozzle, was attached to the headstage. The eyelids were held open by eyeclips. The movement of the NM was monitored by a torque potentiometer attached to the silk thread. (An infrared reflection transducer could also be used to quantify NM extension; Disterhoft *et al.* 1977.) The response of the eyelid muscle was measured by microelectrodes inserted before each session.

The CS was a 350-ms, 1-kHz, 85-dB tone. The US was a 100-ms air puff, presented after an interstimulus interval of 250 ms. The animals were trained in daily sessions of *c.* 120 trials each. A CR in the NM was defined as movement of the membrane 0.5 mm or greater within the 250 ms after the onset of the CS. Learning to criterion was defined as eight CR in nine consecutive trials. In this study, training was continued until criterion was reached; other groups, using similar kinds of US and CS, reported performance of 70–80 per cent CR following 40–500 trials (Gormezano *et al.* 1962; Disterhoft *et al.* 1977). Altering the intertrial and interstimulus intervals was reported to enhance acquisition (Levinthal *et al.* 1985).

The relatively long onset latency of the CR implicates many more synapses than in the system mediating the naïve reflex

The onset latency of the eyelid muscle UR in the stimulated eye was about 7 ms from the time of arrival of the air puff at the cornea. The latency of the NM UR was about 22 ms. The eyelid muscle CR and the NM CR developed with identical kinetics. The mean onset latency of the eyelid muscle CR on the day after criterion was reached was 80 ms, and that of the NM CR was > 100 ms. Essentially similar latencies were found in studies by other groups, using somewhat different conditions but a similar US (Gormezano *et al.* 1962; Disterhoft *et al.* 1977). Note that the onset latency of the CR here is significantly longer than in the paradigm used by Woody and his colleagues to study conditioning of an eyeblink response in the cat. The relatively long onset latency indicates that conditioning of the NM response in the rabbit involves many more synapses than in the system mediating the UR.

The neocortex and hippocampus are not obligatory for an elementary form of NM conditioning

Bihemispheric ablation of the rabbit cortex had no significant effect on acquisition and retention in protocols of delay conditioning (Oakley and Russell 1977; Yeo *et al.* 1983; Mauk and Thompson 1987). Similarly, delay conditioning of eyeblink to tone, by electric shock to the eyelid, was slowed down but not prevented by decerebration in the cat (Norman *et al.* 1977). Circumscribed hippocampal lesions also did not prevent or retard conditioning of the rabbit NM response in a delay paradigm (Solomon and Moore 1975; Powell and Buchanan 1980). Performance in more complex versions of NM conditioning was, however, influenced by hippocampal lesions. The affected parameters included latent inhibition of conditioning by non-reinforced pre-exposure to the CS (Solomon and Moore 1975); blocking of conditioning to the CS_1 component in a compound CS_1+CS_2 stimulus, by previous pairing of CS_2 with the US (Solomon 1977); facilitation of extinction by successive acquisition and extinction sessions (Schmaltz and Theios 1972); and reversal tone discrimination (Berger and Orr 1983). The possible contribution of the hippocampus to NM conditioning is further discussed below.

The cerebellum is obligatory for NM conditioning

Ample information on the role of the cerebellum in motor behaviour, theories describing the cerebellum as a potential 'motor learning machine' (see below), and incidental observations in electrophysiological and neuroanatomical experiments, have all led several research teams to test systematically the role of this brain organ in conditioning of the NM reflex. In brief, it was indeed found that lesions to the cerebellum prevented acquisition of the CR on the ipsilateral but not the contralateral side (Lincoln *et al.* 1982; Lavond *et al.* 1985). Similar lesions abolished retention (McCormick *et al.* 1982*a,b*; Clark *et al.* 1984; McCormick and Thompson 1984*a,b*; Yeo *et al.* 1984, 1985*a,b*; Woodruff-Pak *et al.* 1985; Solomon *et al.* 1986). The UR was not affected.

A few words on the gross anatomy of the cerebellum are pertinent here, because a minimal terminology is needed to follow the experimental results and the model derived from them (Figs. 11.3 and 11.4). A few additional facets of cerebellar connectivity pertinent to learning models will be mentioned in due course below. Structurally the cerebellum consists of a cortex, an internal white mass called the medullary substance, and pairs of deep nuclei, called the fastigial, the interposed, and the dentate nuclei. The cerebellar input, originating in the periphery, brain-stem, and cerebral cortex, synapses onto the cerebellar cortex and the deep nuclei. This input reaches the cerebellum in two fibre systems: mossy fibres and climbing fibres (Figs. 11.3 and 11.4). The mossy fibres originate from a variety of brain-stem nuclei. The climbing fibres originate in the inferior olive complex in the medulla. The sole output of the cerebellar cortex is provided by the Purkinje cells. Most of this output flows via the deep nuclei. The input and output pathways course through three paired cerebellar peduncles, which connect the cerebellum to the brain-stem.

The experiments cited above determined that NM conditioning is disrupted by lesions to the interposed and dentate nuclei, or even to only a small part of the interposed nuclei; to the superior and middle cerebellar peduncles (carrying fibres from the interposed nuclei, and mossy fibre input, respectively); and, in some studies, to the lateral cerebellar cortex. The role of the latter might depend on the stimulus parameters of the CS and US and on

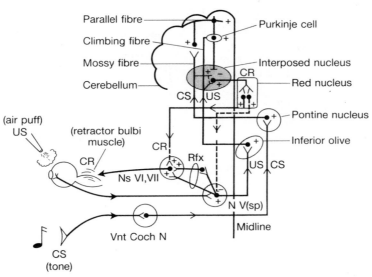

Fig. 11.3. A simplified flowchart of a circuit presumed to be obligatory for conditioning of the NM reflex in the rabbit. The US (corneal air puff) pathway consists of somatosensory projections to the dorsal accessory part of the inferior olive and its climbing fibre projections to the cerebellum. The CS (tone) pathway consists of auditory projections to the pontine nuclei and their mossy fibre projections to the cerebellum. The CR pathway is depicted only in part as projecting from the interposed nucleus in the cerebellum to the red nucleus, and via the rubral pathway to the appropriate motor nuclei. The red nucleus may also inhibit transmission of US information to the inferior olive, so that US activation of the climbing fibres is attenuated when a CR occurs. The interposed nucleus is hypothesized as an essential US–CS convergence and retention site (marked by shading). There might be additional convergence and retention sites, for example in the cerebellar cortex (see Fig. 11.4). For additional components of the circuit, see text. +, excitatory synapses; −, inhibitory synapses, N V (sp), spinal trigeminal (Vth) nucleus; N VI, N VII, VIth and VIIth cranial nuclei; Rfx, minimal naïve reflex pathway; Vnt Coch N, ventral cochlear nucleus. (After Thompson 1986.)

the training protocol (Woodruf-Pak *et al.* 1985). In one study, removal of the cerebellar cortex reduced the CR but did not affect its reacquisition, indicating that the cerebellar cortex is normally involved but is not obligatory for learning (Lavond *et al.* 1987*b*). As we shall see below, the role of the cerebellar cortex in NM conditioning is an important issue which bears on formal learning models suggested for the cerebellum.

Further experiments established that lesions to specific cerebellar input and output paths also disrupt NM conditioning. On the input side, lesions to the dorsal accessory portion of the inferior olive complex prevented acquisition of the NM CR and, in trained animals, gradually extinguished the CR, without affecting the UR. The effect was functionally equivalent to withholding the US (McCormick *et al.* 1985). As to output, an important cerebellar path to the brain-stem runs through the red nucleus in the midbrain. Lesions to the red nucleus and its brain-stem projections disrupted the retention of NM conditioning (Rosenfield and Moore 1983; Rosenfield *et al.* 1985).

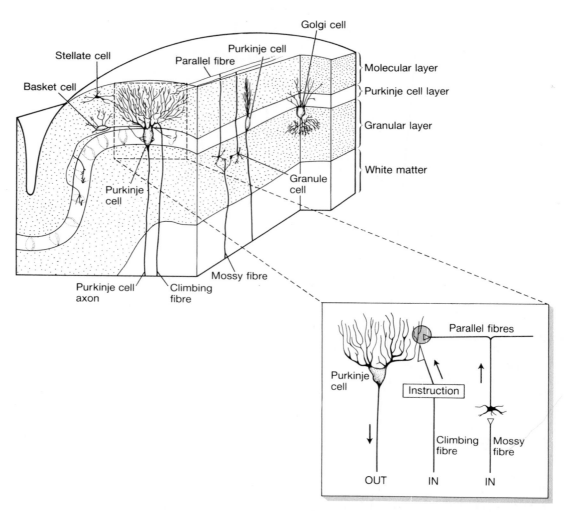

Fig. 11.4. The microstructural simplicity and regularity of the cerebellar cortex attract experimenters and theoreticians alike, and led some of them to suggest that the cerebellum is a 'motor learning machine'. Only five types of neurons—Purkinje, granule, Golgi, stellate, and basket—are organized into the three layers of the cerebellar cortex. The sole output, which is inhibitory, is provided by the Purkinje cells. Input reaches the cerebellum via two excitatory fibre systems, the mossy and climbing fibres. Each Purkinje cell receives converging input from a large number of mossy fibres via *c.* 2×10^5 parallel fibres that are sent by the granule cells. In contrast, each Purkinje cell receives input from only one climbing fibre. See also Fig. 11.3. *Inset:* A simplified, schematic neuronal circuit, showing the convergence of the two major cerebellar inputs on the Purkinje cell. The Marr–Albus models suggested that the climbing fibre instructs the Purkinje cell to respond specifically to the concurrent pattern of parallel fibre activity. The modified synapse is encircled. (Adapted from Ghez and Fahn 1985.)

The role of the cerebellum in NM conditioning was also indicated by electrophysiological studies. Recording of unit activity revealed behaviourally-related responses in the anterior and posterior cerebellar lobes, and in the dentate and interposed nuclei (McCormick *et al.* 1982*a*; McCormick and Thompson 1984*a,b*). For example, in some cases, activity in the dentate and interposed nuclei preceded the NM response by *c.* 40–60 ms. Furthermore, electrical stimulation of the dentate and interposed nuclei yielded, in some cases, eyelid closure and NM extension in both naïve and trained rabbits (McCormick and Thompson 1984*a,b*). In other experiments, electrical microstimulation of the pontine or the lateral reticular nuclei, both of which provide mossy fibre inputs to the cerebellum, was capable of substituting the CS (Steinmetz *et al.* 1985, 1986; Lavond *et al.* 1987*a*). Microstimulation of the dorsal accessory nucleus of the inferior olive was capable of substituting the US (Mauk *et al.* 1986).

A model for circuits subserving NM conditioning

Taken together, the above results led to a schematic model for the involvement of the cerebellum in NM conditioning (Thompson 1986). According to this model (see Fig. 11.3), essential information on the US is provided via the spinal Vth cranial nucleus to the dorsal accessory portion of the inferior olive and its climbing fibre input to the cerebellum. Information on the tone CS is mediated by auditory projections from the ventral cochlear nucleus to the pontine nuclei and their mossy fibre projections to the cerebellum. Information relevant to the CS is also provided by the mossy fibres' projections of the lateral reticular nucleus, but the nature of this information is not clear; it is unlikely to be auditory (Lavond *et al.* 1987*a*). (For pathways mediating visual CS in a visual version of NM conditioning, see Koutalidis *et al.* (1988).) The CR is possibly mediated via multiple pathways (Desmond and Moore 1986). The major one is proposed to project from the interposed nucleus to the red nucleus and then, via the rubral pathway, to motoneurons in the facial, abducens, and accessory abducens nuclei. The red nucleus possibly also attenuates the transmission of somatosensory information about the US to the inferior olive when the CR occurs (Thompson 1986; Donegan *et al.* 1988). The model hypothesizes that US–CS convergence takes place in the interposed nuclei, but does not establish at which cellular site(s) within the nuclei it actually happens, and which microscopic mechanisms are involved.

Hippocampal activity correlates with conditioning of the NM reflex

As noted above, lesion experiments have indicated that the hippocampal formation (see Fig. 6.1) is not essential for elementary forms of NM conditioning. Normal performance in some relatively complex NM conditioning paradigms did, however, depend on an intact hippocampus. Earlier observations on the role of the hippocampal formation in learning in general have led Berger, Thompson, and their colleagues to investigate the role that this brain region plays in conditioning of the NM reflex (Berger *et al.* 1976, 1980, 1983; Berger and Thompson 1978*a,b*). They found that hippocampal unit activity in CA1 and CA3 increased rapidly and markedly during training. This happened first in the US period and later in the CS period. The neuronal activity typically preceded the behavioural activity by *c.* 35–40 ms. Moreover, the pattern of discharge probability appeared to predict the amplitude–time course pattern of the behavioural response (see Fig. 11.5). No such phenomenon was detected in rabbits subjected to control, unpaired training. These results led to the proposal that a 'temporal model' of NM behaviour is established in the hippocampus during training (Berger and Thompson 1978*b*; Berger *et al.* 1980). Note that this did not necessarily imply that hippocampal activity is causally related to the behaviour. One alternative is that the hippocampal activity is secondary to the activity of other brain regions which subserve the behaviour and/or attend the sensory input.

Further analysis revealed heterogeneity in the spontaneous and evoked firing among CR time-locked cell types in, or near, the CA1 and CA3 pyramidal layers (Berger *et al.* 1983). We will return to this point towards the end of this chapter. It was suggested that increased pyramidal activity resulted from enhanced synaptic efficacy of afferent connections, and an LTP mechanism was proposed to underlie learning in this system (Berger and Thompson 1978*b*; Weisz *et al.* 1984).

A set of experiments combining *in vivo* and *in vitro* methods provided further evidence for the correlation of intrinsic changes in the excitability of hippocampal neurons with NM

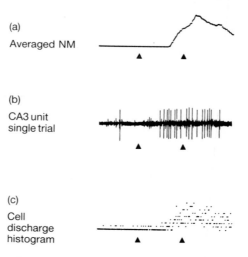

Fig. 11.5. Evidence in support of a 'temporal model' of the NM reflex in the hippocampus. The traces depict the correlation between NM response and discharge patterns of a hippocampal pyramidal neuron in a rabbit, during an advanced stage of training to associate a tone with the NM reflex. (a): The NM response, averaged over a number of trials. (b): Single trial example of discharge pattern of the neuron. (c): Histogram (3-ms bins) of the cell discharge over the same number of trials as for the NM response in the upper trace. First arrow, tone (CS) onset; second arrow, air puff (US) onset. Total trace duration is 0.75 s. The pattern of increased frequency of unit discharge appears to model the amplitude–time course of the behavioural response. 'Temporal models' of NM conditioning were also found in the cerebellum. (From Thompson *et al.* 1984.)

conditioning. Disterhoft *et al.* (1986) trained rabbits with a periorbital shock as the US, sacrificed the animals, prepared slices of their hippocampi, and analysed the properties of CA1 pyramidal neurons in the slice. They found a significant reduction in afterhyperpolarization (i.e. a hyperpolarizing potential succeeding a spike) in cells from conditioned animals. Afterhyperpolarization can attenuate burst activity, and hence its reduction might contribute to an enhanced excitability (a decrease in $I_{K(Ca)}$ may be involved; compare with the *Hermissenda* results, Chapter 5). The same laboratory later also reported enhancement of postsynaptic potentials in CA1 pyramidal neurons, as well as long-lasting translocation of PKC in the CA1 region, following conditioning in a similar protocol (Bank *et al.* 1988; Lo Turco *et al.* 1988). Since the pyramidal neurons in the hippocampal slice were isolated from the rest of the neural system subserving the NM response, it follows that conditioning-related information was stored in these cells.

Tentative conclusions

The notion of a minimal, or essential trace, which emerged in experiments on cat short-latency eyeblink conditioning, reappears here from a different angle and in a different context. The trace normally involved in NM conditioning has multiple components. A candidate essential component involves the cerebellum, and a flowchart model has been proposed for the modifiable pathways there, but the cellular loci of primary change and subsequent retention have not yet been established. The possibility that learning deficits after cerebellar lesions are secondary to subtle sensory and motor deficits, must still be further investigated (Welsh and Harvey 1989). Furthermore, the existence of an essential trace in the brain stem is not excluded by the data. A trace is also present in the hippocampus, which is not essential for motor performance in elementary versions of the paradigm.

More on the cerebellum as a component of learning systems

As noted above, NM conditioning provides but one example of an experimental system that implicates the cerebellum in motor learning. The role of the cerebellum in motor control has long been inferred from clinical observations and ablation experiments. In recent years, the cerebellum and its efferent and afferent pathways have been shown to play a cardinal role in several distinct modifiable reflexes. Several extensively studied examples are mentioned here *en passant*, as a guide for further reading.

Modification of vestibular reflexes

The vestibulo-ocular reflex stabilizes the eye against changes in head position. The signals that initiate this reflex originate in the labyrinth of the inner ear. The reflex is capable of remarkable adaptation. For example, after wearing reversing prisms, eye movements go with, instead of against, head movements (Miles and Fuller 1974; Gonshor and Melvill Jones 1976). Several groups of investigators have studied the involvement of the cerebellum in the reflex (Ito 1972; Robinson 1976). It can be shown that the parallel fibres mediate vestibular signals whereas the climbing fibres mediate ocular signals (see further discussion of the cerebellar cortex below). Masao Ito and his colleagues have reported evidence suggesting that concurrent activation of parallel fibres and climbing fibres induces long-term depression in the parallel fibres synapting on Purkinje cell dendrites, and this contributes to experience-dependent modification of the reflex (Ito *et al.* 1982; Ito 1984, 1987; but see Miles and Lisberger 1981; Bloedel 1987). The long-term depression probably involves glutamate receptor desensitization (Kano and Kato 1985; Ito, 1987; for further studies on long-term depression in Purkinje cells *in vitro*, see Sakurai, 1987; Crepel and Krupa, 1988).

Whereas Ito and his colleagues concluded that the cerebellum is a site of learning and memory in the above reflex, Llinas *et al.* (1975) proposed, on the basis of another paradigm of vestibular-related plasticity, that acquisition and retention take place in a cerebellar input system rather than in the cerebellum. In rats unilateral vestibular damage induces vigorous rolling, which normally disappears a few days after the lesion as the nervous system learns to compensate for the deficit. It has long been known that removal of the cerebellum prevents this motor compensation. Llinas and his colleagues found that a drug-induced lesion of the inferior olive, inflicted prior to a vestibular lesion, prevented the motor compensation. This indicated that the climbing fibre input, originating in the inferior olive, is necessary for acquisition. The question was then posed whether the olivo-cerebellar input is also necessary for retention or, alternatively, whether the modified information is stored in the cerebellum itself. To answer this, the inferior olive was damaged in animals several months after they recuperated from a vestibular lesion. This immediately and permanently reinstated the abnormal motor behaviour characteristic of vestibular damage. The conclusion was, therefore, that retention also requires an intact olivo-cerebellar input, and hence the information is not retained in the cerebellum. (For further arguments in favour of regarding the cerebellum merely as a component in a pathway rather than a memory storage site, see Bloedel 1987.)

Learning of skilled movements

Gilbert and Thach (1977; Thach 1978) recorded from the cerebellar cortex of the monkey while the animal was trained to hold a rod in a fixed position against deflection by an external force. Learning to adjust to a new opposing force was correlated with characteristic alterations in both mossy and climbing fibre activities.

A further note on the red nucleus and conditioning

Finally, since the red nucleus was implicated in NM conditioning, a set of studies that implicate this midbrain nucleus in another type of motor learning might also be mentioned here. Pairing a tone (CS) with an electric shock (US) to the forelimb of a cat led to limb flexion in response to auditory stimulus; conditioning was found to depend on an intact cortico-rubrospinal pathway (Smith 1970). Nakaakira Tsukahara and his colleagues modified the paradigm by replacing the tone with electric stimulation of the cortico-rubral fibres in the cerebral peduncle as the CS, while restricting the cerebral cortifugal outflow to the cortico-rubrospinal pathway by sectioning the cerebral peduncle caudal to the red nucleus. Conditioning resulted in forelimb flexion in the response to previously ineffective stimuli delivered to the cortico-rubral fibres. Tsukahara and his colleagues investigated the cellular correlates of conditioning and suggested that learning is associated with the formation of new functional synaptic contacts by the cortico-rubral fibres on neurons in the red nucleus (Tsukahara 1981, 1984; Tsukahara *et al.* 1981; Murakami *et al.* 1988).

Formal models of the cerebellum as a learning machine

The neuroanatomy and electrophysiology of the cerebellum have given rise to the formulation of detailed theories of cerebellar function (Eccles *et al.* 1967; Marr 1969; Albus 1971; Eccles 1977; Ito 1984; Pellionisz and Llinas 1985). Many of these theories address learning in particular. What attracts theoreticians, besides the microstructural simplicity and regularity of the cerebellar cortex (see Fig. 11.4), is the fact that the Purkinje cell, the sole output of the cerebellar cortex, is a convergence locus for the two cerebellar input systems: the mossy fibres, via the parallel fibres which are axons of the granule cells, and the climbing fibres. The input of many mossy fibres, but of only one climbing fibre, converges on any Purkinje cell. Such an arrangement makes attractive the idea that concurrent activation of a climbing fibre and parallel fibres plays a role in experience-dependent modification of the Purkinje cell.

Marr (1969), for example, suggested a formal learning model in which an inferior olive neuron, via its climbing fibre, somehow instructs a Purkinje cell to respond specifically to the concurrent pattern of parallel fibre activity (see Fig. 11.4). Albus (1971) entertained a closely related model. These models predict modification of the synapses from the parallel fibres to the Purkinje cell during learning, and thus propose the cerebellar cortex as a locus of plasticity. Results obtained with adaptation of the vestibulo-ocular response and learning of skilled movements, mentioned briefly above, are compatible with this proposal. The results on motor compensation of vestibular damage are incompatible. The NM results agree with a scheme in which the climbing fibres 'teach' (US) and the mossy fibres mediate 'learning' input (CS); however, many studies of NM conditioning implicate deep cerebellar nuclei, rather than the cerebellar cortex, as being essential for the memory trace. This requires models in which the interaction between the 'teaching' and 'learning' inputs takes place not on the Purkinje cell but in deep nuclei neurons (see Fig. 11.3). A role for the cerebellar cortex in NM conditioning in the intact animal is, nevertheless, advocated by some authors (Yeo *et al.* 1984, 1985*b*; Yeo and Hardiman 1988), and is not excluded by others (Gellman and Miles 1985; Woodruff-Pak *et al.* 1985; Thompson 1986; Lavond *et al.* 1987*b*).

More on the hippocampus as a component of learning systems

NM conditioning also provides but one example of an experimental system that implicates the hippocampus in learning. Additional findings relevant to the role of the hippocampus in learning in monkeys and humans are addressed in Chapters 14 and 15. Meanwhile the discussion is restricted to some additional evidence that emerges from studies of lower mammals.

Several groups have studied the activity and excitability of hippocampal cells during and after conditioning in tasks ranging from straightforward Pavlovian learning to complex instrumental learning. One example is provided by the extensive set of studies conducted by James Olds and his colleagues in the late 1960s and early 1970s (Olds *et al.* 1972). Their starting point was the expectation that classical conditioning involves rerouting of nerve impulses within well-defined pathways in the brain. For example, after training to associate a CS with a US, excitation in response to the CS should first take place in an 'old' pathway up to a point, and then be routed into a pathway which evokes the CR. Olds regarded the rerouting junction as a 'learning centre', and suggested that it might be identified by comparing electrical activity in the brain of an animal before and after conditioning, and locating novel or altered responses with a short-onset latency following the CS. The argument that short-latency novel responses are necessarily indicators of 'primary' learning changes could be questioned, since if these responses are recorded late in conditioning, they might in fact derive from longer-latency responses which began to appear early in training (Thompson 1976). Nevertheless, this experimental approach may yield information on some temporal and anatomical characteristics of conditioning.

Olds and his colleagues used relatively unrestrained rats, and behavioural tasks somewhat more complex than those encountered so far in this chapter. The animals, chronically implanted with multiple microelectrodes in various brain areas, were typically trained to associate an auditory signal with a food reward. At the preliminary part of the experiment, the rats were instrumentally trained to retrieve the food from a dispenser. Later, the rats were conditioned by repeatedly presenting them with a tone temporally coupled to the food. Another tone was not reinforced and served as a control. The onset latency of the motor CR, i.e. overt movement, was typically 250–350 ms. In certain experiments, extinction was also performed, by following the conditioning session with pseudoconditioning trials.

In the first series of experiments, the existence of conditioning-associated new or greatly enhanced unit responses, whose onset latencies were equal to, or shorter than, those of the auditory sensory responses in the inferior colliculus and the medial geniculate, was considered to be a potential indicator of 'learning centres'. The onset latency of response in these two stations in the central auditory pathway was < 20 ms from the onset of the auditory stimulus. In later experiments, units with longer-onset latencies were also considered.

Putative 'learning centres', defined as above, were mapped in all levels of the brain, but not in all areas (Disterhoft and Olds 1972; Olds *et al.* 1972, 1978; Segal and Olds 1972; Kornblith and Olds 1973; Segal 1973; Disterhoft and Stuart, 1976). Only a small percentage of all the sampled units exhibited short-latency, conditioning-correlated responses. These units differed in their onset latency of response as well as in their degree of CS specificity. The units were present in the pontine reticular formation and the ventral tegmentum, but absent in the dorsal midbrain reticular formation and the tectum. They were present in large proportion in the posterior nucleus of the thalamus, in smaller proportion in the medial geniculate and the lateral thalamic nuclei, and absent in other thalamic areas. In addition, they were present in the hippocampus and in the neocortex. The largest specificity for the conditioned tone was displayed by units in the latter two regions. Extinction led to return of the activity of most units to preconditioning level.

Analysis of large samples of neurons led to partial schemes of the flow of CRs in various brain regions. Results for the hippocampus illustrate the data (Segal 1973). During pseudo-conditioning, the medial septal nucleus and the cingulate cortex exhibited short-latency (12–24 ms) responses to the tone. In contrast, at that stage of the experiment, only a small proportion of units in areas CA1 and CA3 of the hippocampus exhibited a significant response to the tone. On conditioning, medial septal units still displayed the shortest response latencies, but these were no different from the response during pseudoconditioning. Units in area CA3 displayed new conditioning-correlated responses with the shortest-onset latencies. New conditioning-correlated responses with longer-onset latencies were detected in CA1 and the lateral septum, efferents to CA3. The latencies of CA1 units changed, however, with training, and finally became shorter than those of CA3. The dentate and the entorhinal cortex, afferents to CA3, also displayed longer latencies than those of CA3. In addition, conditioning-correlated changes were detected in the cingulate cortex.

On extinction, the responses in the dentate were the first to be extinguished, followed by those in CA3. Extinction-correlated changes were also detected in CA1 and the ventral hippocampus, but could hardly be detected, within the time course of the experiment, in the medial septum and the cingulate cortex. Taken together, the results could be interpreted as suggesting that different hippocampal areas fulfil different, some possibly parallel, functions during conditioning. Area CA3 seemed a major focus of change in the transition from pseudoconditioning to conditioning, and from conditioning to extinction.

Studies along similar lines were later extended. Segal (1977*a*, *b*) probed the response of cells in CA1 by electrical stimulation of the commissural pathway, and found conditioning-dependent enhanced excitability and the emergence of a new type of cellular response, possibly dependent on aminergic transmission. Disterhoft and Segal (1978) trained rats to correlate one (CS$^+$) but not another tone (CS$^-$) with food, and then retrained the animals to reverse their preference. Following retraining, the response rates of hippocampal neurons were more closely related to the increased movement response to the former CS$^-$, which now signalled food. This was in contrast to neurons in the motor cortex, where response was more closely related to the former CS$^+$, which became neutral after reversal. In other words, the hippocampal neurons were more attentive to novelty.

Ample electrophysiological data on units in the hippocampus (as well as in some other brain regions) were generated by the above studies. These results have directed additional attention to possible roles of the hippocampal formation in learning. The difficulties in interpreting the extensive data stemmed mainly from the lack of sufficient information on the properties of identified classes of units and their correlation with specific behavioural acts.

The hippocampus as a cognitive map, and its involvement in perceptual and working memory

Several groups of investigators reasoned that more meaningful information on the role of the hippocampus in behaviour in general, and in learning and memory in particular, could be obtained if the animals were subjected to situations somewhat more 'ecological' than those mentioned so far. The rationale was that the hippocampus may have roles which are not expressed when a restrained animal is required to blink its eyes, or wait in a dull environment for a fixed dispenser to release a reward. The experimental strategy was modelled after the one used in studying the 'receptive fields' of sensory units, except that the hippocampus is many synapses away from sensory neurons, and what stimulus attributes determine the 'receptive field' of its neurons could not be known a priori. Therefore, the way to proceed here was to observe the animal in a variety of spontaneous or imposed situations, and correlate behaviour with unit firing. As Ranck (1973) noted, here the behaviour of the neuron first shapes the behaviour of the experimenter, and only later the experimenter asks what shapes the behaviour of the neuron.

Observation of unrestrained rats in situations which partially resemble natural ones, for example exploration of mazes, proved useful in this context. Some prototypes of the experimental procedures, and the kernels of the rationale underlying them, can clearly be traced back to the signal studies of Willard Small at the turn of the century, Walter Hunter in the 1910s, and Edward Tolman in the 1920s to 1940s (Small 1901; Hunter 1914; Tolman 1938; see also Lashley's work in Chapter 10). These procedures combine operant conditioning with incidental learning. We are, thus, taking one step further towards analysis of potentially more complex representations in the mammalian brain.

Studies conducted with rats and mazes in recent years indicated that an intact hippocampus is indeed critical in, and its activity correlated with, a great variety of learning situations. What the most important parameters common to all these situations are, is still a matter of dispute. Among the major parameters analysed in recent years were spatial environmental cues (O'Keefe and Nadel 1978), the contextual significance of stimuli (Solomon 1980), and the recruitment of working memory (Olton *et al.* 1979, 1980; Becker and Olton 1982; Walker and Olton 1984). These parameters are not mutually exclusive.

We now illustrate selected facets of hippocampal function in non-primate mammals by sampling a set of experiments conducted by John O'Keefe and his colleagues on the role of the hippocampus in spatial learning (O'Keefe and Conway 1980; O'Keefe and Speakman 1987).

The notion that the hippocampus is involved in spatial tasks was based on the observations that hippocampal lesions disrupt such tasks, and that the activity of hippocampal neurons can be correlated with spatial cues. We examine here the latter kind of evidence. Electrophysiological analysis of freely moving rats reveals two major types of units in the hippocampus. Some of the cells display high-firing frequency, short-duration, and simple action potentials, and increase their activity when the animal moves. These cells were termed 'theta units', because they correlate with the slow, 4–12-Hz so-called 'theta waves' of the limbic system's EEG (Ranck 1973). In contrast, the majority of hippocampal units have a complex spike pattern, and could be further divided into subtypes on the basis of spontaneous and evoked activity and behavioural correlations. The activity of many such 'complex spike' units could be correlated with the animals' position in space (O'Keefe and Nadel 1978; O'Keefe 1979; McNaughton *et al.* 1983; Muller *et al.* 1987). They were named, therefore, *place units*. Patches of the environment in which a 'place cell' fires were named *place field(s)* of this cell (Fig. 11.6).

Place units and place fields have intriguing properties. They do not depend on a single sensory modality. Wide variations in behaviour or psychological state were reported by several groups to have only minor effects on firing within the place field. Units may have several place fields. Adjacent units have different fields of the same environment, indicating that they do not form a 'projectional map' (p. 144) of geographical space. One can shuffle spatial cues and show that some place units have place fields correlated with the shifted cues (O'Keefe and Conway 1978; O'Keefe 1979; Muller and Kubie 1987).

If place units indeed map the environment, detect in it novel cues, and adjust to them, then such units could provide a powerful, manipulable tool for studying the neuronal basis of complex internal representations and their modification by experience. In an illustrative study towards that end, O'Keefe and Speakman (1987) recorded single-unit activity in the hippocampal CA1 pyramidal cell layer of the rat while the animal was performing spatial memory tasks. Basically, the task was to choose the correct arm of a four-arm maze in order to obtain food reward (Fig. 11.6). The maze, centred within a black-wall circular enclosure, had elevated detachable arms with low side- and end-lips, so that the rats could easily sense the environment. The enclosure contained six controlled extramaze spatial cues: a dim point source light, a white card, a fan, a black towel, an aromatic marker pen, and a small wire cage containing two spectator rats. The goal arm contained food pellets. The location of the goal arm was rotated in multiples of 90° from trial to trial, and could be identified by the extramaze cues, which were kept in the same spatial relationship to each other and to the goal. The maze itself was left stationary and formed part of the static-background geographical cues.

At the beginning of the experiment, the rats were allowed to explore the maze and retrieve the food reward in the goal arm. The goal and start arms were varied randomly, and, as mentioned above, the controlled extramaze cues rotated with the goal. Later, the rats

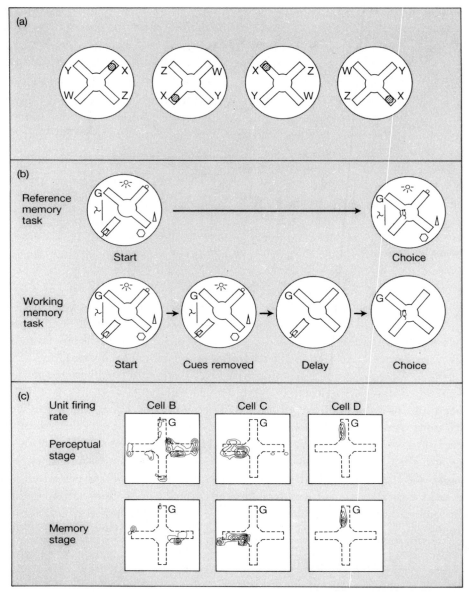

Fig. 11.6. The hippocampus as a cognitive map. (a): An idealized experiment detecting a place unit, correlated with extramaze spatial cue(s). Four controlled extramaze cues, W, X, Y, and Z, are rotated with a fixed intercue orientation, while the maze as well as other salient features of the environment (not shown) are kept stationary. A rat (not shown), chronically implanted with electrodes in the hippocampus, is placed in the maze. The cell fires only when the rat is at a specific spatial location correlated with the controlled cues, irrespective of the 'static' environmental cues. The hypothetical place field, identified by the cell's firing, is schematically depicted by a circle. Its detailed dependence on one or more of the four extramaze cues could be determined by further manipulating these cues. (b): Schematic description of spatial reference and working

were presented with two types of tests. These were spatial reference memory (SRM) and spatial working memory (SWM) tests (Fig. 11.6). In SRM tests, a rat was placed in the start arm, and allowed to choose the goal while the extramaze cues were present throughout the trial. Following successful performance, the rat was further presented with the task, but at this stage it had to spend 30–120 s locked in the start arm before being allowed to run. This was done to yield information on unit activity in a 'perceptual' or 'registration' period. In some experiments, the goal was baited only after it had been chosen, to rule out the sight or smell of food as a cue. Following SRM tasks, the rats were presented with SWM tasks. Here, the cues were available only during the first 30–120 s' 'perceptual' period of each trial, while the animal was locked in the start arm. At the end of this perceptual period, the rat was allowed to choose, with delays of up to 60 s between the disappearance of cues and the beginning of choice. Rats had no difficulty in learning the task, and could actually remember similar tasks for at least 30 min (O'Keefe and Conway 1980).

The rats' behaviour was monitored with an overhead camera, and the activity of CA1 place units recorded with a chronically implanted double-tip microelectrode system ('stereo-electrode'), that allowed simultaneous identification of several neighbouring single units in multiple-unit recordings. Statistical analysis was performed of the correlation between unit activity and the following factors: (1) the spatial location of the animal with respect to the controlled extramaze cues and the goal; (2) the spatial location with respect to the static-background geographical cues; and (3) the spatial location within each arm of the maze.

The firing pattern of approximately 80 per cent of all the units analysed during SRM tasks correlated with at least one of the above spatial factors. The place fields delineated by the activity of approximately 60 per cent of the total units correlated with controlled extramaze cues. The space fields of some cells correlated with combinations of controlled cues and static background.

---

memory tasks performed in a cue-controlled environment. A plus-shaped maze was centred within a black-wall circular enclosure. Six controlled extramaze spatial cues were a dim-point source light, a white card, a fan, a black towel, an aromatic marker pen, and a small wire cage containing two rats. The goal arm (G) contained food pellets. The location of the goal arm was varied from trial to trial, but the extramaze cues and the goal were kept in the same spatial relationship. Two main tasks were used: a spatial reference memory task (SRM), in which the cues were present while the rat was allowed to run to the goal arm; and a spatial working memory task (SWM), in which the extramaze cues were removed after the perceptual stage before the rat was released into the maze. (c): Place fields of three adjacent hippocampal units B–D recorded in an experimental situation similar to that described in (a). The place fields are marked by contours of spike frequency, recorded at the indicated maze positions (contour map calibrations and the number of trials per plot are omitted for simplicity). The fields extend beyond the confines of the maze because the latter was elevated and had no sides, so the rats could lean out beyond the edges of the arms. Perceptual stage: place fields recorded during the entire SRM trials and during the perceptual phase of SWM trials; Memory stage: place fields recorded in the memory phase of SWM trials. The place fields for each unit are similar both in perception and memory. (For further details, see text.) (Modified from O'Keefe and Conway 1978, 1980; O'Keefe and Speakman 1987.)

In the SWM task, 90 per cent of the units with extramaze-correlated place fields maintained their place fields during the retention phase of the trial (Fig. 11.6). When in an SWM task a rat was forced to make a detour and enter a non-goal arm before being allowed to enter the goal arm, the place fields, identified by recordings made in the detour arm, were very similar to those recorded in the same arm when it was the start arm during perceptual periods. Note that the rat could not know in advance which detour arm it would be forced to enter, and that the extramaze cues were no longer available when the detour was made. These results ruled out the possibility that the animal was guided by memory of the specific turn required to reach the goal, and suggested that during the registration period a spatial representation was set out for the entire maze.

• The SWM experiments demonstrated that the place units were not active merely due to perception of sensory information. Unit activity was correlated with the position of the cues in their absence. But what does this correlation mean? Are place units a critical locus of plasticity? Are they integral parts of specific engrams, or only components of an accessory system for attending information stored in other circuits? Some hints concerning possible answers to the above questions, may be obtained by further discussion of the role of the hippocampus in learning.

The many faces of the hippocampus

What is the relation between the activity of hippocampal units in the different tasks discussed so far? Were those the same, or different types of units, that responded in Berger and Thompson's NM conditioning of restrained rabbits, in Olds and Segal's reward-conditioned experiments of freely moving rats, and in O'Keefe and Nadel's maze-exploring rats? This cannot yet be answered, just guessed, because often investigators fall in love with their own paradigms and do not replicate protocols devised by others; this makes the comparison of the data from different laboratories very difficult. Berger *et al.* (1983) did compare their unit data to the Ranck (1973) classification, and identified the neurons time-locked to the NM CR as complex spike cells, and other neurons which burst rhythmically during conditioning as theta cells. Best and Thompson (1984) briefly reported that place units showed cue-correlated changes in activity (but not in the place field) during 'fear conditioning' which associated a tone with a foot shock. In contrast, Christian and Deadwyler (1986) reported that place units did not increase activity during any phase of a tone-discrimination water-reward conditioning, whereas theta cells did.

A straightforward method of determining the relevance of a given type of unit to different behaviours is to monitor continuously the activity of this unit while the animal is engaged in a variety of consecutive behaviours. This is not easy, and is not often done. Ranck (1973) did it systematically in his analysis of the response repertoire of hippocampal cells, but did not correlate the activity with spatial cues, which have since become an important criterion in classification. A recent example of such a multifaceted approach is provided by Eichenbaum *et al.* (1987). They subjected their rats to a quite complex training and testing protocol that even a graduate student might fail. In brief, a water-deprived rat with implanted electrodes was placed in a reasonably spacious chamber which had a small odorant-delivering port ('sniff port') in one of its walls, and a water-reward cup at the opposite end. The rat had to learn to poke the sniff port for 2 s on termination of a tone, and to run to get the water reward only if a particular odorant (S$^+$) was delivered through the port. The experimental contingencies reinforced a stereotyped sequence of behaviours: the rat would sit near the middle of the chamber until the tone stopped. Then it would run to the port, poke its nose, and either maintain it there on delivery of S$^+$, or withdraw it on delivery of the non-reward associated odorant (S$^-$). Following the appropriate response to S$^+$, the rat would run to the reward cup and happily drink its water. Following S$^-$, the rat would either run to the cup but find no water, or stop at the middle of the chamber and wait patiently for the next chance.

About 85 per cent of the units that were held for a sufficient time to allow analysis, displayed behavioural correlates. Three major cell categories were thus identified. 'Cue-sampling' cells fired 0.3 s or later after the onset of odour sampling. Their response increased with training, and was greater during sampling of S$^+$ than during sampling of S$^-$. Moreover, firing was greater when the stimulus on the preceding trial was S$^-$, than if it was S$^+$. 'Goal-approach' cells fired prior to arrival at either the sniff port or the water cup. Both 'cue sampling' and 'goal sampling' fired complex spikes (although with different probabilities). About 40 per cent of all the complex spike cells had place correlates, but their activity could be correlated with the behaviour performed at the same place as well as with the place *per se*. The third category of cells corresponded to theta units. They increased firing during all approach behaviours and decreased it in stationary behaviours, such as odour sampling. Their activity was synchronized with sniffing bouts. Certain behaviourally relevant parameters, such as the trial–initiation signal, or the specific odour identity, were not correlated with unit activity.

Taken together, the different cell types seemed to play complementary roles in the modified behaviour. The cells were classified according to when they fired, and correlated with a behaviour which seemed salient to the experimenter. The question still remains: do we interpret the meaning of the experimental situation as the rat brain does? In other words, are places, or behaviours, or motivation the critical parameter(s) that drives a specific cell to increase its activity? Or are there other stimuli, either environmental or secondary to the activity of other brain centres, that are more important?

Despite the very different tasks used in the different conditioning paradigms described in this chapter, in all of them behaviourally correlated hippocampal units could be detected. This suggests that the activity of hippocampal cells is relevant to global computations, possibly performed on supramodal attributes. In Chapters 14 and 15 we return to the global roles of the hippocampus in learning and memory. In the mean time it becomes clear that whatever the roles of the hippocampus are, some of them become unveiled only when an animal performs tasks that are far more complex than Pavlovian conditioning.

Some conclusions concerning memory traces in the vertebrate brain

• Several general conclusions emerge from our discussion of engram fragments in the vertebrate brain. Even relatively simple learning tasks may involve multiple traces and multiple loci of plasticity. In no case has it been conclusively determined that a specific part of the trace is indeed a primary locus of learning or a permanent site of storage. In some systems, sensory input pathways are modified, and in others motor centres are also modified in learning. In yet other cases, brain systems that are several synapses away from sensory or motor centres show changes with training.

An interesting point is that in most studies, the activity of many sampled cells did correlate saliently with behaviour. This is difficult to reconcile with a learning system composed of exclusive, dedicated circuits, because there would not be enough cells to encode an appreciable number of memories. Rather, it is compatible with the notion that any single neuron subserves a multiplicity of engrams (see Chapter 9). The question is what does the event-related activity of single neurons mean? On the one hand, note that if complex representations are encoded in an ensemble code in large distributed neuronal assemblies, then single-unit activity need not necessarily make sense to the observer. The fact that it does suggests that many units in the assembly serve as nodes in hierarchical subsystems, i.e. windows that disclose meaningful operations of large parts of the system. Moreover, localized encoding may actually coexist with ensemble encoding in the same system (Georgopoulos *et al.* 1988). Therefore, even without yet being able to analyse whole assemblies, the study of complex memories in the mammalian brain appears rewarding. On the other hand, the possibility remains that single-cell activity in complex systems merely reflects a limited repertoire of 'primitive' operations common to many different internal representations, exactly as analysis of biochemical reactions in homogenates reveals 'primitive' molecular operations common to many different types of cells. In such a case, without knowing the global context of this cellular activity, we would not expect to learn interesting things about individual representations. Only analysis of molar activity of circuits and neuronal assemblies will decisively tell us what is the relevance of single-unit activity to the global output of a system encoding complex representations, and what is the role of unit activity in the modification of complex engrams.

Another point, suggested by some of the studies described in this chapter, is that we should probably expect to encounter differentiation in engrams along the following lines: some parts of the trace should reside in brain regions that deal with modality- and task-specific information, whereas other parts of the trace should reside in brain structures that subserve many types of engrams, and could therefore be regarded as basic components of memory systems. The basic components may fulfil two roles. First, they could perform computations obligatory for the generation and/or retrieval of different types of internal representations. Second, they could encode high-order representations, i.e. internal representation of groups of other representations, hence cognitive indices important in the management of memories (Hirsh 1974; Teyler and DiScenna 1986). In reality, of course, all the parts of a trace are intimately interwoven. However, this notion of differentiation does have a methodological merit, because it directs one to look for some parts of the trace in brain structures that are specifically implicated in a specific behaviour, for example sensory association cortex; and for other parts of the trace in brain structures that are implicated in memory by many different tasks. Two candidates for the latter are the cerebellum (for motor aspects of memories) and the hippocampus.

How general are the above conclusions? The answer must await further discussion of memory systems in the vertebrate brain. Before we proceed to deal with more complex memory systems in mammals, we should address two special cases of memory in vertebrates. These special memories, which are acquired only during restricted ontogenetic periods, cast intriguing light on the combined role of genes, development, and individual experience in moulding representations.

12 *The establishment of innately predisposed representations in sensitive periods: imprinting*

Filial and sexual imprinting, biological mechanisms for acquiring social preferences

Imprinting is a process in which an organism learns, during a sensitive period, to restrict its social preferences to a specific class of objects (Hess 1959; Bateson 1966). In many species of vertebrates, the young imprints on its mother by experiencing her visual, olfactory, or auditory qualities (Scott 1962). Imprinting is especially common in precocial birds. A few hours after hatching, their chicks approach and follow a moving object which may, or may not, resemble the adult female. Several hours to a few days later, they show a strong preference for the familiar object, while at the same time actively avoiding other objects.

The first to report systematic observations on imprinting was Douglas Spalding, in the second half of the nineteenth century (Spalding 1873; see Chapter 1). It was, however, Konrad Lorenz who brought the subject to the attention of modern science (Lorenz 1937, 1981). He regarded imprinting as a unique learning process that had evolved to serve two purposes: (1) to form attachment with the mother immediately after hatching; and (2) to restrict, later in life, sexual preference for an adequate mate. The first process is filial imprinting, and the second is sexual imprinting. 'A phylogenetic program determines precisely when the young organism is to learn what. After hatching and becoming able to look around, a greylag gosling utters its lost piping, to which under normal conditions its mother answers with a rhythmic cackle. To this the gosling responds by greeting. The mutually releasing sequence of piping–cackle–greeting is predictable with a high degree of probability and the moment at which it will first take place is equally predictable. The program timing the period of sensibility of that irreversible learning process for this particular moment is obviously adaptive. The inbuilt learning mechanism conveys to the gosling information which, if verbalized, would say: 'When you first feel lonely, utter your lost piping, then look for somebody who moves and says "gang, gang, gang" and never, never forget who that is, because it is your mother' (Lorenz 1981).

This normally adaptive ontogenetic programme can form bizarre social bonds under unusual circumstances. These are illustrated by the famous photographs of Lorenz followed by his ducklings. There was nothing special about Lorenz that made the ducklings follow him; many researchers and farmers have similar experiences.

The relation between filial and sexual imprinting has long been debated in the scientific literature (Bateson 1966, 1978; Hinde 1970; Lorenz 1981). Suffice it to say that in some cases, the same experience that at an early age establishes filial imprinting contributes later in life to the formation of a sexual bond. Sexual imprinting of a bird on a human may result in embarrassing situations indeed. Again, some of the most colourful accounts of what it is like to be courted enthusiastically by a bird, come from Lorenz (1981). In one case, a jackdaw that imprinted on him generalized and 'fell violently in love with a petite, dark-haired girl'. In yet another case, a couple of budgerigars devoted so much energy to courting Lorenz and one of his colleagues that they 'allowed their babies to starve to death'.

The ethological literature is rich in observations on imprinting. To the neurobiologist, imprinting is a potential clue to an interesting type of learning, and especially to the interaction between development and learning. Though the behavioural phenomena are attractive, the neurobiological analysis encounters a major difficulty, since the learned behaviour is complex. The analysis of brain mechanisms of imprinting must, therefore, start with the selection of a suitable experimental system, in which the imprinted behaviour is robust and quantifiable. A favourable system is filial imprinting manifested as an approach behaviour in the domestic chick.

Domestic chicks display both filial and sexual imprinting, which take place during different sensitive periods (Vidal 1980). The approach behaviour wanes in weeks, and ethologists might regard it as a less 'typical' case of imprinting than sexual imprinting, which is essentially irreversible (Lorenz 1981). But for our purpose, it does not really matter whether filial imprinting in the chick typifies imprinting, as long as the phenomenon offers exerimental advantages for the analysis of learning. The approach behaviour of the domestic chick does have such advantages. Approach is a robust, readily quantifiable motor behaviour. The chick can imprint on easily controlled visual stimuli. The sensitive period is measured in hours or a few days at most. The chick brain is rather well studied. And, of course, domestic chicks are readily available.

The approach behaviour in the domestic chick, a manifestation of filial imprinting

A convenient method for demonstrating filial imprinting in the domestic chick is to let the eggs hatch in the dark and then expose the newborn chicks to a visually conspicuous object. This might be a flashing coloured light, a moving toy car, or a university professor. In nature, mother hen not only moves around in the visual field, but also emits calls, which accelerate visual imprinting (Smith and Bird 1963). Therefore, in some training protocols the maternal call of the hen is played in the background, to facilitate learning. After a few hours, the chicks already approach preferentially the familiar object.

Merely observing a freely moving chick as it reacts to a familiar object is not, however, a useful procedure for quantifying imprinting. The visual angle, and hence the apparent size and shape of the object, change as the chick approaches it. This alters the motivation and intensity of response. Behavioural assays were developed that overcome the problem. An extensively used procedure is as follows (Bateson and Wainwright 1972; Fig. 12.1): during training, each chick is individually placed in a wheel-shaped cage with opaque sides, approximately 30 cm in diameter. The curved surfaces on which the chick stands are made of wire mesh and hence are practically transparent. The chick sees a single, visually conspicuous object, such as a rotating flashing cube or a stuffed fowl (Horn 1985). The wheel's axle is mounted on a rod and turns freely, so that when the chick attempts to move towards the object, the wheel rotates but the chick does not advance, and hence its distance to the object does not change.

For testing, the approach counts towards the imprinted and control objects can be measured alternately (McCabe *et al.* 1982). Alternate exposure is problematic, since it may modify the preference with testing. Instead, the chick could be subjected to a choice situation. But a simple choice is also problematic. As noted above, the attractiveness of objects depends, among other qualities, on their apparent size: often the bigger, the better. If the chick were allowed to approach freely the stimulus towards which it first moved, this object would grow bigger in the visual field, become more attractive, and only little reliable information would be obtained on the magnitude of the acquired preference. To solve this difficulty, a receding-wheel choice apparatus was designed (Fig. 12.1; Bateson and Wainwright 1972). In this apparatus, the running wheel is mounted on a trolley which is placed on two rails, approximately 2.5 m in length. The familiar and non-familiar objects are each positioned on an opposite end of the rail. The apparatus is designed so that when the chick walks in one direction, the trolley is carried in the opposite direction. Therefore, when the chick attempts to approach the familiar object, it actually increases the distance to it. The preference intensity is thus titrated; the more the chick prefers an object, the more it travels away from it. At a certain point, the familiar object becomes so small, and the other object so conspicuous, that the chick may start approaching the latter, but then, after a while, reorients itself again.

To neutralize the contribution of general motor activity, a preference score is calculated by dividing the approach count to the familiar stimulus by the total approach counts to both the familiar and non-familiar stimuli (McCabe *et al.* 1982). A disadvantage of the running-wheel simultaneous-choice procedure is the possibility that after a prolonged test the chick would learn that moving in one direction brings it to the opposite direction. This does not happen during the brief time course of standard imprinting experiments.

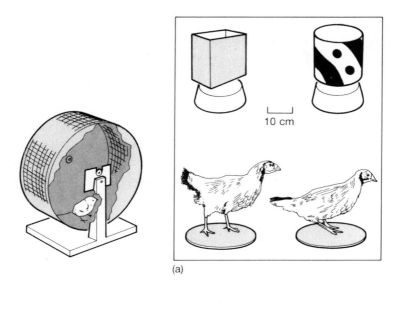

(a)

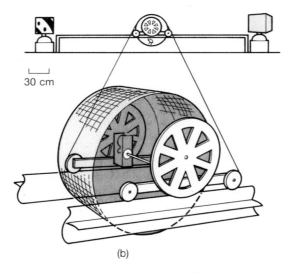

(b)

Fig. 12.1. Running-wheel apparatuses for training and testing filial imprinting in the domestic chick. (a): A chick in a training wheel. The wheel is drawn with one of its opaque sides exposed. The animal runs on the mesh towards a visually conspicuous object, such as those depicted in the inset. Since the wheel rotates on its axle, the distance to the object remains constant. (b): A test-choice apparatus. When the bird runs towards the familiar object, positioned on one side of the rail, the trolley is carried to the other direction. At a certain point the familiar object becomes so small that the chick does not attempt to approach it any more. This procedure permits sensitive titration of preference. (Adapted from Bateson and Wainwright 1972; Horn 1985; Johnson and Horn 1987.)

Identification of a fragment of a brain system that subserves imprinting in the chick

Since the motor output involved in the approach behaviour is complex, it is impractical to trace by neuroanatomical and neurophysiological means the pathways from the effector organs to the relevant brain centres. Patrick Bateson, Gabriel Horn, and Steven Rose initiated, instead, a research programme in which biochemistry was used to identify candidate brain regions (Horn *et al.* 1971, 1973; Bateson *et al.* 1972). Their basic assumption was that in imprinting, some growth processes take place in the brain, and therefore relevant brain centres should display enhanced turnover of nucleic acids and proteins. Bateson and his colleagues searched for such enhanced metabolism by measuring the incorporation of radiolabelled lysine into proteins and radiolabelled uracil into RNA. They found that relatively high amounts of radioactive label were concentrated in the forebrain roof of the imprinted chicks. In the chick, if the dorsal supraoptic commissure is cut, each hemisphere receives visual input only from the contralateral eye. When this 'split-brain' operation was performed before training, and one eye covered, uracil incorporation was detected only in the forebrain roof of the contralateral hemisphere.

One problem with the initial experiments was that the trained chicks were exercised in the light, whereas the control chicks were kept under dim light or in the dark, so that the lower incorporation of label in the control could result from sensory-motor deprivation. The same argument applies to sensory deprivation of the unstimulated hemisphere in the 'split-brain' chicks. An improved behavioural protocol was, therefore, devised, to dissociate the effects of sensory stimulation from those of learning (Bateson *et al.* 1973). The chicks were now trained for different periods in the first day of life, and then retrained for an equal amount of time on the second day. The assumption was that imprinting is accumulative and saturable, so that the more the individual learns on day 1, the less it has to learn on day 2. One would expect, therefore, acquisition-related processes to function more intensely on day 2 in those individuals that did not have enough time to learn on day 1, and vice versa. Indeed, the greater the experience of the chicks on day 1, the lower was the amount of uracil incorporated into the anterior forebrain roof on day 2. Note that all the chicks were exposed to the same stimulus intensity on day 2, so that any difference in brain metabolism on that day could not result from a simple difference in ongoing sensory stimulation. Also, individuals trained for a longer time on day 1, approached the familiar object on day 2 more frequently than did other individuals, implying that the reduced incorporation of label was not due to a reduction in motor activity.

In yet another set of experiments, the correlation between uracil incorporation into RNA and imprinting was approached in another way. Here advantage was taken of the individual variability in learning. All the chicks were trained and tested in an identical protocol. Under these conditions, uracil incorporation into the anterior forebrain roof of different individuals was positively correlated with their intensity of preference for the familiar object (Bateson *et al.* 1975). Taken together, the results did form a reasonably coherent story, suggesting that macromolecular synthesis in the forebrain roof correlates with imprinting, and not with mere sensory or motor experience.

All the above experiments involved the dissection and homogenization of gross parts of the brain. This method was not sufficiently delicate to narrow down the change in activity to discrete parts of the anterior forebrain roof. A finer localization was made possible by *in situ* autoradiography. In a representative set of studies, Horn *et al.* (1979) used the same training protocol described above, in which the chicks are trained for different periods on day 1 but for an equal amount of time on day 2, to control for sensory exposure during administration of the metabolic precursor. The chicks were exposed to a rolling, illuminated yellow slit. On day 1, individuals in the 'undertrained' group gazed at the pattern for 45 min, whereas individuals in the 'overtrained' group gazed at it for 180 min. On day 2, all the birds received an injection of radioactive uracil, followed by an exposure to the imprinting stimulus for an additional hour. The chicks were then sacrificed, and serial sections of their brains subjected to autoradiography. There was a significantly greater incorporation of radiolabelled uracil in the intermediate and medial region of the hyperstriatum ventrale (IMHV) in the 'undertrained' individuals. This could be explained by assuming that the IMHV is somehow involved in imprinting, and the 'undertrained' chicks caught up with their learning on day 2. The IMHV lies within the anterior forebrain roof (Fig. 12.2), the same brain region which was implicated in imprinting by the earlier studies.

The uracil-incorporation experiments tried to identify brain sites active during acquisition. Another method, metabolic mapping with 2-deoxy-D-glucose (2DG), was used to identify sites active during retrieval. This method, already described in Chapter 10, measures the rate of glucose consumption. Koshaka *et al.* (1979) have reasoned that those areas in the

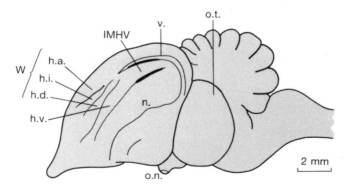

Fig. 12.2. Diagrammatic longitudinal section in the brain of a 2-day-old chick, cut laterally to the midline. h.a., hyperstriatum accessorium; h.d., hyperstriatum dorsale; h.i., hyperstriatum intercalatus; h.v., hyperstriatum ventrale; IMHV, intermediate and medial part of hyperstriatum ventrale, indicated by heavy lines; n., neostriatum; o.n., optic nerve; o.t., optic tectum; W, visual Wulst. The IMHV has been implicated in some aspects of imprinting. (Adapted from Horn 1985; for a comprehensive neuroanatomy of the chick brain, see van Tienhoven and Juhasz 1962.)

chick brain that store the imprinted engram should display enhanced activity and hence consume more energy during testing, i.e. during use of the relevant memory. Koshaka and his colleagues employed the following experimental procedure: chicks were individually kept in the dark for 12 h after hatching, and then placed for 45 min in an illuminated arena, in the presence of a floating red balloon. After training, the chicks were returned to their dark individual cages and kept there for 2 days. On day 3, they were again placed in the illuminated arena with the red balloon. Those chicks that followed the moving object for more than 20 min were considered imprinted, and those that followed the balloon for less than 5 min were considered non-imprinted. Note that the criterion for imprinting was not rigorous: no choice was given between a familiar and non-familiar object. On the subsequent day, both 'imprinted' and 'non-imprinted' birds were injected with [^{14}C]2DG, exposed to the red balloon, and sacrificed. Brain sections were processed for autoradiography. The radioactive label in the lateral neostriatum and in the medial hyperstriatum ventrale of the imprinted individuals was higher than in controls. The label in the medial hyperstriatum ventrale overlapped part of the IMHV.

In acoustically imprinted guinea fowl chicks, Maier and Scheich (1983) found enhanced 2DG utilization in the medial part of the hyperstriatum ventrale; the labelling was interpreted by Horn (1985), though not by the authors themselves, to overlap the most anterior region of the IMHV. In these experiments, the chicks were presented on the first and second day of life with rhythmic tone bursts of either 1.8 kHz or 2.5 kHz, and tested on the following days for their preference of the appropriate tone. The trained chicks, as well as naïve controls, were injected on day 7 with [^{14}C]2DG, placed in a box, and presented with the 1.8-kHz stimulus. The animals were then sacrificed and their brains analysed by autoradiography. High glucose utilization was revealed in all the chicks, whether imprinted or not, in the primary auditory field L. There were, however, differences between the labelling of several areas in the brains of imprinted and control individuals. These areas included the hyperstriatum accessorium, and the lateral and medial neostriatum and hyperstriatum ventrale (Fig. 12.2). The utmost labelling in these areas was found in those chicks that were imprinted on 1.8 kHz, the tone played during administration of the [^{14}C]2DG.

The differential effects of brain lesions on imprinting and operant conditioning

As we already know, metabolic mapping can correlate gross activity of brain regions with behaviour, but cannot establish whether a suspected brain region is obligatory for that behaviour. For the latter purpose, circumscribed ablations are useful.

Salzen *et al.* (1975, 1978) placed a slab of yellow sponge, or a green stuffed cloth ball, in the rearing cage of newborn chicks, and found that lesions in the lateral forebrain prevented acquisition and retention of a preference for either object. In these studies, the hyperstriatum ventrale was left largely intact. Later studies by McCabe *et al.* (1982), using the running-wheel apparatus, found that more restricted lesions to the lateral forebrain had only a small effect on imprinting. In contrast, lesions to the IMHV had marked effects (McCabe *et al.* 1981, 1982; Horn *et al.* 1983). The effect of lesions to the IMHV on *acquisition* was tested first (McCabe *et al.* 1981). Bilateral lesions to this area, placed prior to the completion of training, prevented the development of preference for the training stimulus, but did not affect visuomotor performance in a pecking test. The results suggested, therefore, that the IMHV is necessary for acquisition.

Additional exeriments were performed to determine whether the IMHV is also obligatory for *retention* (McCabe *et al.* 1982). The experimental procedures were essentially similar to those described above, only that here the bilateral lesion in the IMHV was placed after training was completed, approximately 1 day before testing. Sham-operated controls, and chicks with lesions in hyperstriatum accessorium and in the lateral forebrain, showed a strong preference for the training stimulus. Individuals with IMHV lesions still preferred the training stimulus, but less so than the sham controls and the chicks without IMHV lesions. The visuomotor ability of the chicks with IMHV lesions did not differ from that of the other chicks when tested in a pecking test. The conclusion was, therefore, that an intact IMHV is also involved in the retention of imprinting.

Is the effect of IMHV lesions on imprinting due to a global defect in learning? The answer is no. For example, McCabe *et al.* (1982) trained chicks in a maze to prefer one of two geometric patterns by rewarding the approach to the correct pattern. The experiments were performed at 12 °C, and the positive reinforcement was, appropriately, a stream of warm air. Bilateral IMHV lesions had no significant effect on the chicks' performance in this task.

Chicks will work not only for a stream of warm air in a chilly Cambridge laboratory, but also for the mere view of another chick, or even a coloured box (Bateson and Reese 1969). This can be demonstrated using the following procedure: a chick is placed in a box with only one transparent wall. A visually conspicuous object, such as a stuffed jungle fowl or a red box, is placed in front of that wall. Inside the box, on the floor, there are two pedals. If the chick presses one of these pedals, but not the other one, a relay is activated, which illuminates the box or the fowl, and initiates their rotation. Chicks like the show, and learn to step on the correct pedal, while concomitantly imprinting on the illuminated rotating object. In learning to activate the pedal, chicks with bilateral IMHV lesions did not significantly differ from sham-operated controls. However if, following this operant-conditioning training, the same individuals were given a choice between the familiar and the non-familiar objects, the chicks with IMHV lesions performed at chance levels (Johnson and Horn 1986).

• Taken together, the above experiments indicate that brain system(s) obligatory for imprinting are not identical to brain system(s) obligatory for (1) attributing reinforcing value to visually conspicuous objects; and (2) associating responses with reinforcers. The IMHV, according to these results, is necessary for imprinting but not for the latter two capabilities. The data should not, however, be taken to imply that the IMHV has no role whatsoever in associative learning independent of imprinting. For example, Rose and his colleagues (Rose and Harding 1984; Rose and Csillag 1985) trained chicks to avoid pecking at a bitter bead, and reported increased incorporation of 2DG and [³H]fucose in the anterior forebrain roof which overlaps the IMHV.

Possible predispositions, or the individuality of chicks

In many of the experiments described above, two very different types of imprinting stimuli were used. One was a box, the other a fowl. At first the data concerning both types of stimuli were pooled together. However, with time an interesting pattern emerged. The effect of IMHV lesions clearly depended on the nature of the stimulus. These lesions affected imprinting on a rotating coloured box much more than imprinting on a rocking, stuffed jungle fowl. Those chicks with IMHV lesions that were trained to imprint on a jungle fowl, still preferred the fowl to a novel stimulus, whereas those trained to imprint on a box performed at a chance level during testing (Fig. 12.3; Horn and McCabe 1984).

Chicks have an innate predisposition to prefer a stuffed jungle fowl to a box (Johnson *et al.* 1985). The mere existence of such a preference should not, probably, come as a big surprise, since mother hen usually does not resemble a rotating box. However, the visual qualities that constrain the preference are not known. They are clearly not the shape of a bird. A disarticulated stuffed fowl was as attractive to the chick as an intact bird, and conspecifics were not preferred to stuffed ducks, or even to a stuffed polecat! (Horn 1985). Is there a common denominator to the visual qualities of parts of animals that makes them more attractive than an artificial geometrical pattern? It is tempting to suggest that lesions

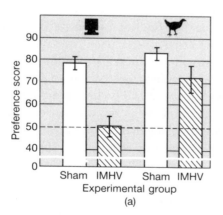 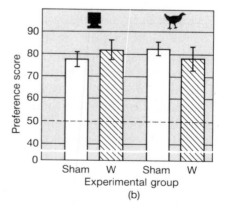

Fig. 12.3. The effect of lesions in the IMHV and the visual Wulst on the formation of preference to a stuffed jungle fowl and a rotating box. Lesions in the IMHV markedly impaired imprinting on the box but not on the fowl. Lesions in the visual Wulst had no effect. Lesions in the IMHV did, however, impair the recognition of individual chicks (not shown; see text). (After Horn 1985.)

to the IMHV do not prevent the preference for a fowl in imprinting experiments, because outside the IMHV there is a neuronal system that anyway directs the chick towards an animal but not towards a box (Horn and McCabe 1984). But before accepting the idea that an innate predisposition for an 'animal' is involved, the Gestalt qualities that make the fowl more attractive than the box must be determined, and their relevance to the shape of an animal convincingly demonstrated.

If there is indeed a predisposition for chicks, then what role does the IMHV play in imprinting on the jungle fowl? Johnson and Horn (1987) hypothesized that the IMHV enables the newborn chick to recognize individual conspecifics. To test this hypothesis, they took 6–10-hour-old dark-reared chicks, and ablated bilaterally the IMHV or, alternatively, the Wulst, which is the analogue of the mammalian primary visual cortex. Approximately 1 day after surgery, the chicks were placed in a running wheel and exposed for a total of 200 minutes to one of two stuffed jungle fowls (see Fig. 12.1(a)). Two hours after the final training session, the chicks were given a choice between the two fowls. The chicks with Wulst lesions displayed a clear preference for the familiar stuffed fowl, whereas the birds with IMHV lesions did not.

• The following *working model* was, therefore, suggested: chicks are born with a disposition to prefer chicks and imprint on them. The neuronal system that endows the chick with this a priori knowledge does not include the IMHV. In addition, another neuronal system, containing the IMHV, is engaged in imprinting and enables the chick to learn the distinctive fine qualities of the particular type of object. In nature, this latter system enables the chick to learn about the individuality of its conspecifics. In a young chick, the main purpose of this system is probably to direct the chick to follow its mother. It may also be useful in maintaining social order (Candland 1969). In the laboratory, the same system enables the chick to learn the characteristics of a rotating red box. Therefore, lesions to the IMHV abolish the ability to acquire and remember red boxes and the individuality of chicks, but do not abolish the innate preference for a platonic chick.

This is an attractive model, but is must be supported by much additional data. Some observations suggest that the picture might not be so simple. First of all, as noted above, the stimulus qualities that supposedly distinguish an animal-like from a non-animal-like object were not determined. The effect of IMHV lesions on imprinting to many more types of artificial as well as natural shapes must still be tested. The relevant stimulus attributes are crucial for understanding the predisposed representation. In addition, as far as IMHV lesions are concerned, they significantly lowered the number of approaches towards the stuffed fowls during training in the running wheel (Johnson and Horn 1987). This was interpreted as resulting from weak learning which did not promote attempts to move towards the object (Bateson and Jaeckel 1974). But suppose the opposite is also correct, and learning depends on activity, so that the intensity of training of the chicks with IMHV lesions is not as strong as that of the controls. If a certain minimal intensity of training is required to establish a distinction between individual fowls, the less-active individuals with IMHV lesions would not learn. Or, suppose a certain motivational threshold is required for learning to imprint on artificial, simple objects; if IMHV lesions reduce motivation, this would explain the failure to learn such objects (Horn and McCabe 1984). A fine analysis of the effect of IMHV lesions on activity and motivation is therefore required.

Some morphological and biochemical correlates of imprinting

The data reviewed so far have implicated the IMHV in normal imprinting. Is there any information on the cellular and molecular processes that take place in this brain region in imprinting?

The data are provocative, but fragmentary. Some hints emerged from ultrastructural studies. Horn *et al.* (1985) found that imprinting training correlated with an increase in the length of the postsynaptic density of spine synapses in the IMHV, but only in the left hemisphere. Other morphological measures, for example the density of synapses, did not significantly change. The number of NMDA-type glutamate receptors also increased in the left IMHV (McCabe and Horn 1988). NMDA receptors are expected to mediate excitatory transmission and subserve LTP (see Chapter 6). All these findings led to the suggestion that imprinting is accompanied by an enhanced efficacy of spine synapses, and that there is hemispheric asymmetry. (Studies of the effect of hemispheric IMHV lesions on imprinting also raise the possibility of left dominance, but the results are still rather inconclusive and will not be detailed here; see Cipolla-Neto *et al.* 1982; Horn *et al.* 1983.)

The data on the functional properties of IMHV neurons are scarce. It is not known how visual qualities of imprinted objects are represented in terms of neuronal activity, and what changes in this representation with learning. This is essential for further understanding of filial imprinting in the chick. Payne and Horn (1984) did record spontaneous multiunit activity from the IMHV after training the chicks to imprint on either a rotating flashing red box or a stuffed jungle fowl. They reported an inverse correlation between the mean firing rate and approach activity in birds trained on the box, but not in birds trained on the fowl. This corroborated the previous results on the different roles of IMHV in imprinting on 'artificial' as opposed to 'animal-like' patterns.

Pharmacological studies also suggested that cellular mechanisms subserving imprinting to different classes of objects, are different. Davies *et al.* (1985) injected chicks with the toxin DSP4, which reduces noradrenaline concentrations in the brain. The chicks were then exposed to either a box or a fowl. The toxin impaired imprinting on the box, but not on the fowl. It did not significantly affect general motor activity, the intensity of approach behaviour, or the accuracy of pecking at a rocking bead. Noradrenergic innervation might, therefore, have a greater role in imprinting on a box than on an animal-like figure. These results are, again, suggestive and correlative. Aminergic systems in the brain are known to affect motivation (Mason 1984), and the behavioural effect of the toxin might, therefore, stem from reduced motivation to imprint on an unnatural stimulus (see above), rather than from defective learning. In another pharmacological experiment, Bolhuis *et al.* (1986) injected newly hatched chicks with testosterone, and trained them to imprint on either a fowl or a box. The hormone enhanced imprinting on the fowl; in addition, in the fowl-trained chicks the level of the hormone in the blood was positively correlated with the preference score. This was not observed in the box-trained chicks. Thus again, a difference emerged between biochemical mechanisms that underlie imprinting on a box and on a fowl.

In addition to the usual problems inherent in attempts to dissociate neural changes related to sensory-motor processes from changes related to learning and memory *per se*, the study of imprinting is confounded by the effects of development. For example, during the first days of life there is anyhow an increase in the length of postsynaptic densities of spine synapses in the IMHV, though in both left and right hemispheres (Horn *et al.* 1985). In addition, mere exposure to light has been proposed as the stimulus establishing a left-hemisphere visual dominance in chicks (Rogers 1982). In spite of the sophisticated training and testing procedures developed over the years, there is always the possibility that changes detected in the brain of an imprinted chick are due to the acceleration of maturation, rather than to the acquisition and storage of a specific representation.

On the generality of imprinting: rats, smells, music

Imprinting is but one special case of a general phenomenon, in which brain maturation is modified by sensory experience during a sensitive period at an early age (e.g. Sherman and Spear 1982; Wiesel 1982; Knudsen 1985; Simons and Land 1987). In many cases, for example in the development of the mammallian visual system, the role of experience is more diffuse and global: it expands, fine tunes, and stabilizes the perceptual potential. In filial imprinting in the domestic chick, early postnatal experience endows the brain also with specific information about the social meaning of one or a few objects. The innately programmed reaction to the imprinted object is well defined: 'follow it'. However, the innate programme that instructs the chick brain what the imprinted object should look like, is rather permissive, since chicks can imprint on very different objects.

Such assignment of a specific social meaning to sensory stimuli perceived early in life, clearly exists in many species. In mammals, the critical sensory modality is olfaction. An example, studied extensively in recent years, is the effect of early experience on odour preference in the rat (Leon 1987). Newborn rats become attracted to the specific odour of their mother. This learning is critical for nursing and other aspects of baby–mother inter-actions. It is associated with enhanced neuronal response to the familiar odours, and enlargement of specific parts of the glomerular layer in the olfactory bulb (Coopersmith *et al.* 1986; Leon 1987; Wilson *et al.* 1987). The trace is initially retained in the brain hemisphere ipsilateral to the stimulated naris, but later, as the interhemispheral commissures develop, becomes accessible to the contralateral side (Kucharski and Hall 1987). The memory for imprinted odours persists into adulthood (Coopersmith and Leon 1986).

Additional processes that can be regarded as filial imprinting in mammals were described but not analysed (Scott 1962; Cairns 1966). The internal representations involved are expected, in most of the cases, to be multimodal and highly complex. Imprinting-like processes may also play a crucial role in both normal and abnormal development of the human infant, and in the formation of the infant–parent bond (Osofsky 1979; Klein 1987; Marks 1987). Studies of imprinting are, therefore, of interest to a vast spectrum of professionals, from molecular neurobiologists to child psychologists.

Some studies of imprinting-like processes have already yielded practical results, though not in humans. One example is the observation that guide dogs for the blind are much better trained if they have been socialized with humans as very young puppies (Pfaffenberger and Scott 1959). Other studies of the effect of early sensory experience on adult behaviour have a smaller practical potential, but do attest to the heterogeneity of innate predispositions, and to the limitless imagination of experimenters. A study by Cross *et al.* (1967) illustrates these kinds of data. These authors have decided to test the effect of early exposure to music on subsequent musical preference. Of all animals, the albino rat was chosen. Newborn rats were raised in acoustic chambers (unfortunately, no adult controls were included, so that the question of the sensitive period remained unsettled). One group was exposed to a selection of Mozart's compositions for several weeks. A second group was exposed to Schoenberg's music. Later, when given an opportunity to select either Mozart's or Schoenberg's music in a choice box, the Mozart-trained group preferred Mozart, but the Schoenberg-trained group did not display any preference for their acquainted composer. Actually, even controls raised in the absence of music displayed some preference for Mozart. The authors concluded that early exposure influences musical preference, but also that there is a predisposition: Mozart's music is innately more rewarding (or less aversive) than Schoenberg's music, at least for albino rats.

13 *The establishment of innately predisposed representations in sensitive periods: bird song*

What is a song?

Although a fascinating example of early postnatal learning in a sensitive period, filial imprinting in the chick does not easily yield to neurobiological analysis. A major reason for that, as noted in the previous chapter, is the complexity of the acquired behaviour. Another type of behaviour which is established and refined by learning in sensitive periods, is bird song. Here, the motor, vocal output can be analysed by sophisticated methods, and the vocal organ permits tracing of efferents from the subserving brain centres.

All birds can emit calls. Many species of birds can also sing. Calls are simple, brief vocalizations, uttered by both sexes in all seasons in response to particular stimuli such as the presence of a predator. *Songs* are complex, stereotyped vocalizations, which are produced spontaneously, predominantly by sexually mature males during the breeding season (Kroodsma and Miller 1982*a,b*; Konishi 1985). Singing is often associated with a characteristic body posture. Males sing to defend their territory and to attract and stimulate females. However, the romantic picture which portrays the male as a troubadour playing a serenade to his beloved one is quite misleading. Some features of songs may be intelligible to males only (Williams and Nottebohm 1985). If this is the case, then the singing male should be primarily depicted as a decisive warrior, attempting to drive his competitor away. Whether a troubadour, or a warrior, or both, one thing is clear: birds sing to communicate.

The song repertoire of a mature bird is, in many species, a beautiful example of an interaction of nature with nurture. As pointed out by Konishi (1985), the plasticity of bird song has been well known to bird fanciers in the Orient since ancient times, and expert manipulation of song was used for aesthetic purposes. This plasticity has also long attracted scientists' attention (Metfessel 1935; Koehler 1951; Thorpe 1954).

Repertoires and dialects

Before attempting to identify the neuronal apparatus that endows birds with musical skill, we should acquaint ourselves with the behaviour and its modification by experience. A song is a series of sounds with silent intervals between them. One can record songs either in the field or in the laboratory with a tape recorder, but for the purpose of quantitative analysis, a frequency vs. time sound spectrogram is preferable (Fig. 13.1). The most elementary sound revealed as a continuous marking in such a spectrogram is a *note*. Notes are commonly 10–100 ms in duration. Notes group together to form *syllables*. Syllables group together to form *phrases*. Phrases group together to form *songs*. Songs are commonly 1–5 s in duration. Different songs form a *repertoire*. The size of a repertoire ranges from one song to many hundreds, depending on the species. Repertoires of geographically distinct populations of the same species often differ, and are then denoted as *dialects* (Baker and Cunningham 1985). Dialects change with time, but may persist in a given population for more than 15 years (Payne *et al.* 1981). Some species perform all or most of their repertoire in cycles that take tens of minutes to be completed (Marler 1984).

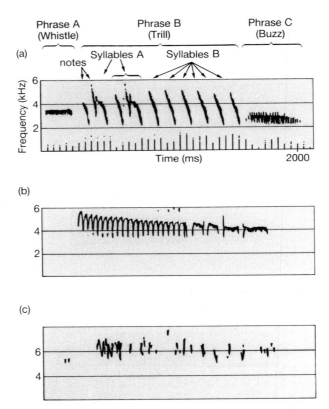

Fig. 13.1. Time–frequency sound spectrograms of the songs of white-crowned sparrows. (a): A song of a normal bird. The terminology of bird song is marked on the recording. (b): A song of a bird raised in acoustic isolation. Structures typical of wild-type song are missing. (c): A song of a deaf bird. The vocalization is highly abnormal because the bird must listen to its own voice to develop even an elementary song. (Adapted from Konishi 1985.)

Why do some species have a very rich song repertoire? Ethologists have offered several possible explanations. One hypothesis is that a large repertoire signals to a female that the male excels in his virility (Kroodsma 1977). Another hypothesis is that a rich repertoire deceives intruders into thinking that they are confronting many males, and this drives them away from the territory of the deceiver (Krebs 1977). Whether any of these hypotheses is valid, is not yet clear. What is clear, though, is that songs contain a core that encodes species-specific cues, but that usually only a small number of song properties are sufficient for that purpose (Konishi 1985). It is also clear that variability is often encountered not only among the repertoires of geographically distinct conspecific populations, but also among the performance of individuals of the same species cohabiting the same ecological niche. It seems as if 'in the evolution of birdsong there is a tension between selection pressures to develop along species-specific lines and pressures to systematically or opportunistically develop novel behaviours, adding more options for invading new communities and sound environments, and thus exploiting new resources. It is this tension that makes song development interesting for epigenetic analysis' (Marler 1984).

Ontogenesis of a song

The interplay between genetic and epigenetic processes is best exemplified by the ontogenesis of song in passerine birds such as canaries, chaffinches, or sparrows. The song develops in these birds in several discrete stages. Let us consider the wild chaffinch, *Fringilla coelebs*, as an example (Nottebohm 1970). The adult chaffinch's song typically consists of two or three phrases and lasts for approximately 2 s. The repertoire varies among individuals, and may reach six different songs. Immediately after hatching, around May, the young chaffinch begins to emit various food-begging calls. Within a few weeks, when summer arrives, the male starts to emit a loose and rambling aggregation of notes of varying complexity and of low volume, sometimes for minutes at a time. This vocal pattern is called a *subsong*. The singing posture typical of the adult is not yet apparent. The subsong keeps changing, and discrete passages, resembling the adult song, gradually emerge. These passages comprise a *plastic song*. During the breeding season, the subsong vanishes and the plastic song crystallizes into the full adult song. The singing posture also matures. The final crystallization takes place within approximately 1 month, usually before the end of March. Thus chaffinches normally acquire a *mature song* at about 10 months of age.

In the chaffinch, no major alterations will be made in the repertoire after song crystallization. However, each spring the bird passes through the subsong and the plastic-song stages, before reaching anew the full, stable song for that season. The degree of vocal plasticity retained by adult songbirds varies among species. For example, the adult canary is endowed with a greater song plasticity than the chaffinch, the zebra finch, or the white-crowned sparrow. A canary male entering its second breeding season retains only one-third or less of the syllable types it sang during the previous year.

Behavioural analysis: how much is learned?

What is the contribution of the genes on the one hand, and of experience on the other hand, to the crystallization of a full adult song? A series of experiments on the white-crowned sparrow, *Zonotrichia leucophrys*, illustrates the combined role of inheritance and environment. Marler and Tamura (1964) have raised white-crowned sparrows in various types of acoustic environments and observed the effect of these experimental manipulations on vocal behaviour. Males taken during the first 2 weeks of life and raised in soundproof chambers, in the absence of adults from the same population, developed songs which had some of the species-specific characteristics but lacked the dialect characteristic of the appropriate population.

In another experiment, isolated males were exposed to recorded sound played into the acoustic chamber through a loudspeaker. It was found that alien dialects could be taught provided the male was exposed to recordings of that particular dialect during the first few months of life, but not later. Song patterns of other sparrow species were not copied. These and other experiments led to the conclusion that male white-crowned sparrows have an innate capacity to develop a basic white-crowned sparrow song pattern, but they must learn their song dialect from older males. Learning takes place during approximately the first two months of life. Indeed, in nature, a young white-crowned sparrow, born in the spring, is exposed to the songs of his father and neighbours up to about three months after fledging; the adults stop singing during the summer and the autumn. Once the song is established, further acoustic experience does not change the pattern. The song dialects are thus transmitted by learning during a sensitive period.

The timing of the sensitive period for song learning in a given species in nature depends on several environmental factors, including social interaction with a tutor, the amount of adult song heard during the hatching year, and the photoperiod (Kroodsma and Pickert 1980; Baptista and Petrinovich 1984; Marler 1984; Konishi 1985). During the sensitive period, songbirds display a predisposition to learn their conspecific song. This was clearly demonstrated by Marler and Peters (1977), who transferred eggs of swamp sparrows from wild nests to nests of canaries. The latter, with some help from Marler and Peters and their assistants, served as foster parents. The transfer prevented any contact with the natural parents. The males were then exposed to recordings of either swamp sparrows or song sparrows. The song syllables learned successfully were all swamp sparrow syllables.

The sensitive period means something to the females, too, in spite of the fact that they do not normally sing. Early experience determines the preferred responsiveness of female white-crowned sparrows to males taken from their home dialect region (Baker *et al.* 1981). Actually, in some cases learning to sing is better described as a joint endeavour of both sexes, since the presence of a female provides the male with instructive cues. For example, when male cowbirds are individually housed with females belonging to different cowbird subspecies, their song dialects are biased towards the preference of their non-singing female companion (King and West 1983). Later observations revealed that one way in which the female cowbird can instruct the male is by responding to appropriate vocalizations with a typical wing stroke. This wing stroke resembles the initial wing movement of a copulatory posture, and causes the male to approach and inspect the female (West and King 1988). The visual display of the female may serve, therefore, as an innately programmed positive reinforcer for the male.

Do young males copy faithfully the dialect of their tutors? Clearly not. Again, an experiment on swamp sparrows can serve as an example. Young males were taken from the field as 2–10-day-old nestlings and reared in the laboratory in acoustic isolation (Marler and Peters 1981). They were exposed to tape-recorded song for 40 days, beginning between 16 and 26 days of age. After training, the birds were isolated in separate soundproof chambers. Their voices were then recorded, beginning at about 95 days of age and continuing up to about 1 year of age. The recordings showed that the first attempts of the males to reproduce memorized syllables begin more than 200 days after they have last heard them. In other words, the sparrows commit the song heard during the first two months of life to memory, and keep the information stored without rehearsal for about 8 months. Analysis of the song during the process of crystallization showed that the birds learned much more from their tape-recorder tutor than they finally manifested in their adult song. In swamp sparrows a crystallized song repertoire typically consists of three syllables per male. In the process of going through subsong and plastic song, experimental birds used as many as 19 learned syllables per male. These extra syllables were discarded at the time of crystallization. In all, the mature song was an outcome of selection, combined with improvisation and invention.

How innate is the song that develops in an acoustically isolated male? Experience plays a role even in this situation. Mark Konishi (1965) deafened young white-crowned sparrows by removing the cochlea. The song developed in the operated birds was rather primitive and broken, lacking species-specific traits, and quite similar to the song of another species subjected to similar surgical treatment. It was much more amorphous than the song developed by birds reared in isolation. Males deafened after exposure to their home dialects during the sensitive period, but before they started to sing themselves, developed songs similar to those of a deafened naïve bird. Once the adult pattern of singing became established, deafening had little or no effect on it.

These results indicated that the bird must be able to listen to its own performance in order to develop normal song. Even the 'innate' song, i.e. the song developed in acoustic isolation, must therefore be 'learned' using auditory feedback. The kernel innate song is, thus, that pattern of vocalization which develops without any auditory feedback. The resemblance (or lack of resemblance) of the innate song so defined to the species song, depends on the species and on the experimental conditions (Marler and Sherman 1983; Marler 1984; Konishi 1985).

• In spite of species and experimental differences, the results of the social isolation and deafening experiments can be generalized to yield the following model: a bird is born with a song motor-control system that needs instructions in order to generate a normal song.

These instructions are provided in two stages, which may partially overlap, depending on the species. First comes a *sensory stage* during which the bird listens to its tutor. There is a genetically determined predisposition to prefer a conspecific tutor. The song thus heard is confined to memory. Later, during the *sensory-motor stage*, which corresponds to subsong and plastic song, the bird must hear itself to match its own vocal output with its innate template as well as with the memory template of the tutor's song. The bird selects the components of its mature song from this inventory of innate and learned instructions. In the absence of the tutor, only the innate information is used. In some species, the central song system can generate some species-specific song structure solely on the basis of an innate template, in the absence of both tutorial instructions and auditory feedback.

The architecture of the song-generating system

A sensible starting point for the analysis of the biological machinery that learns, stores, and performs the song, is tracing the efferents back from the vocal organ, the syrinx, to the brain. The syrinx is a muscle-controlled cartilaginous box, situated at the junction between the bifurcation of the bronchi and the trachea (Fig. 13.2). Two thin elastic membranes, denoted the internal tympaniform membranes, form thin modifiable walls at the rostal end of each bronchus. Two connective tissue folds, the external labia, can gradually protrude into the bronchial bore, thus altering the shape and the dimensions of the bronchotracheal connection.

The sound system of the bird can be thought of as a set of bellows controlled by the respiratory musculature and inducing vocalization during expiration. The major air sacs in the thorax and the abdomen are the bellows. As air flows by the internal tympaniform membranes, as a result of the Bernulli effect, it draws these membranes into the lumen of each bronchus. The pressure of the airflow, the position of the external labia, the membrane tension and elasticity, the shape of the trachea, and the opening of the mouth all set the amplitude and frequency of the sound.

Each side of the syrinx is innervated by the tracheosyringeal branch of the ipsilateral hypoglossus nerve (the XIIth cranial nerve). Sectioning both the left-side and the right-side branches of the nerve renders the bird virtually silent (Nottebohm 1971). On the basis of unilateral hypoglossal lesions in the canary and the white-crowned sparrow, Fernando and Marta Nottebohm suggested that the left hypoglossal nerve has a dominant role in driving the syrinx (Nottebohm and Nottebohm 1976). The syringeal musculature on the left is heavier than that on the right (Nottebohm 1980*a*). Unilateral obstruction of bronchial airflow did not, however, reveal dramatic functional differences between the contribution of left and right bronchi to song production (McCasland 1987).

Nottebohm and his colleagues proceeded, by a combination of neuroanatomical and behavioural techniques, to identify the central neuronal pathways that control the syrinx (Nottebohm *et al.* 1976, 1982; Nottebohm 1980*a*). They utilized the fact that lesions cause

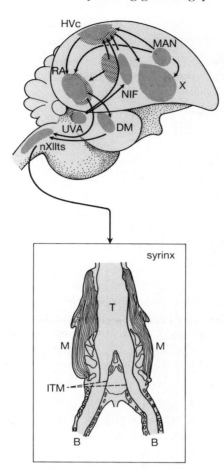

Fig. 13.2. The anatomy and topography of the song-generating system in a passerine bird. The diagrammatic representation of the brain depicts the song-control system and its interconnections. DM, dorsomedial nucleus of nucleus intercollicularis; HVc, hyperstriatum ventrale, pars caudale; MAN, magnocellular nucleus of the anterior neostriatum; nXIIts, nucleus hypoglossus, pars tracheosyringealis; NIF, nucleus interfacialis; RA, nucleus robustus archistriatalis; UVA, thalamic nucleus uva; X, area X. Hatched areas represent auditory pathways. Motor commands flow from NIF to HVc to RA. The latter two nuclei contain sensory-motor circuits for auditory-vocal interactions; for their role in learning, see text. *Inset:* The syrinx. B, Bronchi; ITM, internal tympaniform membranes; M, muscles; T, lumen of trachea. (Adapted from McCasland 1987.)

anterograde and retrograde neuronal degenerations, which could be traced by appropriate staining methods. In addition they used tracing techniques based on the neuronal transport of dyes and radioactive compounds. Unilateral sectioning of the tracheosyringealis branch of the hypoglossal nerve resulted in degeneration of cells in the caudal part of the ipsilateral hypoglossal (XIIth) nucleus. Further lesions demonstrated that this nucleus is connected either directly or indirectly to a system of discrete brain nuclei (Fig. 13.2). The functional connections between most of these nuclei were later corroborated by electrophysiological studies (see below). In brief, the nucleus tracheosyringealis receives input from the nucleus robustus archistriatalis (RA), both directly and indirectly via the nucleus dorsomedialis (DM). The RA receives input from the caudal part of the hyperstriatum ventrale (HVc) and the magnocellular nucleus (MAN). The HVc in turn receives input from the MAN as well as from the nucleus interfacialis (NIF), and in addition projects to area X. The latter is also innervated by the MAN. The NIF receives input from the nucleus uva in the thalamus (UVA).

The effect of lesions in most of these nuclei has been investigated. Unilateral lesions in area X of the adult had no significant effect on song (Nottebohm *et al.* 1976), nor did bilateral lesions in the UVA (McCasland 1987). Bilateral lesions in the MAN disrupted the development of song in juvenile birds but did not affect maintenance of song in adult birds (Bottjer *et al.* 1984). Severe song deficits were observed following damage inflicted in the adult to the HVc and the RA (Nottebohm *et al.* 1976), and to the connections between the NIF and the HVc (McCasland 1987).

HVc lesions usually had the most dramatic effects. Some descriptions of birds with HVc lesions almost invite anthropomorphic interpretations. For example, a canary with a bilateral lesion in the HVc remained in moderately good physical shape but became melancholic and would not sing. Only episodes of intense sneezing interrupted his otherwise motionless perching. On one occasion a female canary was placed in a contiguous cage. The operated male suddenly became excited. He soon adopted an upright position, inflating the gular area, raising his wing joint while slightly dropping his wings' primary feathers, keeping his legs flexed and performing pumping motions with his head. The throat and the sternum, however, were not pulsating and the bill remained closed. The bird was thus engaged in a silent song. The motivation for singing was there, but the motor output system was not recruited. He looked like a very sad bird indeed.

Lesions placed in the left HVc and RA of the canary had a more dramatic effect on song than did lesions in the right part of these nuclei. This led to the proposal that in addition to peripheral asymmetry (see above), there is also a central lateralization of song control (Nottebohm *et al.* 1976; Nottebohm 1980*a*). This suggested an intriguing analogy between bird song and lateralization of human cerebral functions, including speech (Bradshaw and Nettleton 1983; Damasio and Geschwind 1984; Benson and Zaidel 1985). However, the interpretation of the effects of unilateral lesions on bird song in terms of functional laterality has been questioned (Konishi 1985; McCasland 1987), especially on the basis of electrophysiological experiments (see below).

The correlation of brain volume with song

Whereas ablations identified specific brain areas as obligatory for normal song, neuro-anatomical observations unaccompanied by lesions correlated the morphology of these areas with the behaviour. In bird song, such a correlative approach is especially feasible because only the male sings, male song develops with age, and song is a function of season. The female, young male, and male in a silent season can, therefore, serve as intraspecies controls for the actively singing male. In addition, closely related species sometimes differ significantly in their song repertoire. Comparison of the relevant brain nuclei among such species may also cast light on the relevance of specific nuclei to song production.

Several interesting conclusions emerge from studies that correlated brain neuroanatomy with bird song.

Brain nuclei implicated in song by lesion and behavioural studies are larger in males

There is a clear-cut sexual dimorphism in candidate vocal control areas of the brain of songbirds. In canaries and zebra finches, the HVc and RA are three to six times larger in males than in females. This male/female ratio in the volume of these nuclei is much larger than the ratio between total brain volume of males and females or between the size of brain areas which are not implicated in vocalization (Nottebohm and Arnold 1976; Nottebohm *et al.* 1981; Bottjer *et al.* 1985).

Ontogenesis of song-related brain nuclei parallels that of song

For example, the HVc, RA, and area X increase markedly in male zebra finches between the second and the eighth week of life, a sensitive period for song learning and development, whereas the increase in brain regions not involved in vocalization is much smaller (Bottjer *et al.* 1985). The massive neuronal projection from the HVc to the RA is also established in the zebra finch during this sensitive period (Nordeen and Nordeen 1988*a*).

The size of song-related brain nuclei is correlated with seasonal variations in song repertoire

Each year male canaries learn anew a substantial part of their song repertoire (see above). The HVc and RA in such males were found to be almost twice as large in the spring, when the male produces stable adult song, than in the autumn, when song starts to emerge again following several months of silence. The increase in the size of other brain regions was much smaller (Nottebohm 1981).

The size of song-related brain nuclei is correlated with the complexity of intraspecies dialects and interspecies repertoires

Geographically distinct populations of marsh wrens differ markedly in the complexity of their song repertoire. Canady *et al.* (1984) measured the size of the HVc and RA in two such populations, one from New York and the other from California. The New York wrens learn about 50 different songs, whereas the Californian wrens learn about 150 different songs. The HVc and RA in the Californian wrens were about 30–40 per cent larger than the corresponding brain nuclei in the New York birds.

An analogous phenomenon was found by Brenowitz and Arnold (1986) in their studies of the rufous-and-white wren and the bay wren. Here males of the two species have similar song repertoire sizes and do not differ in the size of their song-control nuclei. However, in these species females sing in duets with the males. Female rufous-and-white wrens have a much smaller song repertoire than do female bay wrens. The volume of the HVc, RA, area X, and the hypoglossal nucleus in the latter females was found to be significantly larger than that in the former.

● In conclusion, the amount of tissue in brain nuclei implicated in song by lesion and behavioural studies is positively correlated, in several passerine species, with the amount of song as a function of sex, age, season, and population.

The interpretation and possible implications of the above finding deserve a few comments. It is attractive to propose that tissue in song-related nuclei is the physical substrate of engrams necessary for song production. Four possibilities could be considered in this respect. (1) The nuclei grow because they have to store more memory, and the added cells expand rudimentary networks which operate at near-saturation capacity. For example, adult female canaries have approximately 15 000 neurons per HVc per hemisphere, males in the autumn have 25 000, and in the spring 41 000 (Nottebohm 1985). The dendritic trees of cells in the masculine HVc expand, too, and the changes in the density of connections can in principle be estimated (Gurney 1981). This imposes potential quantitative constraints on the information-encoding capacity of the neuronal network, since it means that 15 000–25 000 neurons per hemisphere, with a certain density of connectivity, are not suffcient to store the appropriate part of an engram contributing to the canary vocal repertoire, whereas 41 000 neurons with another density of connectivity are sufficient. Of course, these calculations would become meaningful only on identification of the nature of the relevant engram in the HVc. Similar speculative calculations could be performed in the case of other species, e.g. the zebra finch, in which about 18 000 neurons are added to the HVc in each hemisphere in the sensitive period (Nordeen and Nordeen 1988*a,b*). (2) The nuclei grow because the relevant network is generated *de novo* from a new subtype of cells. Thus the size of the relevant network is even smaller than noted above. (3) The extra cells do not encode song-related representations but are needed for functions auxiliary to song learning. And finally, (4) the enlargement of brain tissue is not a prerequisite for learning and memory, but rather a by-product, somehow resulting from the extensive use of networks.

As we shall see, at least part of those neurons that supposedly participate in encoding enduring song memories, may themselves have a surprisingly limited life span. The relevance of the changes in the size and composition of neuronal populations in the passerine brain to memory and behaviour is further discussed below.

Sex hormones and musical skill

Since in most songbird species only the male sings, and since sex hormones turn an embryo into a male, the role of sex hormones in the development, learning, and maintenance of song was long suspected. Treatment with the male hormone testosterone was shown in the late 1930s to induce male-like song in females of several species of songbirds (Leonard 1939; Shoemaker 1939; Baldwin *et al.* 1940). Chaffinches castrated before song development did not sing as adults unless injected with testosterone (Nottebohm 1969). Moreover, testosterone administered to normal chaffinches in the winter, a period in which they are normally silent, induced song (Thorpe 1958). In swamp sparrows and in song sparrows, castration delayed the development of subsong and plastic song, and prevented song crystallization, which was accomplished only upon administration of exogenous testosterone (Marler *et al.* 1988).

Much of what we currently know about the effects of sex hormones on songbirds' brain and song, stems from studies conducted in recent years on the zebra finch and the canary. Similar to that of the chaffinch, the male zebra finch's song is androgen dependent; castration of an adult male abolished the song but administration of testosterone reinstated the behaviour (Prove 1974). Gurney and Konishi (1980; Gurney 1981, 1982) have systematically investigated the effect of an early postnatal administration of androgens and oestrogens on the morphology of the song-control nuclei and on the expression of song in the adult. They implanted zebra finch chicks, immediately after hatching, with slow-releasing pellets of the oestrogen 17β-oestradiol (E_2), or the androgen 5α-dihydrotestosterone (DHT). Testosterone itself is not very useful in distinguishing the effect of androgen from oestrogen because, similarly to many other vertebrate organs, bird brains can aromatize testosterone into E_2, whereas DHT cannot be aromatized.

The hormones did not significantly affect the morphology of the song system and the expression of song in the male. In contrast, they did have a marked effect on the female. The oestrogen led to a dramatic change in the morphology of the HVc and the RA: the volume of these nuclei increased, mainly due to an increase in cell size. In other words, the oestrogen masculinized the song nuclei. The androgen DHT increased cell number, but had only a small effect on the volume of the nuclei. In those cases where testosterone was used, it induced masculinization, but this was interpreted as resulting from its *in vivo* conversion into oestradiol.

However, in spite of the morphological changes in their brain, the females that were treated as chicks with the steroid hormones did not sing as adults. In other words, morphological masculinization at an early age was not sufficient. Gurney and Konishi reasoned that, as in the male, song in the masculinized zebra finch female must also depend on the availability of androgens. To test this hypothesis, they implanted pellets of testosterone or DHT into adult zebra finches that had been treated as chicks with either E_2 or DHT. These adults were then housed individually in soundproof chambers, and their behaviour observed. Both testosterone and DHT induced song in adult females that had been treated as chicks with E_2, but not in adult females that had been treated as chicks with DHT. The adult females treated with E_2 began singing within 24 hours after exposure to the androgen. At first the song was fragmentary and unstable, but following a month of continuous exposure to the androgen it became highly stereotyped and similar to the male's song. The emergence of song induced by the androgen was paralleled by an increase in the volume of the HVc, RA, MAN, and area X. These areas were not increased in normal adult females treated with androgen.

Taken together with additional data, these experiments have led to the suggestion that the default pattern of brain development in the zebra finch is female: increasing the oestrogen level can defeminize, but not demasculinize, the song system. A similar situation, in which the brain develops in a female pattern unless perturbed by increasing the availability of oestradiol, or androgens which can be aromatized to oestrogen, at the appropriate brain nuclei, is encountered in mammals (Arnold and Gorski 1984; Kelley 1986). In the zebra finch, in the first post-hatching days, the male and female have the same number and size of cells in their homologous brain nuclei, but then neurons in the female forebrain song nuclei shrink or die, whereas those in the male nuclei increase in size and number (Bottjer *et al.* 1985; Konishi 1985; Konishi and Akutagawa 1988; Nordeen and Nordeen 1988*b*). This means that the appropriate hormonal balance that defeminizes the brain prevents an ontogenetic programme of cell atrophy and death. This defeminization during a critical period 'organizes' the brain for male-like action, but subsequent expression, or 'activation', of normal masculine behaviour requires a constant supply of androgen in the adult.

The sex hormones seem to act directly on cells in the syrinx and the forebrain song system. Injections of radioactively labelled testosterone into adult zebra finches have revealed fewer hormone-concentrating cells in females than in males in song-control nuclei but not in some other brain regions (Arnold *et al.* 1976; Arnold and Saltiel 1979; Arnold 1980). Similar results were obtained in the chaffinch (Zigmund *et al.* 1980). Steroid hormones are known to increase protein synthesis in many systems, and indeed testosterone was found to increase protein synthesis in most song-control nuclei of female white-crowned sparrows, prior to induction of song by this treatment (Konishi and Akutagawa 1981).

Brain reinnovation?

What happens to neurons in the forebrain song nuclei of the adult brain when these nuclei expand following hormonal manipulations in the female and seasonal variations in the male? We already know that the soma and dendrites of individual neurons expand; but is there also direct evidence for neurogenesis in the adult? A series of studies by Nottebohm and his colleagues has addressed the question of neurogenesis, and yielded intriguing results (Goldman and Nottebohm 1983; Paton and Nottebohm 1984; Burd and Nottebohm 1985; Nottebohm 1985; Paton *et al.* 1985; Alvarez-Buylla and Nottebohm 1988).

The species used in these studies was the canary. In contrast to female zebra finches, normal female canaries will sing when treated with an androgen as adults, though the song is simpler than that of adult males. Appearance of song is accompanied here, too, by a marked expansion of the HVc (Nottebohm 1980*b*). It is, therefore, a convenient system for manipulating the size of the song nuclei. The method used for the detection of neurogenesis was autoradiographic monitoring of the incorporation of radioactively labelled thymidine, which serves as an index of DNA synthesis in the S phase of the cell cycle, preceding mitosis. Goldman and Nottebohm (1983) chronically implanted 1-year-old females with testosterone, or cholesterol as a control, and then injected them with radioactively labelled thymidine. About 5 weeks later, the birds were sacrificed and their brain sectioned and subjected to autoradiography.

The surprise was that all the birds, including those that received cholesterol, showed a large number of labelled cells in their forebrain, including in and around the HVc. These included ependymal and subependymal cells in the ventricular zone, endothelial cells, glia, and neurons. It was estimated, on the basis of incorporation of radioactivity, that in both control and experimental birds neurons were added to the HVc at an average rate of *c.* 1.5 per cent per day of total HVc neuronal pool! Since in normal adult female canaries the size of the HVc is practically constant, this implies that there must be a massive cell death to offset the large numbers of added neurons. Follow-up experiments after cessation of [³H]thymidine administration, showed that about half of the labelled cells survived less than 1 month, and virtually no labelled cells were detected after 1 year (Nottebohm 1985). Later experiments, using a similar methodology, detected thymidine incorporation in the adult male (Paton and Nottebohm 1984; Paton *et al.* 1985).

Where did the neurons come from? When the canaries were sacrificed shortly after injection of radioactively labelled thymidine, autoradiography revealed no label in neurons, but very heavy labelling in the ventricular zone overlying the HVc. It was, therefore, suggested that neuronal precursors exist in the adult canary brain in the HVc ventricular zone, and that these cells divide (and hence incorporate the DNA precursor) and then migrate into the HVc.

Are the cells indeed neurons? Paton and Nottebohm (1984) injected normal adult males and testosterone-treated adult females with radioactive thymidine twice a day for 14 days, and 1 month later recorded intracellularly from neurons in the HVc. The same cells were then filled with horseradish peroxidase (HRP) and the birds sacrificed. Of 74 neurons thus stained with HRP, seven were found also to contain silver grains, indicating incorporation of radioactive thymidine. Six of these double-labelled cells were found in females, and only one in a male. Subsequent studies provided evidence that the thymidine-incorporating neurons are probably interneurons, since they were not labelled by retrograde transport of HRP from nuclei X and RA, which project to the HVc (Paton *et al.* 1985). Electron microscopical analysis revealed on these neurons synapses with different morphological characteristics, indicating different types of input (Burd and Nottebohm 1985).

What is the physiological significance of the apparent massive neuronal replacement in the adult canary brain? Neurogenesis in the adult brain is not an idiosyncrasy of the canary; it has also been recently detected in the HVc of zebra finches (Nordeen and Nordeen

1988*a*), and was previously detected in some other vertebrates (Altman and Das 1965; Johns 1977; Kaplan and Hinds 1977; Graziadei and Monti Graziadei 1979; Bayer *et al.* 1982). Thymidine incorporation studies usually involve chronic administration of rather large doses of radiolabelled compound, and the possibility cannot yet be utterly excluded that these doses affect neuronal physiology. However, assuming that thymidine incorporation faithfully reflects natural neurogenesis, three questions could be raised:

1. What role does neuronal replacement fulfil in the adult brain? If the brain invests so much energy in cellular turnover, this process should be expected to be essential for brain function. One possibility is that the replaced cells store dispensable memories (see below). This is somewhat counterintuitive: one could imagine more parsimonious ways of altering memories, especially if they are stored in a distributed manner (see Chapter 9). Another possibility is that the cells must be replaced because they deteriorate in the course of fulfilling exhausting metabolic functions, auxiliary to memory.

2. Does neuronal replacement fulfil a role in the learning of bird song? Clearly, even if it does, the process is not exclusive to song learning. Thymidine labelling occurs in forebrain areas which are not considered to be involved in the motor control of song. In addition, neuronal turnover is very active in females, who do not sing. One suggestion was that the females must still renew their brains to comprehend the seasonally altered song of the males (Goldman and Nottebohm 1983). Again, as mentioned above, it is possible that those interneurons that are replaced in the forebrain song nuclei are not involved in memory storage, but have supporting roles.

3. What role, if any, does cell proliferation play in the masculinization of the song system in the adult female canary? The results of Goldman and Nottebohm (1983) did not show a significant difference between the number of labelled neurons in testosterone- and cholesterol-treated birds, but did show a marked effect of the hormone on the proliferation of glia and endothelia. It is not clear to what extent masculinization in the canary requires massive remodelling and establishment of new internuclei pathways, and what is the role of non-neuronal cells in this process. It is possible that in normal female canaries all the necessary pathways do exist, at least in rudimentary forms, and that this is why post-hatching hormonal priming is not necessary. In other species the song pathways in the female may be incomplete. For example, in normal adult zebra finch females the RA is not innervated by the HVc (Konishi 1985). Female zebra finches do not sing if treated with androgen in adulthood; they must be primed with oestradiol as chicks. Such priming probably establishes the otherwise missing pathway(s).

Neuronal activity correlated with song and learning

The identification of a system of brain nuclei whose presence is necessary for song, and whose morphology correlates with it, made it possible to address the electrophysiological properties of neurons that are related to the internal representation of song. The major findings are given below.

Activity of neurons in the NIF, HVc, and RA is correlated with sound production

Both single- and multiunit recordings from HVc neurons in several species of songbirds showed activity which was correlated in time with the production of song elements (McCasland and Konishi 1981; McCasland 1987). This activity was found in the HVc in both right and left hemispheres. Deafening did not affect it, thus indicating that it was due not to sensory neurons but rather to neurons with motor function.

Additional information on the properties of such neurons was obtained from single-unit recordings performed in freely behaving mocking-birds (McCasland 1987). Several types of neurons with apparently specific roles in song production were thus revealed. Some cells preceded the initiation of song with very long (*c.* 0.5 s) bursts. They were termed 'anticipatory cells'. Other cells produced briefer stereotyped bursts in correlation with all song syllables. The activity of yet other cells was correlated only with specific syllables. Some such cells fired well in advance of sound onset. The neurons that demonstrated activity correlated with the production of song syllables did not respond to the same syllables when presented as an auditory stimulus, confirming their role as neurons with motor function.

Multiunit recordings revealed neural activity correlated with song elements in the NIF and the RA, too (McCasland 1987). A burst of activity in NIF neurons preceded each syllable and an inactive period preceded each silent interval. In the RA, activity was correlated with syllables similarly to activity in the HVc, and was markedly inhibited for up to a few seconds at the termination of song. Neuronal activity correlated with vocalization was not found in the MAN, UVA, and area X.

Motor commands flow from the NIF to the HVc to the RA

About 80 per cent of the spikes in the NIF, 55 per cent of those in the HVc, and 35 per cent of those in the RA occurred before sound onset. Determination of the time relative to sound onset at which 50 per cent of the spikes have occurred in each nucleus, revealed that 50 per cent of the spikes in the NIF had already occurred about 30 ms before sound onset, whereas the corresponding time points were about 0 ms for the HVc and about 20 ms after sound onset for the RA. These results led McCasland (1987) to suggest that the motor pathway leading to the hypoglossal nucleus is hierarchically organized, with the commands flowing from the NIF to the HVc and then to the RA. Since recordings from the UVA, which relays input to the NIF, as well as recordings from other potential input regions in the vicinity of the NIF, did not show song-related neuronal activity, it was concluded that the NIF is a source of timing cues for song.

Neurons in the HVc, which is a motor nucleus, respond to acoustic stimuli

The results reviewed above have demonstrated that the NIF, HVc, and RA contain neurons that send motor commands for vocalization. However, the HVc also contains neurons that respond to auditory stimuli. Katz and Gurney (1981) recorded intracellularly from the HVc in the zebra finch's brain and found that many of the neurons showed alterations in electrical activity in response to tone bursts. The same neurons were injected, after the recording, with HRP and their position in the HVc verified by histological analysis. Auditory responses were later recorded from single units in the HVc of other birds as well (Margoliash 1983; Paton and Nottebohm 1984; Muller and Leppelsack 1985). Moreover, HVc neurons were found to display enhanced metabolic activity correlated with acoustic stimuli, as indicated by the accumulation of radioactive 2-deoxy-D-glucose (Scheich *et al.* 1979). Auditory information is thus accessible to the motor system that controls song. This might actually be expected: we already know that the bird learns to sing by matching its vocal output with a song model (see above); cells that control the vocal organ must therefore interact with cells that listen.

Auditory units in the HVc indeed know when the motor neurons in the same nucleus fire: activity in auditory units was absent when the bird sang and seconds afterwards, whereas activity in motor units was not affected by playback of the song. On the other hand, canaries were found to terminate their song prematurely in response to the playback (McCasland and Konishi 1981). Thus, in spite of inactivity of HVc auditory units, the bird can still listen and respond behaviourally to song.

HVc neurons respond specifically to the bird's own song

Recordings made from single units in the HVc of the white-crowned sparrow identified neurons that responded only to combinations of temporal and spectral features which characterize the individual's own (autogenous) song (Fig. 13.3; Margoliash 1983). Such units were denoted 'song-specific' neurons. No units were identified that responded solely to the entire song; still, the results show that in the bird's song system, quite complex representational combinations are integrated in single units. The probability of identifying single units which integrate more complex representations, if such neurons exist at all, might be expected to be low. We do not, therefore, know yet whether 'grandmother cells', or 'andante cells', exist in the white-crowned sparrow brain.

Multiunit recordings, sampling a large population of neurons, have later suggested that the entire population of HVc auditory neurons is selective for autogenous song (Margoliash and Konishi 1985; Margoliash 1986). No such selectivity was found in field L, an auditory region not committed to song. The selectivity of the auditory units in the HVc depended on the rate and order of the acoustic information, since playing the song in reverse or modifying its speed diminished its stimulatory efficacy.

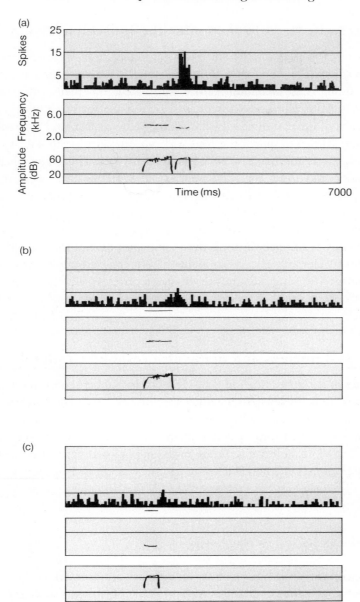

Fig. 13.3. Auditory responses of a unit in the HVc of the white-crowned sparrow, selective for a double-whistle component of the bird's own song. (a): The upper plot is a spike frequency histogram. The horizontal lines directly beneath the plot are timing marks. The two lower plots are of the bird's double whistle. They depict frequency (*middle*) and amplitude (*bottom*) vs. time. (b): There is only a weak offset response, or no response at all, to the first phrase in isolation. (c): No response to the second phrase in isolation. HVc auditory neurons learn to respond to a specific song pattern during the sensory-motor phase of song acquisition. (From Margoliash 1983.)

HVc neurons must learn to recognize the individual's song

The specificity of HVc neurons for autogenous song is high. Birds that were isolated as chicks, and have therefore developed abnormal song (see above), contained in their HVc 'song-specific' units that were optimally stimulated by their own abnormal song (Margoliash 1983). Multiunit responses showed preference for the adult's own song when compared to songs of other individuals sharing the same dialect, and even when compared to the song model (tutor) to which the bird was exposed during the sensory critical period early in life (Margoliash and Konishi 1985; Margoliash 1986). This means that HVc auditory neurons are plastic and must learn, by interacting with the bird's own vocal output, to respond to a specific song pattern during the sensory-motor phase of song acquisition. These neurons probably lose their plasticity after song crystallization; this was deduced from the observation that in adult white-crowned sparrows chronically implanted with electrodes, responses to autogenous song were stable over several months (Margoliash 1986). Therefore, in this species, other, as yet unidentified sites of plasticity are involved in learning to recognize conspecific song later in life. It is not yet clear whether HVc neurons retain their plasticity in adults of other species, especially those which acquire new song elements every year.

The hypoglossal nucleus and nerve also respond to sound

Recordings made in the zebra finch in the tracheosyringeal branch of the hypoglossal nerve and in the hypoglossal nucleus from which this branch originates, revealed long-latency auditory responses, which were eliminated by lesions in the HVc or RA (Williams and Nottebohm 1985). Different song syllables elicited response in different neurons within the hypoglossal nucleus. The auditory response in the hypoglossal nerve was preceded by 12–18 ms by auditory responses in the HVc. No response was elicited by sound in the tracheo-syringeal nerve of females.

Motor theories of song perception

Why is it important for adult birds to store an auditory representation of their own song in their song-generating system? Maintenance of song, involving auditory feedback and matching with a template, is not a satisfactory explanation; we already know that birds deafened as adults continue to sing, and that HVc auditory responses are actually inhibited by the bird's own singing (see above). A role in song perception seems therefore more plausible.

Postulated perceptual roles of autogenous song representations invoke interesting heuristic analogies between birdsong and other communicative vocalizations, including human speech. There are two major classes of theories of speech perception (reviewed in

Peterson and Jusczyk 1984). One class of theories proposes special perceptual filters for speech. Another class of theories, denoted 'motor theories', proposes that we perceive speech in terms of the way that we produce it, i.e. that speech sounds are perceived and distinguished by tacit knowledge of the vocal gestures used in their production (Liberman *et al.* 1967). Motor theories of speech thus propose an inherent linkage between production and perception. The finding that song-control nuclei in the bird's brain can also specifically respond to the bird's own song, has recently led to the proposal of two versions of motor theories of birdsong production.

One proposal suggests that the system that produces song in the adult also serves as an autogenous reference for classifying songs of other birds. Such a reference enables the bird to compare its own song with that of neighbours and strangers, and to determine, by the magnitude of the difference between the songs, its territorial distance from the other bird. This in turn might contribute to reproductive success (Margoliash and Konishi 1985; Margoliash 1986). In other words, when a young bird learns what its own song is like and memorizes it, it also learns how to quantify self-identity.

Another proposal is more radical and suggests that actual activity in the forebrain motor nuclei is an integral part of perception. According to this suggestion, when a bird hears a song it deciphers the message by converting the sensory data into the motor commands necessary to reproduce the same sounds (Williams and Nottebohm 1985). If this is the case, then one should expect the information from the hypoglossal nerve to be transmitted back to the forebrain. Williams and Nottebohm (1985) reported that this was indeed the case, and that multi-unit responses in the MAN could be recorded with latencies longer than those in the hypoglossal nerve. The flow of information during song perception may therefore include backwards activity in parts of the song-generating system.

If song is only intelligible to a neural system that is capable of producing the same vocalizations, then the female, who does not sing, may not understand a large proportion of what the male says. It might still understand part of the message, such as the love calls, but miss much of the excitement inherent in the vocal exchange between two males.

A tight genetic envelope, with a touch of novelty

• In conclusion, the study of birdsong has established that a set of interconnected brain nuclei serves as part of the physical substrate of the bird's culture (see Fig. 13.2). Ethological and behavioural studies first determined that song indeed includes elements of transmitted culture, anchored in a genetic programme and interwoven with individual innovation. Neuroanatomical studies then identified parts of the brain system involved, and suggested intimate links between the morphology of this system and the acquisition and performance of song. Pharmacological manipulations established that the behavioural programme and its modifications are controlled by sex hormones. Finally, electrophysiological analysis identified motor and sensory correlates of plastic and stable song, and cast some light on the possible properties of neuronal representations in this system. The stimulus selectivity of neurons in song-generating centres, e.g. in the HVc, is modified by sensory-motor experience. These neurons are, therefore, cellular substrates for learning. We also know what kind of sensory input affects the behaviour, and what is the ontogenetic time window during which such input is effective. Note, however, that in spite of the rich data available on the song system, we do not really know where the song template itself is located and how song is represented there.

The genetic envelope which imposes constraints on the mature song is rather tight. In almost all birds, even if raised beside a record player, the adult would end up singing a repertoire very similar to that of its conspecifics rather than an aria from *La Bohème*. And yet, within this genetic envelope, an intricate process of selection among many possible combinations takes place. From the behavioural data we know that this selection passes through unstable conformations before the system stabilizes. These phenotypic unstable states are expected to be a manifestation of the activity of quasi-stable neuronal assemblies, encoding prerepresentations (see Chapter 9). Brain substrates on which the selective mechanisms operate are provided by the growth of song nuclei, and selection probably occurs already in the course of this growth process.

Interestingly, a mathematical model for learning temporal sequences, by selection among spontaneously generated activity patterns in a network of neural-like elements, was recently proposed (Dehaene *et al.* 1987). The connectivity and Hebbian modification rules described in this model will not be described here but may, in principle, form a basis for simple song learning. In nature, the number of possible combinations of song elements is so large that no two individuals end up with exactly the same set of stable representations. This means that although each bird relies heavily on a well-established species experience of millions of years, the selection that its nervous system finally makes adds a touch of novelty.

14 *The generation of complex internal representations by sensory impressions, and their use: lessons from monkeys and vision*

Vision as an example of the formation of complex internal representations by sensory impressions

In most examples discussed so far, experience modified innate reflexes or stabilized innate pre-representations. In this chapter, we explore how sensory information generates novel, relatively complex internal representations, including those that are less dependent on innate templates. The models discussed in this chapter describe anatomical topography and neurophysiological flowcharts. Fine-grain analysis, incorporating data about the neuronal encoding of the relevant internal representations, must await further conceptual and technological developments.

A substantial part of what we know about the biology of the memory of complex representations stems from studies of the primate visual system. This is not surprising. About 50 per cent of the primate cortex is devoted to vision. Ample information is available on the properties of neurons and brain areas engaged in vision, and sophisticated, sensitive methodologies are available for studying visual behaviour. Moreover, much clinical data exist on disorders of visual recognition in humans.

Multiplicity of visual pathways and brain maps of the visual space

In mammals, visual information is transmitted from the retina to the cortex by two major pathways. The optico-geniculo-cortical pathway, primarily devoted to pattern and motion analysis, involves the projection from the retina to the dorsal lateral geniculate nucleus (LGN), from the LGN to the striate cortex in the occipital lobe, and from the striate cortex to the extrastriate cortex. The optico–colliculo–cortical pathway, primarily devoted to visual attention and the control of eye movements, involves the projection from the retina to the superior colliculus, from the colliculus to the pulvinar or the lateral posterior nucleus of the thalamus, and from the latter to the extrastriate cortex. Both pathways interact at various levels, including the superior colliculus, thalamus, and cortex.

In the present discussion, we confine our attention to cortico–cortical visual pathways in the adult monkey. It should be emphasized that the mammalian visual system also provides a most striking example of developmental plasticity, and ample data are available on the alteration of functional properties of visual neurons by combinations of retinal and extra-retinal inputs (Hubel *et al.* 1977; Sherman and Spear 1982; Wiesel 1982; Fregnac and Imbert 1984; Kleinschmidt *et al.* 1987; Fregnac *et al.* 1988; see also p. 106). However, this developmental plasticity, which most probably shares cellular mechanisms with experience-dependent modification in the adult brain, will not be further discussed here.

It is now well established that the visual world is multiply represented in the mammalian cortex. About 20 distinct visual areas, interconnected by more than 80 pathways, exist in the cerebral cortex of the macaque. Visual attributes are processed in the cortex both in parallel and serially, with analysis becoming gradually more global. Channels in which information about form, colour, binocular disparity and motion is segregated and recombined, have been intensively investigated in recent years (Hubel and Livingstone 1987; Maunsell and Newsome 1987; DeYoe and Van Essen 1988; Zeki and Shipp 1988). Only very limited aspects of findings on the neuronal systems involved in high-level visual processing will be discussed below. Of special interest to us in this respect is the finding that in the infero-temporal cortex, the last station in the striate-temporal cortical visual pathway (Fig. 14.1), neurons have large receptive fields and respond to complex stimulus qualities. Sometimes, preferential response is elicited only by highly differentiated patterns such as hands or faces (Gross *et al.* 1969, 1972; Desimone and Gross 1979; Perrett *et al.* 1982; 1987; see also the complex patterns in Optican and Richmond 1987; Richmond *et al.* 1987). These findings clearly demonstrate that the temporal cortex processes complex visual representations.

The role of the temporal lobe in visual learning

It was already known in the nineteenth century that temporal lobe lesions lead to dramatic behavioural deficits (Brown and Schafer 1888). The deficits observed in monkeys following bilateral temporal lobectomy were later termed the Klüver–Bucy syndrome (Klüver and Bucy 1938). The animal becomes tame, fearless, sexually hyperactive, and apparently unable to use vision to assess the significance of objects. It displays a compulsive tendency to react to visual stimuli and to place objects in its mouth. It is now known that the emotional, social, and sexual changes result from damage to the limbic system (Figs 14.1 and 14.2), but the abnormal visual behaviour is also caused by damage to the connections of the limbic system with the inferotemporal cortex, and to the inferotemporal cortex itself (Fig. 14.1).

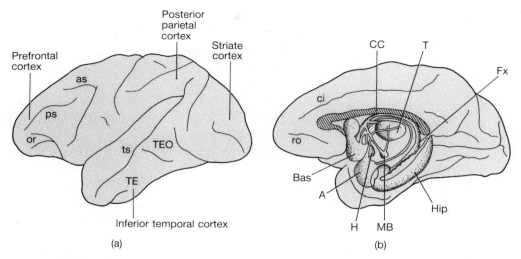

Fig. 14.1. Lateral (a) and sagittal (b) schematic views of the macaque brain, depicting structures mentioned in the text. A, amygdala; as, superior arcuate sulcus; Bas, basal ganglia; CC, corpus callosum; ci, cingulate sulcus; Fx, fornix; Hip, hippocampus; H, hypothalamus; MB, mammillary bodies; or, orbital sulcus; ps, principal sulcus; TE, anterior inferotemporal cortex; TEO, posterior inferotemporal cortex; T, thalamus; ts, superior temporal sulcus. The amygdala, hippocampus, and interconnected diencephalic nuclei are parts of the so-called 'limbic system' (see Fig. 14.2). (After Mishkin and Appenzeller 1987.)

The role of the latter in vision only became apparent in the 1950s, when the effect of inferotemporal neocortical lesions was dissociated from that of limbic lesions (Chow 1951; Mishkin 1954). The role of the inferotemporal cortex in visual processing was later confirmed and extended by a variety of invasive techniques and behavioural assays (Cowey and Gross 1970; Mishkin 1972; Bender 1973; Blake *et al.* 1977; Delacour 1977; Sahgal and Iversen 1978; Fuster *et al.* 1981; Moss *et al.* 1981; Iwai 1985). The role of the temporal lobe in vision was also confirmed by clinical observations in humans (Geschwind 1965*a*, *b*; Horel 1978; Cowey 1982).

● The effects of bilateral inferotemporal lesions could be summarized as follows. (1) They impair some aspects of vision but not other sensory modalities. (2) They do not cause 'simple' visual deficits. Their effects can be easily dissociated from those of primary visual cortex lesions, in that they do not markedly impair acuity and object detection in the visual field, but do impair pattern discrimination and some forms of visual learning. For example, monkeys with restricted inferotemporal lesions can still pick up a raisin or catch a cockroach with no apparent difficulties. (3) There is a functional differentiation in the inferotemporal area, and lesions in the posterior inferotemporal area (area TEO; see Fig. 14.1) produce behavioural deficits different from those of lesions in the anterior inferotemporal area (area TE).

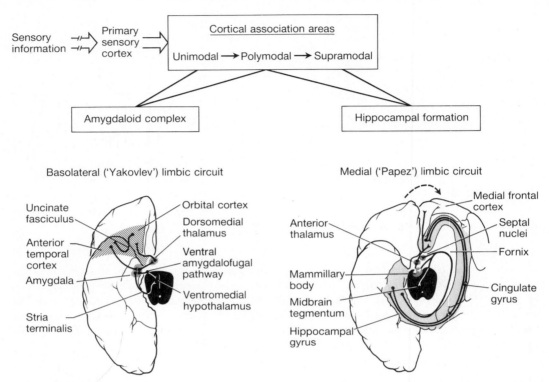

Fig. 14.2. The limbic system can be depicted as composed of two major, interconnected circuits, one centred on the amygdaloid complex, the other on the hippocampal formation. The amygdaloid complex interconnects with unimodal and polymodal association cortices. The hippocampal formation interconnects with polymodal and supramodal association cortices, and hence receives higher-level information than the amygdala. Evidence for the involvement of limbic circuits in learning and memory is discussed in this chapter. For further details as well as explanation of the terms uni-, poly- and supramodal, see text. The limbic circuits are depicted on a horizontal ventral view of a hemisphere; in the scheme of the Papez circuit, the medial hemispheric wall was rotated downwards (broken arrow) to the horizontal plane. (Adapted from Livingston and Escobar 1971; LeDoux 1987.)

Differentiation in the effect of inferotemporal lesions on visual learning

Two types of learning tests were instrumental in demonstrating the differential contribution of posterior and anterior inferotemporal areas to visual learning. These tests are conventionally termed *discrimination* and *recognition* tests, respectively, but actually they tax faculties more complex than their name implies.

Visual discrimination is the ability to distinguish between visual attributes of concurrently presented objects. In laboratory animals, discrimination training involves the association of discriminanda with a reinforcer. Therefore, the task taxes perceptual abilities as well as the ability to associate percepts with reinforcements. In a typical visual discrimination task, a monkey is placed in the Wisconsin General Testing Apparatus (see p. 31), and confronted with a sequence of 10–20 pairs of visually conspicuous 'junk' objects. One object in each pair conceals a food reward, such as a raisin, half a peanut, or a dried banana slice. The same sequence is repeated at a fixed interval, usually 24 hours, until the animal learns to choose consistently the baited object in each pair. Lesions in area TEO impair success in the above test more than lesions in TE.

In contrast, an intact TE is more critical than an intact TEO for normal performance in recognition tests. Recognition is the ability to distinguish familiar from novel stimuli. Commonly used visual recognition tasks are based on delayed response (see Chapter 2). Tasks involving delayed response tax not only perceptual and learning abilities, but also working memory. A delayed-response task which was especially informative in elucidating the function of area TE is the trial-unique, delayed non-matching to sample task (Mishkin and Delacour 1975; Fig. 14.3). In brief, the monkey is presented with a visually conspicuous 'junk' object over a central, baited food well. The animal uncovers the well and retrieves a food reward. After a delay (usually 10 s) the same object is paired with another, novel 'junk' object, each presented over a lateral well (Fig. 14.3). To retrieve the food treat, the monkey must now avoid the familiar object and displace the novel one. The procedure is repeated, with new 'junk' objects in each trial. A typical session consists of 20 trials separated by 30-s intertrial intervals.

Note the following. (1) The monkey is required to retain at least some characteristic visual attributes of an object following a brief exposure to that object. (2) The information from one trial is irrelevant to the next, hence it is stored in, and retrieved from, working memory (see Chapter 2). (3) The familiar object is not rewarded in the test. (4) No information is available in memory on association of reward with the correct choice object, because this object is novel. Reward is only an incentive for participation in the test. The test is, therefore, considered to measure relatively 'pure' visual recognition. Success in the trial-unique, delayed non-matching to sample task requires mastering the principle of selecting novelty—for which monkeys have a natural inclination.

Furthermore, delayed matching and non-matching to sample tests can be tailored to quantify additional aspects of visual performance and memory. After the monkey masters the basic principle, the task can be made more difficult in two ways (Gaffan 1974). One is to increase gradually the delay between the sample and the choice. This titrates memory duration. The other way is to give the animal not one but several objects to remember, in lists of various length (usually up to ten). In each list, several sample trials are presented successively, followed by their respective matching in the same order. For example, in a list of three, the presentation would be A, B, C, then A vs. D, B vs. E, C vs. F (select A if it is a matching to sample task, D if non-matching to sample, etc.). This incorporates in the test measures of memory span, distractibility, and also of discrimination, because when the list is made sufficiently long, the burden placed on discrimination ability becomes heavy and the subject may err even if recognition memory *per se* is intact.

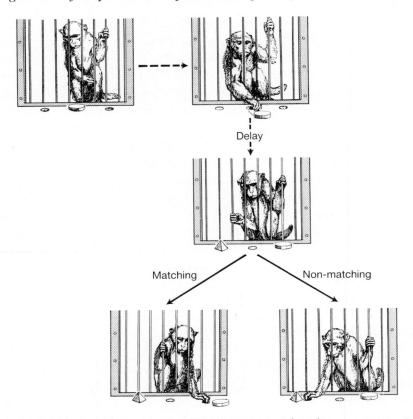

Fig. 14.3. Delayed-response tasks that measure object recognition and working memory are informative in analysing the function of brain structures involved in visual memory. The diagram illustrates two related tasks. The monkey is placed in a Wisconsin General Testing Apparatus (see Fig. 2.4) and is first presented with a visually conspicuous object over a central baited well. The animal displaces the object to obtain the food reward. After a delay of a few seconds, the monkey is confronted with the same object paired with a new one. In the delayed matching to sample task (*left*), displacement of the familiar object is rewarded. This test measures visual recognition and association of an object with reward. In the delayed non-matching to sample task (*right*), displacement of the non-familiar object is rewarded. In the trial-unique version of this task, each object is seen in only one trial during the whole series of tests. The trial-unique, non-matching to sample test is especially appropriate for quantifying recognition memory, because reward is only an incentive for participation in the test, and the ability to associate an object with the reward does not contribute to success. Monkeys master the non-matching tasks quicker than the matching tasks, because they have an innate tendency to explore novelty. (Modified from Mishkin and Appenzeller 1987.)

Mortimer Mishkin and his colleagues (Mishkin 1982) trained monkeys in these procedures, and then lesioned selected portions of the temporal lobe. Following recovery, the monkeys were retrained on the same task. The effects were quite dramatic: animals with TE lesions required many more trials to reach a criterion, and, furthermore, their performance deteriorated if the delay between sample presentation and choice was increased. In contrast, the performance of animals with TEO lesions deteriorated only when the number of unique objects in each trial was increased.

• The discrimination and recognition tests thus succeeded in dissociating some roles of areas TE and TEO. The recognition tasks are considered to place a heavier burden on mnemonic faculties and on high-level processing than the concurrent discrimination tasks. This is because in concurrent discrimination, the animal reacts to familiar stimuli which contain all the information necessary to prompt a correct response. In contrast, in trial-unique, delayed non-matching to sample, the animal must retain over a delay information about an object seen only once in the preceding phase of the test, use this information in a more 'abstract' way, as a comparator in categorizing the choice object as novel, and then repeat the procedure with another pair of objects. The observation that area TE is more important than area TEO for visual recognition and working memory, fits into a scheme in which visual information is hierarchically processed in the striate-inferotemporal pathway, and is accessed in the more anterior stations by more global mental operations. In all probability, there is no sharp functional transition between areas TE and TEO, but rather a graded continuum of increasing globality in the computations made over the visual representation.

The above results do not exclude the possibility that the inferotemporal cortex is just a relay station and memory is stored further away from the striate-temporal visual pathway. However, since the TE is the last station in this hierarchical pathway, the assumption that complex visual representations are formed there, and that TE circuits are an integral part of the neuronal system that stores these representations, does make sense. Further evidence in support of this assumption is provided below.

Retention of task-specific information in inferotemporal neurons

Can the activity of inferotemporal neurons be correlated with learning? Apparently it can. In recognition tasks, some TE neurons differentiate visual stimuli by their contextual significance, e.g. if they signify matching or non-matching to a sample (Gross *et al.* 1979; Mikami and Kubota 1980; Baylis and Rolls 1987). Moreover, inferotemporal units retain task-specific information during task performance. A study by Fuster and Jervey (1982) illustrates this point. They have investigated the activity of temporal cortex units, mostly in area TE, during the performance of a delayed matching to sample task. Each monkey sat in a chair facing a white panel with translucent buttons. A trial began with presentation of a coloured light in a centrally located button. The animal turned it off by pressing the button. After a delay of 16–20 s, two or four colours appeared simultaneously in a horizontal row of buttons under the sample button. One of the colours matched the sample. Pressing the matched button delivered juice into the monkey's mouth. After an intertrial period, the sequence was repeated. Although the task may seem simple, it took up to six months to train the monkeys to perform it reliably. This emphasizes that matching to sample is not an easy task for monkeys to perform, and is clearly more difficult than non-matching to sample.

Most cells in this study showed an excitatory reaction to the sample opening a trial, with an onset latency of > 70 ms. The cells could be classified into two categories, depending on whether they responded differentially to colours. Naturally, the differential cells were more interesting. Most of them were excited differentially by different colours, others were excited by some colours and inhibited by others, and some reacted to only one colour. The majority of the cells exhibited a different frequency of firing during the delay than during the intertrial period; and some of the cells showed different levels of activity during the delay, depending on the sample colour (Fig. 14.4). Differential sample response, though, was not a necessary condition for differential delay activity. On presentation of the choice, the cells usually reacted in a manner qualitatively similar to their reaction to the sample. After the choice, unit activity returned to the pretrial baseline.

An intriguing possibility is that the retention of an elevated firing level during the delay between sample and choice, reflects retention of task-related information. Other explanations should, however, be considered. Sensory afterdischarge was not very likely, since some

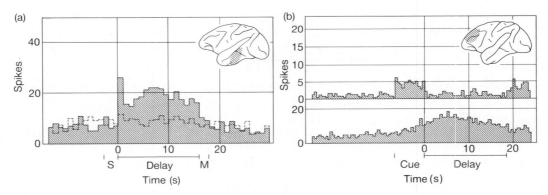

Fig. 14.4. The activity of many neurons in the inferotemporal and prefrontal cortices is correlated with discrete events in delayed-response trials. In some cases, activity is specifically correlated with the delay period of the task. (a): Spike frequency histogram of a single inferotemporal cell during a delayed matching to sample task. In this version of the task, the animal selected one of several colours to match a sample presented before the delay. The cell increased its activity only during the delay if the sample was red (solid line, shaded histogram) but not green (broken line, superimposed on the red-response histogram). S, sample; M, match. (After Fuster and Jervey 1982.) (b): Spike frequency histograms of two units in the dorsolateral prefrontal cortex during a visuospatial delayed-response task, in which the monkey learned which of two identical wooden blocks covers a baited food well. The upper unit increased firing in the cue period, during which the monkey observed the baiting and covering of the wells, and in the choice period following the delay. In contrast, the lower unit increased activity during the delay (after Fuster 1973). One interpretation of the results is that neurons in the inferotemporal and prefrontal cortices retain task-specific visual information. *Insets:* The shaded areas depict the brain areas from which recordings were made.

cells displayed sustained activation during the delay which was not preceded by a reaction to the sample (Fig. 14.4). Alternatively, the increased firing during the delay may have reflected alteration in non-specific alertness, or attention to background stimuli during the delay. The latter explanations were again difficult to reconcile with the data, in this case with a differential, colour-specific sustained activation during the delay. Finally, delay-locked discharge may have something to do with preparation of motor responses. This could not be excluded. Nevertheless, the possibility that the sustained activity, terminated on completion of the choice, indeed reflected a working-memory process, is a plausible one.

We have already noted that the anterior part of the inferotemporal cortex, area TE, differs from its adjacent posterior cortical area in being involved in recognition and working memory more than in concurrent discrimination. The above experiments, and similar ones, have led Fuster and Jervey (1981, 1982) to suggest a further topographical differentiation in the temporal cortex: delay-activated cells were more common in the lower bank of the superior temporal sulcus than in the middle temporal gyrus. It was, therefore, concluded that cortex in the superior temporal sulcus (see Fig. 14.1) specializes in the temporary storage of visual information. In later experiments, using an essentially similar experimental approach, Miyashita and Chang (1988) found delay-activated neurons in the anterior ventral part of the temporal cortex of the macaque. In this case, rather complex patterns were used, and the delay-activated neuronal activity was selective for the shape irrespective of size, orientation, or colour, implying a categorized percept of the picture.

Note that the above data set only a lower limit for the duration of retention of hypo-thetical visual representations in inferotemporal cortex or circuits accessed by it: information can be retained for at least many seconds. These electrophysiological data do not set an upper limit for long-term retention in this region.

The limbic connection

What causes the representation of an object to be stored in the inferotemporal cortex and its interconnected circuits? We return here to a problem discussed earlier in the context of cellular mechanisms of learning: why is some information only transiently processed whereas other information is stored? In *Aplysia*, sensory information was 'printed' if a modulating pathway was concurrently active. Similarly, in LTP, only if depolarization was concurrent with transmitter release did the synapse become potentiated. In the mammalian visual system, too, mechanisms must exist to tell the cortex that a stimulus is worth remembering.

These mechanisms involve distinct allocortical as well as subcortical structures. A major role is fulfilled by the limbic system (see Figs 14.1 and 14.2), to which the inferotemporal cortex projects.

The limbic system

A short note concerning the limbic system is pertinent here, because this system will occupy much of our discussion in this and the following chapter. *Limbus* means 'border' in Latin; the term was used by Broca (1878) to describe the ring of tissue surrounding the anterior brain-stem, and later used to describe an interconnected system of primitive cortex subserving emotion (MacLean 1970). As knowledge grew, it became clear that the composition, connections, and functions of the 'limbic system' are more complex than was thought at first (Swanson 1987). Our only concern here is the role of limbic structures in learning and memory.

From a methodological point of view, it is advantageous to consider the limbic system as being composed of two interconnected circuits (Livingston and Escobar 1971; see Fig. 14.2). One circuit, which could be described as being centred on the *hippocampal formation*, is called the medial or 'Papez' circuit (after the American anatomist James Papez; Papez 1937). In this circuit, information flows from the entorhinal cortex to the hippocampal formation, from there through the fornix to the anterior thalamus (directly or via the mammillary bodies), from the anterior thalamus to the cingulate gyrus, and from there back to the entorhinal cortex via the cingulum bundle. The other major limbic circuit is described as being centred on the *amygdaloid complex*; this is the basolateral circuit which, in analogy with the Papez circuit, might be dubbed the 'Yakovlev' circuit (after the American neurologist Paul Yakovlev; Yakovlev 1948). This circuit includes orbitofrontal, insular, and anterior temporal cortical areas, together with their connections to the amygdala and the dorso-medial nucleus of the thalamus.

The hippocampal formation and the amygdaloid complex both interact with many cortical and subcortical areas; some of these interactions will be mentioned in due course (for a comprehensive neuroanatomical review, see Amaral 1987). One striking difference in the connectivity of these two 'limbic centres' should be noted right away. It concerns the processing of sensory information by the neocortex. As mentioned earlier, cortical association areas, i.e. cortical areas receiving information from the primary sensory cortex, process information at increasing levels of globality. A convenient classification is the following (see Fig. 14.2): the unimodal association cortex is practically committed to a single sensory modality. Unimodal areas feed to polymodal areas, in which cross-modal convergence among sensory modalities occurs. Polymodal areas feed in turn to the supramodal association cortex, in which more global processing is performed (Mesulam *et al.* 1977). The amygdaloid complex interconnects directly with the unimodal and polymodal association cortex. In contrast, the hippocampal formation interconnects directly with the polymodal and supramodal cortex. This implies that whereas both the hippocampus and the amygdala access information from all sensory modalities, the hippocampus receives information which is more highly processed than that received by the amygdala (Amaral 1987). This also indicates that the different circuits of the limbic system perform different computations on cortical information.

Hippocampus, amygdala, or both?

Let us now return to the role of limbic structures in learning and memory. As noted above, the role of medial temporal structures in associative learning has long been suggested by clinical data, and supported by gross ablations in monkeys. Orbach *et al.* (1960) demonstrated that bilateral resections, including the hippocampus and amygdala, led to impairment in the acquisition and retention of visual discriminations, independent of temporal neocortical damage. This and additional studies (Mishkin 1954; Mishkin and Pribram 1954; Weiskrantz 1956; Scoville and Milner 1957; Penfield and Milner 1958) further established the contribution to cognitive deficits of lesions in both brain parts implicated by the Klüver–Bucy syndrome, namely the phylogentically older medial temporal structures, and the temporal neocortex.

Whether both the amygdala and the hippocampus, or only one of them, are necessary for normal visual learning and memory, has been a matter of dispute, because some of the early lesions damaged both structures conjointly. Independent bilateral hippocampal lesions in monkeys had a clear effect on visual learning in some experiments (Jones and Mishkin 1972; Mahut 1972; Gaffan 1974; Moss *et al.* 1981; Mahut *et al.* 1982; Parkinson *et al.* 1988) but not in others (Orbach *et al.* 1960; Correll and Scoville 1965; Mahut 1971; Jones and Mishkin 1972; Spiegler and Mishkin 1981; Zola-Morgan and Squire 1986). Some of these studies used identical behavioural procedures, such as recognition in delayed non-matching to sample. A difference in the preoperative training of the monkeys was partly blamed for the discrepancy in the results; preoperative experience might have established strategies that facilitated postoperative learning (Zola-Morgan and Squire 1986).

In some experiments, lesions which included large parts of the amygdala, but spared the hippocampus, markedly suppressed performance in object and place discrimination tasks (Jones and Mishkin 1972). On the other hand, disruption of the temporal stem, which carries connections of the temporal cortex and amygdala but not the hippocampus, did not significantly impair performance in trial-unique, delayed non-matching to sample tasks, whereas conjoint amygdala–hippocampal lesions did (Zola-Morgan *et al.* 1982). The latter lesions severely impaired performance in other visual learning and memory tasks as well, and their effect was retained for at least 18 months (Zola-Morgan and Squire 1985*a*). However, only experiments that explicitly compared the effect of conjoint vs. independent amygdala and hippocampal lesions on performance in the same tasks, established that both the amygdala and the hippocampus are indeed important for visual memory (Mishkin 1978).

The involvement of the hippocampus in memory is already known to us from previous discussions. Hippocampal activity was especially important in tasks involving spatial and working memory (see Chapter 11). The relevance to hippocampal function of the same two parameters is revealed in studies on visual learning in monkeys. Many of these studies derive their conclusions from the effect of hippocampal ablations on behaviour (e.g. Parkinson *et al.* 1988), but some studies are based on electrophysiology and metabolic mapping. For example, in a delayed-response task involving spatial cues, units in the monkey's hippo-campus displayed activity locked to temporal and spatial parameters of the task. Watanabe and Niki (1985) trained monkeys to hold a lever for 4–5 s in order to turn on a red cue light at a right or left position. Thereafter, the red light was turned off and a 3-s delay period began. If the animal continued to hold the lever during the delay, two white choice lights, right and left, were turned on. In order to get a juice reward, the animal had to choose the side illuminated during the cue period, by releasing the lever and pressing the appropriate button. About 15 per cent of the hippocampal units showed alteration in activity correlated with some event in this task, and of these, more than 40 per cent increased or decreased their firing preferentially during the delay period. This was taken to indicate that the hippocampus was engaged in the retention of provisional memory. Some units had a differ-ential response to right or left, and were suggested to engage in place mapping (see Chapter 11).

Additional evidence for the role of the hippocampal formation in working memory was obtained by Friedman and Goldman-Rakic (1988). They used the 2-deoxy-D-glucose method (see p. 160) to examine metabolic activity in the hippocampal formation and the amygdaloid complex of rhesus monkeys during performance of a battery of visual and visuospatial delayed-response tasks. Control animals performed other sensory–motor and learning tasks not involving working memory. Utilization of glucose in the hippocampal formation, but not in the amygdaloid complex, was higher in animals performing the working memory tasks compared to controls.

A limbic structure that we do encounter here for the first time is the amygdala (see Figs 14.1 and 14.2). It plays a critical role in stimulus–reward association. Note again that such association does not play a direct role in the trial-unique, delayed non-matching to sample task, since the chosen object was not previously encountered and rewarded. Amygdalectomy (affecting also the periamygdaloid cortex) impaired performance on a visual object–food reward association task as severely as did TE lesions. However, in contrast to TE lesions, amygdalectomy had only a small effect on object recognition in the trial-unique, delayed non-matching to sample task (Spiegler and Mishkin 1981). The role of the amygdaloid complex in attributing reward value to objects was emphasized by other studies as well (Gaffan and Harrison 1987; Gaffan *et al.* 1988; Nishijo *et al.* 1988), and may account for part of the bizarre oral behaviour observed in the Klüver-Bucy syndrome.

The function of the amygdala is not, however, confined to the association of reward with visual stimuli. As noted above, it is connected to different unimodal as well as polymodal cortical sensory areas, and hence is in a position to play a role in cross-modal associations. Evidence for this emerges from an experiment in which monkeys were trained on visual and tactual versions of the delayed non-matching to sample task (Murray and Mishkin 1985). In this case, the visually and tactually conspicuous 'junk' objects were mounted on corks that fitted snugly into the food wells, thus forcing the monkeys to palpate them to obtain a reward. After reaching criterion, the light was gradually reduced and the animals trained so that they could perform the task in total darkness. After reaching criterion again, the monkeys received bilateral ablations of either the amygdaloid coplex or the hippo-campal formation. The injured animals relearned the intramodal versions of the task, and were then subjected to an intermodal version of it. Here each sample was presented in the dark, but the choice had to be made in the light, and a response was scored the moment the monkey touched an object, to ensure use of visual cues only. The hippocampectomized monkeys performed well, but the amygdalectomized monkeys failed. This failure could not have been attributed to sensory or motivational deficits, since the animals performed well on each of the intramodal versions of the task. The conclusion was, therefore, that cross-modal association was impaired.

The diencephalic connection

The amygdala and hippocampus each send fibres to diencephalic nuclei (see Fig. 14.2). Transection of these pathways, the amygdalofugal and the fornix, respectively, impaired visual recognition memory. Again, as in the case of the amygdala and the hippocampus, the effect of a combined lesion was much more severe than the effect of separate lesions (Bachevalier *et al.* 1985). A similar picture has emerged from lesions inflicted on the corresponding diencephalic targets, namely the *anterior thalamic nuclei* and the *mammillary bodies*, to which the hippocampus projects, and the magnocellular part of the *dorsomedial nucleus* in the posterior thalamus, to which the amygdala projects. A lesion in the medial portions of the posterior thalamus impaired object recognition (Aggleton and Mishkin 1983*a*; Zola-Morgan and Squire 1985*b*), so did a lesion in the medial portion of the anterior thalamus. However, conjoint removal of both the anterior and posterior thalamic regions caused an even more severe defect (Aggleton and Mishkin 1983*a, b*). The same lesions did not significantly affect pattern discrimination—a distinction already familiar to us from our discussion of area TE in the inferotemporal cortex. A separate lesion in the mammillary bodies, receiving hippocampal input, also had some deleterious effect on object recognition, although the extent of the effect has still to be determined (Saunders 1983; Aggleton and Mishkin 1985).

● At a first glance, the data on the accumulative effects of conjoint limbic and dien-cephalic lesions may seem to fit into the 'mass action' and 'equipotentiality' principles (p. 156). But this is an erroneous interpretation. From what we have already said, it is clear that the hippocampus and the amygdala have different roles: the hippocampus is instru-mental in working memory and in cognitive mapping, the amygdala in cross-modal and object–reward associations. The fact that in some instances lesions in the hippocampus, the amygdala, and their respective diencephalic interconnections, have additive disruptive effects on learning, simply means that these circuits subserve faculties which contribute together to the appropriate overall behavioural response in the test. Given the appropriate test (e.g. a demanding working-memory situation in the case of the hippocampus), the role of the above two limbic circuits could be dissociated. Later in this chapter we will return to the roles of limbic circuits in learning.

A note on modulatory transmitters and pathways, arousal, and attention

Limbic activity is but one factor influencing the conversion of a primary percept into an enduring memory. Another factor is the activity of diffusely projecting fibre systems which regulate arousal by releasing modulatory transmitters in the forebrain. Among these are projecting fibres from the brain-stem that release norepinephrine (Moore and Bloom 1979), dopamine (Moore and Bloom 1978), and serotonin (Azmitia 1978), as well as cholinergic projections from the basal forebrain (Mesulam *et al.* 1983) (Fig. 14.5). Arousal is also regulated by regionally released as well as circulating hormones. All these modulatory signals are expected to affect the excitability of neocortical and limbic circuits, and hence the sensitivity in which these circuits attend to ongoing sensory stimuli in general. Neuromodulatory and hormonal systems have been implicated in all types of learning. This conclusion is based on an extensive body of pharmacological and pathological evidence (for representative references, see De Wied and Gispen 1977; Matthies 1980; Coyle *et al.* 1983; McGaugh 1983; Mason 1984; Overstreet 1984; Flood *et al.* 1985; Ogren 1985; Altman *et al.* 1987; Arendt *et al.* 1988; Hagan and Morris 1988; Mastropaolo *et al.* 1988). The specific contribution to learning and memory is often difficult to determine, because of the multiplicity of the behavioural effects of neuromodulators and hormones, and the dependence of their effect on the inner connections and on the other inputs of the network. The prevalent view is that neuromodulatory transmitters and hormones can facilitate or inhibit synaptic connections and cellular excitability in circuits subserving complex representations, similarly to the way in which facilitatory or inhibitory transmitters modulate synaptic efficacy and cellular excitability in networks subserving simple reflexive memories (see Chapter 4).

As everyday experience tells us, learning can benefit not only from an appropriate level of general arousal, but also from attention, i.e. selective analysis of inputs (Glass and Holyoak 1986). In visual learning, attention can be experimentally increased by training the animal to expect a stimulus at a certain spatial location, or by increasing the difficulty of the task. (For methods of manipulating facets of attention in various sensory modalities, see Hillyard 1985.) For example, in a recent study of visual discrimination in the rhesus monkey, task difficulty, and hence presumably attention, were increased by increasing the similarity between the discriminanda. This resulted in improved behavioural discriminative abilities, as well as enhanced response and selectivity of neurons in extrastriate cortex (Spitzer *et al.* 1988). Attention may be regulated by several brain systems: the diffused modulatory transmitter systems mentioned above; the intricate ascending projections from the brain-stem reticular formation to the neocortex via the dorsomedial and intralaminar thalamic

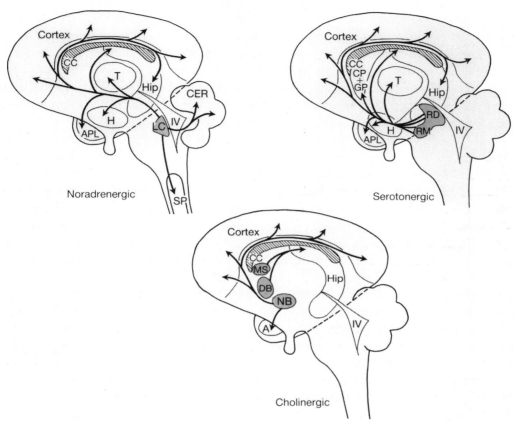

Fig. 14.5. Examples of diffuse subcortical–cortical neurotransmitter systems that play a role in learning and memory. A, amygdala; APL, amygdala-piriform lobe; CC, corpus callosum; CER, cerebellum; CP, caudate-putamen; DB, diagonal band of Broca; GP, globus pallidus; H, hypothalamus; Hip, hippocampus; IV, fourth ventricle; LC, locus coeruleus; MS, medial septal nucleus; NB, nucleus basalis of Meynert; RD, dorsal raphe nucleus; RM, medial raphe nucleus; SP, spinal cord; T, thalamus. The model of visual memory discussed in this chapter proposes specifically that the basal forebrain cholinergic system plays a central role in stabilization of representations (see Fig. 14.7 below). Deterioration of this cholinergic system is specifically linked to memory dysfunction in Alzheimer's disease (Coyle *et al.* 1983; for other brain systems involved in the aetiology of Alzheimer's disease, see Chapter 15). (Adapted from Emson and Lindvall 1986.)

nuclei (a system that is also critical in general arousal); and by other thalamocortical systems (Crick 1984*b*; Mountcastle *et al.* 1987; Rafal and Posner 1987).

The striate–parietal pathway: a complementary visual processing line

So far we have dealt with visual recognition processed by the striate–inferotemporal–limbic–diencephalic system. Another system, involving the *posterior parietal cortex*, processes information on the location of objects in the visual space.

The posterior parietal cortex (see Fig. 14.1) has been implicated, on the basis of clinical observations in humans and ablations in primates, in visual and tactual processing, sensory-motor integration, attention, and commanding movement (Hyvärinen 1982; R. A. Andersen *et al.* 1987). Of special interest to us here are those many studies that have assigned to this region a role in visuospatial learning. The involvement of the posterior parietal cortex in spatial perception, and the distinction between its function and that of the inferotemporal cortex, can be illustrated by a set of experiments performed by Pohl (1973).

Monkeys with lesions in various cortical regions, including the posterior parietal and the inferotemporal cortex, were subjected to object discrimination and landmark discrimination tests. The object discrimination test was similar to the ones described above, and involved displacement of either a red cylinder or a silver cube to obtain a concealed half-peanut reward. In the landmark discrimination task, two wells, *c.* 35 cm apart, were covered with identical matboard plaques; the monkey was rewarded for choosing the food well closer to a salient 'landmark', a red-and-white cylinder, which was positioned randomly from trial to trial close to either the left or the right well (Fig. 14.6). The monkey had thus to respond on the basis of the proximity of the external landmark to the goal. This is termed 'allocentric' or 'extrapersonal' perception, and is usually contrasted with 'egocentric' perception in which spatial position is determined relative to the observer (e.g. left or right; see the test used above in studying the role of hippocampal units in visual memory). How much such tasks indeed distinguish between 'egocentric' and 'allocentric' spatial perception is a matter of dispute; these terms are introduced here merely because they are often used in the literature describing visuospatial tasks. For our purpose, suffice it to remember that performance in the landmark test is clearly dependent on visuospatial abilities.

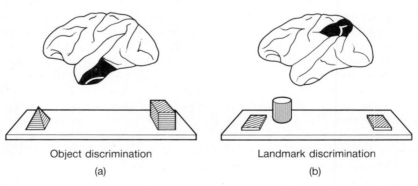

Object discrimination   Landmark discrimination
(a)   (b)

Fig. 14.6. Visual form recognition and visuospatial abilities are differentially affected by lesions in the inferotemporal and posterior parietal cortices (a): Object identification and discrimination in delayed response tasks (see Fig. 14.3) was impaired by lesions to area TE in the inferotemporal cortex. (b): In landmark discrimination, a visuospatial task, the monkey learns to identify one of two identically covered wells by its proximity to a visually conspicuous landmark, such as a coloured cylinder. Lesions in the posterior parietal cortex impaired performance in this task. The lesioned areas are depicted in black. (Adapted from Pohl 1973; Mishkin and Ungerleider, 1982.)

The parietal lesions severely impaired the performance in the landmark discrimination task, but not in the object discrimination task. This was in contrast to the effect of the inferotemporal lesions, which severely impaired object discrimination, but had only a relatively mild effect on performance in the landmark task. Similar effects of parietal lesions on 'allocentric' spatial tasks have been obtained in other studies (Milner *et al.* 1977; Brody and Pribram 1978; Sugishita *et al.* 1978; Petrides and Iversen 1979; Mishkin *et al.* 1982; Lawler and Cowey 1987). Disconnecting the posterior parietal cortex from the primary visual cortex established that the former indeed receives from the latter the input necessary for normal visuospatial performance (Mishkin and Ungerleider 1982).

● Taken together, the data indicate that in addition to the striate-inferotemporal visual processing pathway, dealing with object identification (i.e. *what* is the object?), there is a striate-parietal pathway involved in perception of the object's position relative to other objects in the visual space (*where* is the object?). It is now clear that this dichotomy between object identification on one hand, and analysis of its spatial relationship on the other, emerges already at low levels of visual analysis (Maunsell and Newsome 1987; DeYoe and Van Essen 1988; Zeki and Shipp 1988).

The prefrontal cortex: *ad hoc* processing of internal representations and their use for guiding behaviour

The prefrontal cortex, that anterior pole of the mammalian brain so highly developed in primates, has long been regarded as subserving higher cognitive functions, including some that differentiate *Homo sapiens* from other mammals (Luria 1966; Fuster 1980; Milner and Petrides 1984; Stuss and Benson 1986; Goldman-Rakic 1987). In spite of the impressive information accumulated over the years, and especially in the past decade, the functional neuroanatomy of the prefrontal cortex has still not been worked out in detail, and in many respects this is '*the*' *terra incognita* of the primate brain. Clearly, the prefrontal cortex is engaged in supramodal, global information-processing, and in the regulation of complex behaviours. Several theories have been proposed to explain how it fulfils these functions (reviewed in Stuss and Benson 1986; Goldman-Rakic 1987). Within the scope of our discussion, we will briefly consider only a limited aspect of prefrontal function relevant to the processing of visual information.

All the brain regions that we have so far implicated, either directly or indirectly, in processing visual representations, i.e. the inferotemporal cortex, the posterior parietal cortex, the amygdala, the hippocampus, and the diencephalic nuclei, are interconnected to the prefrontal cortex (Jones and Powell 1970; Fuster 1980; Macko *et al.* 1982; Maioli *et al.* 1983; Van Hoesen 1985; Stuss and Benson 1986; Goldman-Rakic 1987, 1988). It is thus only natural to ask, in the context of our discussion, what role the prefrontal cortex plays in visual learning. More than 50 years ago, monkeys with resections of the anterior frontal cortex were shown to fail in delayed-response tasks (Jacobsen 1936; Jacobsen and Nissen 1937). Since then, many studies have addressed the reason for this behavioural failure. Heuristic interpretations blamed deficits in spatial perception, attention, short-term memory, and operation of command lines to motor centres. Although all the above shortcomings possibly contribute to the abnormal performance of a prefrontally ablated monkey in visual memory tasks, recent data and theories place the emphasis on the con-tribution of prefrontal function to working memory, and on the role of prefrontal cortex in *ad hoc* processing of internal representations and their use for guidance of behaviour (Fuster 1980; Passingham 1985; Goldman-Rakic 1987).

When visual memory is considered, two gross anatomical divisions can be discerned in the prefrontal cortex: dorsal and ventral. The dorsal area includes the fundus and banks of the principal sulcus and, in many studies, also the dorsolateral convexity above that sulcus (Fig. 14.1). The ventral area is often meant to include the inferior convexity, ventral to the lower bank of the principal sulcus, but the borders of the area vary in different studies, sometimes extending posteriorly as far as the arcuate sulcus, and ventromedially to include the ventromedial cortex, with large portions of the orbital and the cingulate cortex. Dorsal and ventral prefrontal lesions have different effects on visual learning in delayed-response tasks. Whereas dorsal lesions severely impede visuospatial functions, ventral lesions have a more deleterious effect on visual recognition (Pohl 1973; Passingham 1975; Brody and Pribram 1978; Mishkin and Manning 1978; Fuster 1980; Bachevalier and Mishkin 1986; Goldman-Rakic 1987; Lawler and Cowey 1987). For example, ventromedial lesions had a greater effect on delayed non-matching to sample than did dorsolateral lesions. In contrast, dorsolateral lesions impeded the classical spatial delayed response, in which the animal had to remember which of two food wells covered with identical covers was baited (see p. 32). Both prefrontal lesions did not markedly affect learning on concurrent visual discrimina-tion (Bachevalier and Mishkin 1986).

Unit recording supplemented the lesion data in implicating prefrontal regions in the performance of visual tasks (Fuster and Alexander 1971; Fuster 1973; Sakai 1974; Niki and Watanabe 1976; Suzuki and Azuma 1977; Kojima and Goldman-Rakic 1984; Fuster *et al.* 1985; Suzuki 1985; Watanabe 1986; Joseph and Barone 1987). The classic study of Fuster (1973) illustrates the data obtained in such experiments. Single-unit recordings were made from the dorsolateral prefrontal cortex while the monkey performed a visuospatial delayed-response task, involving retrieval of an apple piece under one of two identical wooden blocks. The majority of units displayed alteration in firing frequency correlated to events in the trial (Fig. 14.4). Similarly to the inferotemporal units described previously, here, too, some units fired during the delay period at a level above the intertrial baseline. This could be interpreted as retention of information during the delay. Again, other interpretations should not be excluded, for example, setting up a forthcoming motor response (Kubota and Funahashi 1982).

Moreover, reciprocal functional interactions could be demonstrated between units in the dorsolateral, prefrontal, and inferotemporal cortices. Decreasing neuronal activity in each of these regions, by implanted microthermoelectric coolers, led to reversible changes in both spontaneous and task-related activity in the other region. The monkey's performance in a delayed matching to sample task was also impaired by cooling each of the areas (Fuster *et al.* 1985).

●   So, does the prefrontal cortex store visual memories? Yes, and no: the crucial issues are for how long, and for what purpose. The accumulated evidence argues against long-term storage of visual memories in this brain part. The prefrontal cortex is not critical for various visual tasks that require long-term retention, such as concurrent discrimination. Its role becomes apparent only in situations that involve working memory and/or necessitate deviation from behavioural routines (see also the example concerning human frontal lobe lesions, p. 263). Further, it is always good practice to consider the function of brain regions in conjunction with that of other brain regions. This is especially pertinent in the case of the neocortex, and all the more so the prefrontal cortex, whose functions do not seem to make sense in isolation from the rest of the brain. Taken together, the above and additional data and arguments led to the notion that the prefrontal cortex is involved in working memory, or *ad hoc* integration of internal representations in guiding novel behavioural responses (Fuster 1980; Passingham 1985; Goldman-Rakic 1987). It might also be expected to take part in generating novel, enduring internal representations, by the recombination of existing representations. In other words, the prefrontal cortex plays a pivotal role in what we would call, in everyday language, 'thinking'.

A model of a visual memory system

From what has been said above it is evident that multiple brain systems deal with visual information, and those systems are differentially recruited in different learning and memory tasks. One major processing stream is the striate-temporal, which feeds into the limbic and diencephalic systems, and via them, as well as via cortico-cortical fibres, to the ventromedial prefrontal cortex, especially the inferior convexity. This system specializes in processing visual form. Another processing stream is the striate-parietal system, which projects to the principal sulcus and adjacent dorsal prefrontal cortex. This system specializes in visuospatial processing.

   The form and the visuospatial analysis systems had probably evolved as separate modules under different phylogenetic pressures. As we already know, modularity of function, though along different lines, is further revealed when the neocortical processing systems are pursued to their limbic connections. Of course, there is an intimate cross-talk between the form and visuospatial systems at various levels of analysis. A region in which much 'high-level' cross-talk is expected to take place is the prefrontal cortex.

   The results of Fuster *et al.* (1985), mentioned above, on the functional interaction of temporal and prefrontal cortices, as well as the extensive neuroanatomical evidence on reciprocal pathways between the prefrontal cortex, other neocortex areas, and the limbic system (Goldman-Rakic *et al.* 1984; Goldman-Rakic 1988), all caution us against regarding the frontal lobe as the 'command post' in the brain. It is, indeed, tempting to regard the frontal lobe as a site of the mental faculty that Aristotle termed the 'common sense' (Sorabji 1972). But supramodal integration takes place also in the hippocampus, and 'common sense' might be better described as residing in networks distributed over large parts of the brain. This is because there is an intimate interconnectivity and feedback between all the brain stations and levels mentioned so far. These functional loops are a critical ingredient of a visual learning model proposed by Mishkin and his colleagues (Mishkin 1982; Mishkin and Appenzeller 1987; see Fig. 14.7).

According to this model, information encoding visual percepts flows from the striate cortex to extrastriate processing centres, and is recombined to yield visual representations in the visual association areas of the temporal cortex. These representations are stabilized only if limbic circuits are sufficiently activated concurrently with, or within a brief time window after the activation of sensory cortex by the visual stimulus. Under these conditions, the limbic circuits feed back into the inferotemporal cortex to reinforce the stabilization of the representation. It is proposed that the feedback loop involves the basal forebrain cholinergic system (Fibiger 1982; Mesulam *et al.* 1983; Aggleton *et al.* 1987; Fig. 14.5). In addition, projections from other modulatory fibre systems, especially catecholaminergic systems, may globally facilitate neocortical and limbic circuits. Finally, the assemblies can communicate with the prefrontal cortex, which again feeds back to the basal forebrain and to the candidate visual storage areas in the temporal lobe. It is suggested that the prefrontal cortex accesses visual association cortex for 'on-line' processing of transient

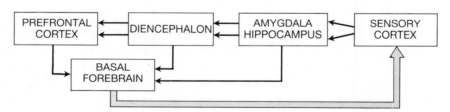

Fig. 14.7. A flowchart model for a visual memory system. The model is for the striate-temporal-prefrontal system that specialises in learning and memory of form. Information on the visual attributes of an object flows from the primary visual cortex to the sensory association cortex in the prestriate and inferotemporal cortex (denoted 'sensory cortex'). From there, the processed information reaches the limbic system in two parallel lines, one to the amygdala, the other to the hippocampus. Further stations of the system are located in the diencephalon, i.e. in the thalamus and mammillary bodies, and in the prefrontal cortex. The limbic, diencephalic, and prefrontal structures feed back into the basal forebrain, which in turn modulates neuronal assemblies so that the visual representation is stored. Critical parts of these assemblies reside in the inferotemporal cortex. Only appropriate activation of the stations downstream from the sensory cortex would result in 'printing' enduring visual representations. The structures in this diagram are known or expected to be reciprocally connected. (From Mishkin and Appenzeller 1987.)

representations, and the hippocampal formation for the retrieval of 'off-line', more enduring representations; the prefrontal cortex uses both kinds of visual representations for *ad hoc* guidance of behavioural responses (Goldman-Rakic 1987).

It is important to reiterate the hypothetical functions of the limbic system in the above model, because these functions are highly relevant to the further discussion of complex memory systems in the next chapter. Three points should be noted. (1) Visual representations can endure for a few seconds without involvement of the limbic system. For example, monkeys with temporal lobe lesions did succeed in delayed-response tasks, provided that the delay was less than *c.* 8 s (Zola-Morgan and Squire 1985*a*). Further evidence for this conclusion is provided later (p. 263). (2) The limbic system plays a role in the consolidation of short- to long-term memory. (3) Since limbic neurons must interact with inferotemporal neurons to consolidate the representation, limbic neurons may themselves become part of the assembly that encodes the representation. If this is so, then sufficient activation of limbic circuits should retrieve the appropriate representation. The latter conclusion is derived from theoretical considerations concerning distributed memories (see Chapter 9). It is also supported by experimental data, demonstrating the retrieval of visual representations by limbic stimulation (see Chapter 10), as well as the selective response of hippocampal and amygdaloid neurons to complex visual representations, such as faces (Heit *et al.* 1988).

*   *   *

In Chapter 9 we summarized a few principles of neuronal encoding in central sensory systems. The idea was that modality-specific strategies of information-processing are guided by some global principles which the brain uses in handling sensory data. Similarly, a few global strategies appear to be used in the experience-dependent modification of complex internal representations. Critical components of neuronal circuits that encode internal representations are assumed to be stored in cortical areas bordering with the appropriate primary sensory areas: visual memories in the inferotemporal, auditory memories in the superior-temporal, somatic memories in the posterior-parietal, olfactory and gustatory memories in the orbital-prefrontal cortex. All these memories have access to each other via cortico-cortical and subcortical pathways. In addition, supramodal memories are presumably stored at temporoparietal and dorsolateral prefrontal areas, and in humans these include linguistic memories. In all cases, the prefrontal cortex can integrate and use the internal representations for the *ad hoc* guidance of behaviour, and probably for generating 'higher-order' representations in processes of implicit learning (insight; see Chapter 2). And finally, the limbic system is critical for the consolidation and retention of different types of uni-, poly-, and supramodal representations. Is the limbic system critical for *all* memories in the mammalian brain? The answer to that will now be provided.

15 *Amnesias, further clues to our own memories*

Direct information on the neurobiology of human memory has so far been derived primarily from pathologies. (Future non-invasive imaging techniques, with appropriate spatial and temporal resolution, will permit real-time visualization of normal brains in learning, and will revolutionize the field; see Chapter 10.) Pathologies, in this context, are amnesias. These are devastating afflictions. Memories mould our personality and link us to society. Therefore, when memory deteriorates so do many aspects of mental and social life. Both practical and theoretical aspects of amnesia have long fascinated physicians and scientists alike. It is mostly the desire to understand human amnesia that has served as an incentive for the study of brain lesions in monkeys, described in the previous chapter.

The study of human amnesia has unique advantages but also severe disadvantages. On the one hand, *Homo* is the most sophisticated experimental subject for neuropsychological research, uniquely capable of verbal communication. On the other hand, human amnesias are unfortunate events that cannot, of course, be intentionally replicated. This inherent irreproducibility makes it difficult to establish the relevance of brain damage in amnesics to their behavioural deficits. To this one should add the basic problems involved in all research which infers function from dysfunction. Nevertheless, as we shall see, when combined with data from neuropsychological research in monkeys, the studies of amnesia do contribute to an internally consistent picture of memory systems in the primate brain.

The literature on amnesia is voluminous, covering and debating in great detail issues ranging from clinical neurology to theories of the human mind. Only limited aspects of this fascinating interface between the clinic and the laboratory are discussed below. (For selected reviews of the neuropathology and neuropsychology of acquired amnesia, see Angelergues 1969; Squire 1982, 1987; Parkin 1984; Signoret 1985; Weiskrantz 1985.)

The taxonomy of amnesia, and some operational definitions

Amnesia is a non-unitary syndrome which can be classified according to several criteria. Among these are the aetiology, the specificity, the span, and the persistence of the amnesia.

The aetiology of amnesia

Amnesia may be *organic* or *psychogenic*. Psychogenic amnesia, characterized by the absence of an identified organic brain pathology, is sometimes part of the clinical picture in hysterical neuroses (Rowe 1980). It is not discussed here. Organic amnesia may result from a variety of causes, including physical trauma and brain injury, brain tumours, surgical ablations performed to remove tumours or alleviate intractable epilepsy, strokes, anoxia, toxins, electroconvulsive shock therapy, and degenerative processes due to heredity, mal-nutrition, infection, intoxication, or ageing (Angelergues 1969; Signoret 1985). Aetiological classification is crucial from a clinical point of view, but not very informative from a theoretical point of view, because different aetiologies may involve the same brain structures.

Non-dissociated vs. dissociated amnesia

Brain damage may cause memory disorders associated with a gamut of perceptual and cognitive deficits. Students of memory are, naturally, most interested in amnesia occurring in the absence of marked perceptual and cognitive abnormalities, or clearly dissociated from them. A traditional method for evaluating the specificity of amnesia is to compare the ranking of the patient on a general intelligence scale, such as the Wechsler Adult Intelligence Scale (WAIS) (Wechsler 1958), with that on a memory scale, such as the Wechsler Memory Scale (WMS) (Wechsler 1945; for a discussion of memory tests, see Erickson and Scott 1977). In normal subjects, WAIS and WMS scores are essentially equivalent. Ideally (from the limited point of view of the experimenter, of course) the amnesic patient should retain a normal score on the WAIS but have a subnormal score on the WMS.

At this point, when the clinical assessment of amnesics is considered, it is pertinent to readdress one facet of the terminology of memory, namely that of memory phases. In previous chapters, the differentiation of memory into short- and long-term phase(s) was based either on theoretical grounds, or on pharmacological evidence, or both. Memory phases were not accurately defined in terms of time windows, partly because phase duration depends on the behavioural task and the species, and partly because phases are intermingled. In general, the theoretical consideration was that long-term memory should outlast the limited life span of proteins which are post-translationally modified in learning. The pharmacological results showed that there is a period, during and immediately after learning, in which the consolidation of memory is sensitive to certain drugs. The data clearly indicated the existence of multiple, some probably parallel, phases of memory (Chapter 8).

However, a clinician attempting to test memory deficits in a patient must be more pragmatic. In the clinic, memory is conventionally broken into immediate memory, recent memory, and remote memory. Immediate memory, or immediate recall, is measured by immediate repetition tasks, in which the patient is requested to repeat information presented by the examiner. For example, the patient is requested to repeat a string of digits, either forwards or backwards; the normal range, as we already know, is from the magical-number 7 to 10 for forward repetition (and 4–7 for backward repetition). Some authors refer to these tests as a measure of 'attention span'. 'Recent' memory is clinically measured by recall of information in the range of minutes. In a routine test, the subject is requested to remember a list of four to five verbal and/or visual items (e.g. ball–point pen, green apple, red tie, key holder), and recall it following 5 minutes of distracting activities. Remote memory, as the term implies, covers more remote events, and can of course be further subdivided into less remote and more remote periods. Routine tests of remote memory are based on biographical or general factual knowledge. More sophisticated tests are used for fine-grain analysis of memory deficits, and selected examples are given below.

Global vs. modality-specific deficits

On the basis of what we already know about the functional topography of memory in the mammalian brain, we might expect some brain lesions in humans to impair multimodal memory, and other lesions to impair only modality-specific information. In addition, we may expect the extent of the lesion in a given brain region to determine the extent of the behavioural deficit. This is indeed the case. Damage to limbic structures, and in some respects to the frontal lobes, impairs facets of multimodal, 'global' memory. The location and magnitude of the lesion determine the severity, the endurance, and the degree of multimodality of the amnesia.

• Note, however, already at this stage, that there is no such thing as a complete, global amnesia, and that even the most severe amnesics do learn and remember some things. Examples of 'global' amnesics, and what they do learn, are given below.

In contrast to lesions in the afore-mentioned brain structures, lesions in appropriate neocortical association areas could lead to modality-specific performance deficits, which may be interpreted as involving damage to a modality-specific memory. One group of such deficits is agnosia: an acquired impairment in recognizing a previously known sensory stimulus, unaccompanied by perceptual deficits (Geschwind 1965a,b; Teuber 1965; Heilman and Valenstein 1979; Rubens 1979; for colourful narrative descriptions, see Sacks 1985). For example, lesions to temporal and parietal cortical areas may produce visual agnosia. In this case, circumscribed unilateral right lesions are usually more disruptive than unilateral left lesions, indicating hemispheric specialization of the cortex (Milner 1968; De Renzi 1982; Warrington 1982; Cummings 1985). Agnosia may furthermore be restricted to a category of stimuli. For example, topographical amnesia, the inability to recall the spatial layout of the environment, is associated with right posterocortical lesions (Paterson and Zangwill 1945; De Renzi 1982). Another striking example of a category-specific agnosia is prosopagnosia, an impaired recognition of faces but not of many other types of visual objects; this deficit is associated with bilateral lesions in the occipitotemporal cortex (Bruyer *et al.* 1983; Assal *et al.* 1984; Damasio 1985).

A caveat is pertinent here. It is not at all clear how much mnemonic deficits contribute to agnosias. The cortical loci of the lesions are in line with the notion that modality-specific memory traces are damaged. But the defect may as well be in execution of some computations over normally registered and stored representations. This caveat will surface again below in considering the meaning of 'global' amnesias; however, in 'global' amnesias the experimental data and the 'globality' support a more intimate relevance of the deficit to memory *per se*. The neuropsychology of modality-specific behavioural deficits will not be further discussed here; we will restrict our attention to 'global' amnesics.

The span of amnesia

Two terms are used to describe the time window of the affected memory, relative to the onset of amnesia. Retrograde, or premorbid amnesia, is loss of memory for events that have occurred prior to the onset of amnesia. Anterograde, or postmorbid amnesia, is loss of memory for events that have occurred after the onset of the amnesia. The term 'onset of amnesia' is problematic if the cause of dysfunction is a disease of slow onset, such as chronic alcoholism (see below). Although cases of 'pure' anterograde or retrograde amnesia have been reported (e.g. Andrews *et al.* 1982), amnesias usually display both elements. The extent of premorbid amnesia is sometimes practically difficult to establish (Sanders and Warrington 1971; Marslen-Wilson and Teuber 1975; Squire 1982; Weiskrantz 1985). In most cases, recent premorbid memory is more severely affected; the older the memory, the more it is resistant to amnesia.

The persistence of the amnesic condition

Some cases of amnesia, caused, for example, by electroconvulsive shock therapy (ECT) for severe depression, are essentially transient, although partial permanent deficits may remain. Other cases, caused by accidental or surgical ablations or by degenerative brain disease, are essentially stable. But even following ablations, partial recovery is possible (Teuber *et al.* 1968; Weiskrantz 1985).

Examples of 'global' amnesics

Hundreds of amnesics are described in the clinical and neuropsychological literature. In recent years, the introduction of computerized tomography has permitted localization of the lesions, and neuropsychological tests have standardized the assessment of memory deficits and their specificity. Only a few examples of noted cases are mentioned here, as a prelude for further discussion of the amnesias.

H.M.

H.M. is probably the most noted and studied amnesic in history (Scoville and Milner 1957; Milner *et al.* 1968; Corkin 1984). He became amnesic in 1953, at the age of 27, following bilateral removal of his medial-temporal lobe, including the anterior two-thirds of the hippocampus, the posthippocampal gyrus, and the amygdala (Fig. 15.1). The operation was performed to alleviate intractable epilepsy. The resulting amnesia was anterograde, accompanied by partial premorbid loss of memory for events in the decade preceding the operation. More remote events were not affected. H.M. scored above average on the WAIS, but his rank on the WMS was more than 3 S.D. below normal. Both verbal and non-verbal memory were affected. Performance in immediate memory tests did not decline following surgery. In contrast, longer-term memory was most severely impaired. Similarly to other amnesics, H.M. had additional neurological complications, which were not considered to contribute to his amnesia, and will not be detailed here. The temporal lobe ablation did alleviate his epilepsy.

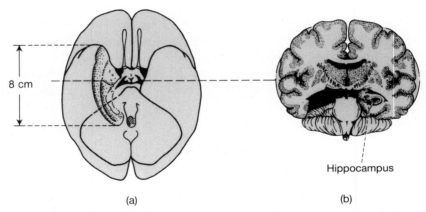

(a)   (b)

Fig. 15.1. The estimated extent of removal of tissue in the medial-temporal lobe of H.M. Surgery was bilateral but one hemisphere is shown intact for comparison. (a): Ventral view of the brain. (b): Cross-section at the indicated plane. A computerized tomography (CT) scan performed in 1984, 31 years after the operation, essentially confirmed the extent of bilateral medial-temporal lobe lesions; it also revealed atrophies in the cerebellum, possibly related to chronic anticonvulsive medications, and a mild cortical atrophy, considered unrelated to the amnesic syndrome (Corkin 1984). (Adapted from Scoville and Milner 1957.)

The following are excerpts from a description of H.M.'s condition 14 years after the onset of his amnesia:

He still fails to recognize people who are close neighbours or family friends but who got to know him only after the operation . . . although he gives his date of birth unhesitatingly and accurately,

he always underestimates his own age and can only make wild guesses as to the date . . . During . . .
nights at the Clinical Research Center, the patient rang for the night nurse, asking her, with many
apologies, if she would tell him where he was and how he came to be there. He clearly realized
that he was in a hospital but seemed unable to reconstruct any of the events of the previous day.
On another occasion he remarked 'Every day is alone by itself, whatever enjoyment I've had, and
whatever sorrow I've had.' Our own impression is that many events fade for him long before the
day is over. He often volunteers stereotyped descriptions of his own state, by saying that it is 'like
waking from a dream'. His experience seems to be that of a person who is just becoming aware of
his surroundings without fully comprehending the situation, because he does not remember what
went before . . . H.M. was given protected employment in a state rehabilitation center . . . partici-
pating in rather monotonous work . . . A typical task is the mounting of cigarette-lighters on cardboard frames for display. It is characteristic that he cannot give us any description of his place of
work, the nature of his job, or the route along which he is driven each day, to and from the center.
(Milner *et al.* 1968)

The performance of H.M. on just one type of recognition task, is illustrated in Fig. 15.2.
We will later use the same task to compare H.M.'s memory to that of another amnesic
patient. The task had two versions, verbal and non-verbal (Teuber *et al.* 1968). For the non-
verbal version, 160 cards bearing nonsense designs of unfamiliar geometric figures were
used. For the verbal version, the cards bore nonsense syllables or three-digit numbers. The
subject was shown one card at a time, for 3 s each. The test was presented in successive
blocks of 20 cards. In each block, eight figures recurred, whereas the rest were shown only
once. The subject had to identify the recurring figure (or word, or number). H.M. clearly
failed in both the verbal and non-verbal versions of the task (Fig. 15.2).

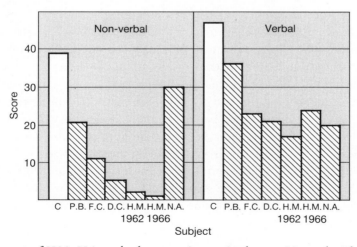

Fig. 15.2. Performance of H.M., N.A., and other amnesics on visual recognition tasks. The subjects had to
recognize nonsense designs (non-verbal task), or nonsense syllables, or three-digit numbers (verbal task). For
details of the test, see text. C, controls; P.B. and F.C. sustained unilateral left temporal lobectomies; D.C. had a
bilateral medial-temporal resection combined with orbitofrontal undercutting. H.M. was tested twice, 9 and
13 years post-surgery. N.A. was tested 4 years post-injury. (From Teuber *et al.* 1968.)

H.M. did succeed in some learning tasks which were based on routine perceptual-motor responses. This preserved learning, and its significance, are further discussed below. As for other tasks, H.M.'s severe anterograde amnesia persisted for more than 30 years, though some slight improvement was detected.

R.B.

H.M. attracted so much interest because the bilateral medial-temporal damage was so extensive, and the neuropsychological follow-up so comprehensive. Many cases have been described in which less extensive medial-temporal lesions resulted in memory loss (Scoville and Milner 1957; Penfield and Milner 1958; Milner 1972). In these cases, the extent of hippocampal lesion was often correlated with the severity of the amnesia. The behavioural deficit, however, could not be correlated solely with the hippocampus, much less with its subfields. This is why one case, that of patient R.B., raised much attention: here post-mortem analysis indicated that the amnesia could have been attributed to a bilateral lesion confined to hippocampal field CA1 (Zola-Morgan *et al.* 1986).

R.B. developed amnesia following an ischaemic episode associated with heart surgery at the age of 52. His score on the WAIS remained normal but memory tests revealed a marked anterograde amnesia in both verbal and non-verbal domains. This was accompanied by only a slight retrograde amnesia. R.B. was well aware of his condition, which clearly interfered with his daily activity. His memory deficit was, however, significantly milder than that of H.M.

R.B. died from cardiac arrest 5 years after the onset of his amnesia. With the consent of his family, his brain was subjected to a detailed histological post-mortem analysis. The major finding was a bilateral lesion involving the entire CA1 field of the hippocampus. There were also smaller pathological findings in other brain areas, including the left globus pallidus, the left internal medullary lamina of the thalamus, the anterior amygdaloid area, the cerebellum, and the cerebral cortex. In the latter, scores of small foci of cell loss were discovered. Since cognitive functions besides memory were unimpaired, these distributed minor cortical lesions were considered to be of little clinical significance. The authors proposed, in accordance with the earlier findings concerning the role of the hippocampus in memory in humans and other mammals, that the bilateral CA1 lesion was the sole, or at least the major, cause of R.B.'s amnesia (Zola-Morgan *et al.* 1986). Such a lesion is expected to interrupt the processing of information in limbic circuits (see p. 238).

N.A.

N.A. was a 22-year-old radar technician in the US air force who, while in his barracks, was accidentally stabbed by a room-mate with a miniature fencing foil (Teuber *et al.*

1968). The foil penetrated the right nostril, and punctuated the base of the brain while taking an obliquely upward course slightly to the left. Computerized tomography, performed 19 years later, disclosed a small lesion in the left dorsomedial thalamic nucleus, but additional small lesions or damage to connected brain regions could not be excluded (Squire and Moore 1979). Following recovery from the acute traumatic consequence of the tragic mock-duel, N.A. showed severe anterograde amnesia but also retrograde amnesia for about 2 years immediately preceding the accident. His performance on intelligence tests was within normal range, but his performance on a standard memory scale was about 2 S.D. below normal (Wetzel and Squire 1982). In all, N.A.'s amnesia was significantly milder than that of H.M., showed greater improvement with time, and was much more severe on verbal than on non-verbal tasks (e.g. see Fig. 15.2).

B.Y.

N.A.'s lesion was mostly unilateral. B.Y., in contrast, sustained a bilateral thalamic lesion (Winocur *et al.* 1984). He was 38 when stricken by infarction, which left him amnesic, though intellectually normal and without apparent sensory or perceptual disorders. At first he had both a severe retrograde and anterograde amnesia. The retrograde amnesia markedly, though not completely, improved within weeks, but the anterograde amnesia persisted. Computerized tomography revealed a bilateral lesion in the medial thalamus, involving parts of the dorsomedial, centromedian, and ventro-posteromedial nuclei. The mammilothalamic tract may have also been involved. The amnesic picture was quite similar to that of N.A., except that performance on non-verbal tasks, such as copying a figure from memory, was much more impaired. This was consistent with the bilaterality of the lesion, involving not only the left, 'linguistic', but also the right, 'visual' part of the brain.

Korsakoff's syndrome

Korsakoff's syndrome, also called Korsakoff's psychosis or the chronic phase of the Wernicke–Korsakoff syndrome, is the most common organic brain disease causing amnesia while sparing, relatively speaking, other intellectual functions (Victor 1976; Butters 1984). Korsakoff's syndrome is commonly associated with chronic alcoholism. In such a case, the disease may begin with ataxia, oculomotor abnormalities, and mental confusion. This is the Wernicke, acute stage, which results from metabolic disorder and responds well to thiamine treatment. Wernicke's disease may develop into Korsakoff's syndrome, which is associated with structural brain damage and is, therefore, irreversible. A similar syndrome may develop in non-alcoholics following viral encephalitidies, infarctions, and tumours of the third ventricle. The patient displays severe retrograde and anterograde amnesia. The retrograde memory loss usually shows a temporal gradient and remote memories are less affected. In one study, the severity of anterograde amnesia was uncorrelated with the severity of amnesia of far-remote events, indicating that the mechanisms required for recalling far-remote events are distinct from those required for new learning and for memory of more recent events (Shimamura and Squire 1986). General intelligence deficits may also occur in Korsakoff patients; still, the rank on the WMS ranges 1–3 S.D. below that on the WAIS (Weiskrantz 1985).

Typically, severe Korsakoff patients fail in both verbal and non-verbal memory tasks. They do not remember simple everyday facts, may confabulate, and cannot plan even the simplest of tasks. They may perform quite well on an immediate memory test, but not on recent memory tests. For example, they can repeat the examiner's name, date, and time of day, but forget them within 1–3 minutes. Patients become socially helpless, fixated at a certain age and date, apathetic, and only slightly aware of their condition.

It has been suggested that in some tasks, Korsakoff patients are affected in learning more than in memory. Given the appropriate time, they may learn and retain information that otherwise they fail to remember. The following experiment illustrates this point (Huppert and Piercy 1978): Korsakoff patients and normal subjects were shown 120 pictures from magazines, and asked to remember them. Ten minutes, 1 day, or 1 week later, a sample was presented of 40 of the original pictures, randomly interspersed with 40 new pictures, and the subjects were requested to identify whether the picture was familiar or new. If Korsakoff patients were allotted enough time to reach a recognition criterion similar to that of normal subjects at 10 minutes, their memory after 1 day or 1 week also did not significantly differ from that of controls. But to reach a normal recognition at 10 minutes, each picture had to be presented to the Korsakoff patients for a period four to eight times longer than that given to controls. Patient H.M., tested similarly, was reported to forget more rapidly than both control and Korsakoff patients (Huppert and Piercy 1979).

The neuropathology of Korsakoff patients is heterogeneous, but in all cases there is damage to the dorsomedial thalamic nucleus (Victor 1976), or the mammillary bodies, or both (Butters 1984). Some symptoms of Korsakoff's syndrome, including failure to rearrange information and personality change, may be contributed by cortical lesions, e.g. in the frontal lobes (Butters 1984; Shimamura *et al.* 1988).

The common denominator: damage to limbic circuits

• The common denominator of brain pathology in all the amnesics decribed above is damage to limbic circuits (see Fig. 14.2). (This pathology is also shared by many other cases of amnesias of different aetiologies, including amnesias resulting from viral encephaliti- dies, strokes in the posterior cerebral artery territory, closed-head trauma, and Alzheimer's disease; Signoret 1985.) The common denominator of the behavioural pathology in the afore-mentioned 'global' amnesics is defects in recent and remote memory, but sparing of immediate memory. This implies that limbic circuits are not mandatory for the initial stage(s) in the formation of experience-dependent representations, but are mandatory for consolidation of at least some kinds of such representations into a long-lasting form. As we already know, the role of limbic circuits in the consolidation and retention of memory was clearly verified by testing the effect of controlled circumscribed lesions on behavioural plasticity in monkeys, described in Chapter 14. The data on global amnesias and on models of temporal-lobe amnesia in monkeys, thus provide decisive neuro-anatomical support to the fundamental distinction between short- and long-term memory, which was made in earlier chapters on theoretical and pharmacological grounds. (For ample relevant evidence from psychology, see Klatzky 1980; Baddeley 1976, 1986.) However, the terms short and long term as used in behavioural, clinical, pharmacological, and theoretical contexts, respectively, do not necessarily refer to similar underlying mechanisms. For example, in amnesia short term implies seconds to minutes; in pharmacological studies, it implies seconds to hours; from a theoretical point of view, it may mean seconds to weeks.

The conventional 'rule of thumb' of clinical neurology is that the extent and location of the limbic lesion are critical determinants in the behavioural profile of the resulting amnesic profile: bilateral limbic lesions cause multimodal, relatively permanent amnesias, whereas unilateral limbic lesions lead to more transient amnesias, which are hemispheric dependent, i.e. left lesions affect primarily verbal memories whereas right lesions affect primarily visual memories. (The general issue of hemispheric specialization and its role in linguistic and spatial faculties will not be further discussed here. For selected reviews, see Bradshaw and Nettleton 1983; Damasio and Geschwind 1984; Geschwind and Galaburda 1984, 1985; Benson and Zaidel 1985.)

In recent years, a further differentiation in amnesias has been offered. It is between 'medial-temporal' and 'diencephalic' amnesias. H.M. and R.B. were dubbed 'medial-temporal amnesics', whereas N.A., B.Y., and Korsakoff patients were dubbed 'diencephalic amnesics'. It was proposed that they differ not only in their damaged brain regions, but also in their behavioural profile. In some tasks, diencephalic amnesics learn slowly, but their memory, once established, is not markedly feeble (see the earlier example of Korsakoff patients). In contrast, in the same tasks medial-temporal amnesics (as well as ECT-induced amnesics) forget rapidly (Huppert and Piercy 1979; Squire 1981, 1982). The proposal that amnesias have differential effects on the rate of forgetting bears on a cardinal issue: what is disrupted in 'global' amnesics—consolidation proper, retention, or retrieval? This issue is not yet resolved (Weiskrantz 1985; Squire 1987). From our earlier discussion of the roles of limbic circuits in memory, it appears that all three functions may be subserved by these circuits, and therefore could be affected by limbic damage. It is likely that the extent and location of the damage may affect one of the functions more than the others. Indeed, Weiskrantz (1985) has questioned the physiological justification for the diencephalic vs. medial-temporal dichotomy, arguing more in line with the conventional clinical view that there is only one core form of 'global' amnesia, and that the different cases represent different degrees of damage to the same functional system.

What can 'global' amnesics learn?

As noted above, even severe amnesics retain some postmorbid memory. In the laboratory, they perform well in tasks ranging from simple, reflexive eyeblink conditioning (Weiskrantz and Warrington 1979), learning of motor tracking, simple mazes, and jigsaw puzzles (Milner *et al.* 1968; Brooks and Baddeley 1976), recall of 'word lists' following priming with fragments of these words (Graf *et al.* 1984), and up to microcomputer operation (Glisky *et al.* 1986; for a review of additional tasks, see Parkin 1982). The patients do not remember that they have been trained in the relevant task. These tasks, though differing in complexity and in the domain of knowledge, have been interpreted by most authors as requiring the acquisition of some *procedures*. The procedures may be either motor, or perceptual-motor, or even perceptual-cognitive.

The performance of 'global' amnesics in a task requiring mirror-reading skill illustrates both their preserved learning capacity and their memory deficit. Cohen and Squire (1980) presented N.A., Korsakoff amnesics, ECT amnesics, and control subjects with cards projected by mirror-reflection in a tachistoscope. Each card displayed three low-frequency English nouns with 8–10 letters (Fig. 15.3). The subjects were requested to read each word triad aloud, and press a button when finished, thus registering the reading time per word triad. They read five blocks of ten word triads on each of three consecutive days and then on a fourth day, approximately 3 months later. For each block of ten triads, half were repeats in all blocks and half were unique. The reading time of the unique triads was used to evaluate the ability to acquire the procedure of mirror-reading; the performance on the repeated triads reflected, in addition, the ability to remember specific words.

All the amnesic patients learned the mirror-reading skill, as indicated by a decrease, with training, in the time required to read the unique word triads. The skill was clearly preserved after more than 3 months. The performance of N.A. and Korsakoff patients in this part of the task was equivalent to that of their matched controls (Fig. 15.3). The learning curves for the repeated word triads were steeper, but the facilitatory effect of the word repetition was smaller in the amnesic patients. This indicated that despite being able to learn and remember the skill normally, the amnesics were poor at remembering which words they had seen. Using a similar procedure, Squire *et al.* (1984) demonstrated that patients who learned mirror-reading before receiving ECT retained their skill after the treatment, despite retrograde amnesia for the pretreatment sessions and words.

The sparing of motor and some perceptual-motor and cognitive skills in medial-temporal amnesia, was also demonstrated in monkeys (Mishkin 1954; Orbach *et al.* 1960;

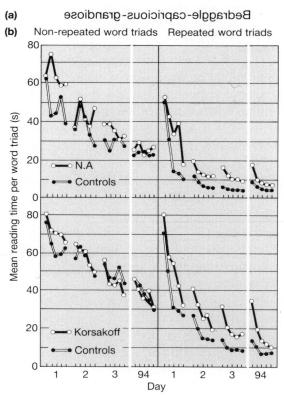

Fig. 15.3. Preserved learning and memory of a mirror-reading skill in amnesic and non-amnesic subjects. (a): Example of a word triad, as presented to the subjects by mirror-reflection in the tachistoscope. For further details of the test, see text. (b): Acquisition of the mirror reading skill in three daily sessions, and its retention 3 months later. The amnesics acquired the ability to read unique word triads at a normal rate. Repeated presentation of words facilitated acquisition in both amnesics and non-amnesics, but the performance of the amnesics in this part of the test was inferior, due to impaired memory of the specific words. They thus learned the procedure, but had difficulties in learning the content. N.A., patient N.A. Controls were matched for age, intelligence, and educational background. The controls for Korsakoff patients were also alcoholics. (Adapted from Cohen and Squire 1980.)

Zola–Morgan *et al.* 1982; Zola–Morgan and Squire 1984; Salmon *et al.* 1987) and even in rodents (Eichenbaum *et al.* 1986). For example, in a set of experiments described below, monkeys with a conjoint, bilateral lesion of the hippocampus and amygdala (intended to simulate the pathology of patient H.M.) were severely impaired in a visual pattern discrimination task, which was interpreted as taxing factual knowledge. By contrast, they were only slightly impaired in another visual discrimination task, which was interpreted as taxing acquisition of skill.

The assumption that some pattern discrimination tasks measure primarily skill-like responses, as opposed to factual knowledge, needs to be elaborated. The argument is that if in a discrimination task, the same piece of information is repeated over and over and over again, the animal acquires the correct response more as a skill, or a procedural response, than as factual knowledge (Iversen 1976; Zola-Morgan and Squire 1984). If, on the other hand, the task is 'easy' and visual discrimination could be learned quickly, then the animal probably does not rely on a skill-like response but rather on facts concerning the discriminanda. One way of achieving rapid discrimination learning is to use simple, three-dimensional discriminanda which themselves contain a reward, e.g. coloured food (Jarvik 1953). Thus, normal monkeys require hundreds of trials to learn to discriminate successfully between patterns of a square and a cross, but can learn after only tens of trials to discriminate between two differently coloured peanut half-shells provided that one of the shells contains raisins. A conjoint hippocampal-amygdaloid lesion severely impaired learning of the 'easy', supposedly 'factual' task, but impaired only mildly learning of the more 'difficult', supposedly 'procedural' task (Zola-Morgan and Squire 1984). The same lesion also did not affect performance in a motor task, requiring the monkey to retrieve bread by manipulating its hand through several rows of vertical sticks; nor did it affect another motor task, requiring the monkey to retrieve a piece of candy threaded onto a bent metal rod.

A dissociation between performance on a recognition task assumed to tax factual knowledge (delayed non-matching to sample) and on a supposedly 'procedural' perceptual task (concurrent discrimination) was also detected in monkeys during ontogenesis. Infant monkeys learned about as quickly as adult monkeys to discriminate long lists of object pairs, but were still incapable of mastering the recognition task (Bachevalier and Mishkin 1984). Note that delayed non-matching to sample also requires working memory, hence the proposal that 'procedural' memory can function independently of working memory, whereas brain systems for working memory and factual knowledge are intimately related.

The multiplicity of memory systems: knowing *how* vs. knowing *that*, and further subdivisions of acquired knowledge

• Taken together, the data on human amnesics and on monkey 'models' indicate that there are at least two major types of memory systems in primates: (1) memory of 'how', also called 'procedural', 'skill', and 'habit' memory; and (2) memory of 'that', also called 'declarative' memory (N. J. Cohen 1984; Mishkin and Petri 1984; Mishkin *et al.* 1984; Mishkin and Appenzeller 1987; Squire 1987). The neural system for 'declarative' memories encompasses the relevant sensory association cortex and the medial-temporal and diencephalic structures mentioned above, and is connected to the prefrontal cortex (see p. 248 for a description of this system in relation to processing of visual information). The prefrontal cortex, in conjunction with the hippocampus, subserves working memory, i.e. temporary use of 'declarative' knowledge for *ad hoc* guidance of behaviour.

Indeed, a major deficit in patients with prefrontal cortex lesions is reduced performance in problem-solving tasks that require flexibility in response strategies, and are assumed to tax working memory (Milner and Petrides 1984). A classic example is provided by the Wisconsin Card Sorting Test. The subject is presented with a series of stimulus cards and a deck of response cards. The cards bear coloured geometric patterns (e.g. a single blue star, three red circles), and can be matched by categories (e.g. colour, form, number). The examiner selects a sorting category (e.g. colour), but does not inform the subject. The latter is instructed to place a response card in front of a stimulus card, wherever he or she thinks it should go. The examiner then informs the subject if the response was right or wrong, and the subject uses this information to obtain correct responses in the following matches. After ten consecutive responses, the examiner shifts the sorting category without warning, and the subject must unveil it again to obtain correct matches. The procedure is then repeated with other sorting categories. Patients with prefrontal lesions find this task abnormally difficult (Milner 1963). The interpretation is that they have difficulties in using temporarily stored information to regulate their actions.

We are now in a position to readdress a question posed earlier in this chapter: in which parts of the above-mentioned 'declarative' memory system are engrams retained? Arguments raised earlier favoured storage in the relevant association cortices. In the absence of intact limbic circuits, the representation can be retained and retrieved for a few seconds only (hence normal immediate recall but not recent memory in amnesics). One may speculate that later, on consolidation, limbic neurons themselves become nodes in networks that encode specific representations. If so, then stimulation of limbic neurons should retrieve representations. This may explain the mental images in stimulation experiments à la Penfield (see Chapter 10). In addition, recording from limbic areas should access specific representations. This appears to be the case. Recording from hippocampal and amygdaloid neurons in epileptic patients (in the process of preparation for surgical removal of epileptic foci) revealed specific responses to faces and words (Heit *et al.* 1988). These data indicate that limbic neurons are intimately associated with neuronal assemblies that encode information on faces and words. The data also show that the limbic system has evolved to deal with semantic representations. It has indeed been suggested that the hippocampus maps semantic attributes and spatial attributes in an analogous fashion (O'Keefe and Nadel 1978). The role of limbic systems in language requires, however, further clarification, since even the severe limbic damage in H.M. produced only relatively minor language deficits (Corkin 1984).

The data on global amnesics also indicate that the role of the limbic system in retention and/or retrieval is time dependent. Even severe amnesics retain declarative representations that at one time, long before the onset of amnesia, required limbic activity for their consolidation. A possible explanation for this phenomenon is that with time, some assemblies associate extensively with many other assemblies. Therefore, the former degrade more gracefully (see Chapter 9), so that damage to the limbic contribution does not suffice to erase the memory. In addition, these stabilized assemblies are more easily accessed via routes other than the limbic connection.

The neural systems subserving habit formation are less well understood. One suggestion is that it is subserved by cortico-striatal pathways (Mishkin *et al.* 1984). The nuclei of the corpus striatum receive information from sensory cortical areas and project to the pre-motor and motor cortex. This may provide a mechanism for the experience-dependent generation and modification of connections between sensory stimuli and motor responses. The data supporting the role of the striatum in skill knowledge stem from several lines of evidence. One consists of experiments testing the effect of anatomical and chemical lesions in the striatum on the performance of laboratory animals in learning tasks which presumably require 'procedural' rather than 'declarative' knowledge (Phillips and Carr 1987). Another piece of suggestive evidence is the observation that memory deficits can occur early in Huntington's disease, which affects the striatum (Signoret 1985). There were also reports that Parkinson's disease, in which the nigrostriatal pathway is degenerated, may lead to deficits in the performance of 'procedural' tasks such as learning a tracking response (Phillips and Carr 1987). However, in the above cases the lesions were not confined solely to the striatum, and in many instances the contribution of memory rather than motor deficits was not convincingly demonstrated. Additional brain structures which should be considered in habit formation are the cerebellum and associated nuclei; these were discussed in Chapter 11.

A point deserving proper attention is the rudimentary nature of the current categorization of memory into 'declarative' and 'procedural', or 'that' and 'how' systems. 'Declarative' memory is not a unitary process. It is conventionally classified into episodic and semantic memory (Tulving 1972, 1983). Episodic memory is the memory of events in an individual's life. Semantic memory (also called reference memory) is knowledge that is independent of events in the individual's life. For example, memory of a visit to Jerusalem is episodic, but the fact that Jerusalem is the capital of Israel is semantic. Both episodic and semantic memories are damaged in severe 'global' amnesia and hence both are subserved by the limbic system. (Some authors claim that semantic memory is less affected; see the discussion in Tulving (1984).) It has been proposed that facets of episodic and semantic knowledge may further be stored and/or accessed in a dual code, one a propositional code, the other imagery (Kosslyn 1980; Paivio 1986). This is expected to induce an additional differentiation in the neuronal systems that subserve declarative knowledge. And as if this is not enough, McCarthy and Warrington (1988) recently suggested that semantic knowledge may be multiply represented in the brain, each representation being linked to the input modality by which the knowledge was gained. Their suggestion was based on behavioural analysis of a patient who suffered from focal left temporal damage. The patient displayed a selective deficit in defining characteristics of animals and plants, but not of inanimate objects, in response to their spoken names. In contrast, he was able to characterize the same animals or plants when presented with their pictures.

How the above-mentioned categorizations of knowledge are realized in neuronal substrates of declarative memories is still a mystery. The situation is no less elusive when so-called 'procedural' memories are considered. There are many different types of learning and memory tasks which are currently dubbed together as 'procedural', including classical conditioning, motor and perceptual skill acquisition by operant and incidental learning, and others. The proposed common denominator of these learning tasks is a certain acquired 'automaticity' in response; in all probability they represent the output of different types of brain systems.

In conclusion, as far as brain systems are concerned, all one can say at this stage is that there is a cluster of cognitive functions, which are involved in the acquisition of factual knowledge and require an intact limbic system; and there are other learning functions which do not depend on an intact limbic system. Factual knowledge that has sedimented into a 'stable knowledge base' of the organism resembles 'automatic' knowledge in that it is less dependent on limbic circuits. Deposition in a 'stable knowledge base' is a time-dependent process, but 'time' is not necessarily absolute time; some factual knowledge may become stable as the years go by, whereas some may be stabilized rather quickly by repetitive training.

The important point is not the actual number of memory systems in the primate brain, but rather the 'organology' of memory that the data suggest. It is likely that multiple learning and memory systems in our brain have evolved in response to different phylogenetic drives. It is also plausible to assume that the first 'how' systems preceded the first 'that' systems in evolution, because knowing how is more basic for survival than knowing that. Reflexive memory systems, such as modifiable defensive reflexes in invertebrates and vertebrates, are primitive 'learning how' systems. It is not clear at what phylogenetic stage declarative memory emerged; what is clear, though, is that knowing 'that' is a key to mammalian intelligence and its pinnacle, human intelligence. 'Learning that' and 'learning how' systems may share molecular and cellular mechanisms and operational rules. 'That' systems, with their extensive use of working memory, may have developed novel neuronal mechanisms to deal with strings of temporarily retrieved representations. Note, however, that although intuitively a 'how' system might look more primitive and simple than a 'that' system, the cognitive function cannot be used as a reliable indication for the complexity of the neuronal apparatus that subserves it in the mammalian brain.

Epilogue

The neurobiology of learning has undergone an impressive revolution. The achievements of recent years, as well as the vastness of the unknown lying ahead of us, can be sampled by readdressing the selected set of fundamental questions posed in Chapter 1. These questions were: (1) How are internal representations encoded in the nervous system? (2) What differentiates between a 'learning' and a 'no learning' situation? (3) Is 'teaching' instructive or selective? (4) Is memory retained in the system that learns? (5) What structural and functional modifications take place in neuronal systems during learning and memory or, in other words, how are the processes referred to in (1)–(4) realized in the biological hardware? We will address the last question first.

The biological hardware

These are the frontiers which have been pushed forward most in recent years. Several types of molecular and cellular processes which correlate with learning were described, and they can be considered as candidate mechanisms for realizing learning and memory *in situ* (Chapters 3–8). In some cases, evidence was provided for the necessity of specific molecular and cellular processes in certain learning situations (Chapters 4–8). Whether these processes also suffice to establish a memory, and further, whether they are exclusive in that respect *in vivo*, is for future research to determine. Very likely, exclusiveness is the exception rather than the rule, since in the systems studied so far, several cellular and molecular mechanisms operate in parallel. Constitutive reductionism (Chapter 1) was thus rewarding; and heuristic mechanistic models of simple learning (Chapters 4 and 5) are already within the realm of explanatory reductionism.

The important point is not that a current particular mechanistic model of learning may reflect facets of reality, but rather that we can now tell what types of molecular and cellular mechanisms could be expected to function in acquisition, short-, and long-term memory. Note, however, that molecular models *per se* describe mechanisms and procedures, but do not reveal the global design and product of the neuronal information stored as a particular memory. Unveiling this 'semantics' requires deciphering the cellular and multi-cellular context in which the above mechanisms operate, and ultimately the code used by the neuronal system to retain and manipulate the relevant internal representation (see below).

Loci of learning vs. loci of memory

In simple systems, the loci that change in learning are also the ones that retain the memory (Chapters 4 and 5). The available data already suggest that in the above systems, long-term memory involves additional loci, which are not modified in learning (see Chapter 8). In the mammalian brain, the anatomical sites in which learning occurs are also parts of the enduring engram (e.g. association cortices), but memory can be processed *ad hoc* in loci that do not store it permanently, i.e. the prefrontal cortex (Chapter 14).

Instructive, or selective?

Some current theories propose that in the vertebrate brain, sensory input does not directly generate an enduring internal representation, but rather indirectly selects and stabilizes a representation among a multiplicity of 'pre-representations', which are provided by internal diversity in the neuronal system (Chapter 9). The question of whether learning is selective or instructive was only referred to, mostly implicitly, in the context of a few experimental systems, e.g. bird song (Chapter 13). It is an important theoretical issue. Learning by selection is intellectually appealing because it unifies, under Darwinian rules, species phylogeny and mental ontogeny. It is of interest from a philosophical point of view because it depicts the brain partly as an autonomous system that has evolved, and hence is innately programmed, to generate spontaneously potential representations of the external world. Molecular and synaptic mechanisms, capable of realizing selection at the molecular and cellular level, were identified, e.g. Hebbian-like synapses (Chapter 6). Behavioural evidence supporting the notion of stabilization of pre-representations in learning was provided (e.g. in Chapters 12 and 13). Furthermore, some formal computational models support the notion that categorization can readily take place by input-induced selection of self-generated states in networks of neuron-like units (Chapter 9). It is also difficult to reconcile the extensive intrinsic activity of neuronal populations (Chapter 10), with a stringent interpretation of 'instruction'; these intricately connected neuronal populations interact dynamically with the input, and do not simply succumb to its instructions. However, three points should be noted. First, so far there is no direct experimental evidence for selection in neuronal assemblies in learning. The experimental evidence supporting selective mechanisms in neuronal systems is derived from studies on neuronal development. Second, in primitive reflexive learning systems the internal diversity, essential for selection, is non-existent or at most very limited. Third, when the behavioural data implicate selective mechanisms, the presumptive selection is intimately associated with programmed neuronal growth (bird song, Chapter 13). Therefore, in this case learning cannot be portrayed merely as the selection of alternatives from an existing internal diversity of pre-representations. This programmed growth places constraints on the selection, while at the same time generating internal diversity on which selection might operate. The role of selection in learning is bound to become clearer as more becomes known about complex learning systems.

Differentiation between 'learning' and 'no-learning' situations

Four conditions may influence the ability of sensory inputs to generate or modify internal representations: (1) the general arousal state of the organism, which determines to what extent the organism attends to inputs in general (Chapters 2, 4, 11, and 14); (2) selective attention to an input, which determines how effectively the organism attends to that specific input (Chapter 14); (3) persistence of the input; and (4) association of an input with a reinforcer. The reinforcer may be extrinsic and clearly identifiable (e.g. in classical or operant conditioning), or intrinsic and conjectural (in implicit learning; Chapter 2). The role of both persistency and associativity in learning is detected at all levels: molecular (Chapters 4–8), cellular (Chapters 4 to 6), and global (Chapter 14).

Neuronal encoding of internal representations

This question brings us back to the definition of learning, with which we started our discussion (Chapter 1). Learning was defined in this book as experience-dependent generation and/or modification of enduring internal representations. The reason for founding this definition on the notion of internal representations was twofold: first, to unify conceptually the phenomena of learning on the basis of a fundamental and inherent property of nervous systems. Of course, experience-dependent representations vary tremendously in their complexity, in the extent that their activation depends upon on-line sensory input, and in their ability to guide action rather than mere reaction (Chapters 2, 4, 5, 11, and 14). In some reflexive memories the representations may be simple and even trivial, whereas in some complex memories the representations are highly complex and based on a richer repertoire of coding principles. Still, the above definition emphasizes that in all cases nervous systems learn by modifying encoded versions of the world. The treatment of representation in a most general sense, and as an inherent and fundamental property of nervous systems, radically extends previous treatments of memories as representational systems (e.g. Roitblat 1982). Second, the definition was so formulated specifically to focus attention on the cardinal problem in the neurobiology of memory, namely the representational codes. A leit-motif of this book is the notion that ultimate understanding of learning and memory demands an account of how internal representations are encoded in neuronal systems. At present, we can provide only partial answers in a few reflexive systems (Chapters 4 and 5), and only very rudimentary answers in complex systems (Chapters 11–15). This is the level of neurobiological analysis which addresses the content of memory. This is also the level of analysis in which breakthroughs are badly needed, and eagerly expected. Advances in these frontiers of research require novel theories on brain function, and methods for monitoring and analysing the concerted activity of multineuron systems. Such advances also require concepts and data from the cognitive sciences, including psychophysics and neuro-computing (Chapter 9). A trend towards such interdisciplinary research programmes is already evident.

Combining all of the above, and taking more of our ignorance into account, it is still tempting to offer a few selected global conclusions about learning. The main conclusion is that *there is no master solution to the mechanisms of learning*. This conclusion is supported by experimental findings at all levels of analysis, as well as by theoretical arguments: (*a*) There are strong phylogenetic pressures to develop and refine learning capabilities, because learning has an immense adaptive value in many ecological niches. In addition, evolution appears opportunistic and eclectic. Therefore, we should expect evolution to make use of whatever mechanism possible in the phylogeny of learning systems, at every level of biological organization. These arguments predict that in addition to elementary 'universal' building blocks in biological learning machines (see Chapters 3 and 8), we should find novel biological technologies of learning at more advanced phylogenetic echelons (Chapter 6), and rich combinations of both primitive and novel components in different memory systems. (*b*) The assumption that different learning systems evolved under different phylogenetic pressures also argues for the coexistence, especially in higher organisms, of radically different types of memory systems geared towards different goals (Chapters 11–15). (*c*) Multiple learning and memory mechanisms ensure greater data security and flexibility, and parallel mechanisms speed up the process. Examples of the coexistence of multiple, sometimes parallel learning mechanisms at molecular as well as molar levels were provided in Chapters 4–8, 11, 14, and 15.

In conclusion, we should expect to see, in the coming years, a surge of findings on a multiplicity of molecular and cellular mechanisms which subserve learning. We should hope to gain insight into the way complex internal representations are encoded in the brain. And finally, we should not forget that the neurobiology of memory, in addition to being an exciting branch of the basic sciences, is also endowed with an immense applied potential. Understanding the biology of learning and memory could result in alleviation of the suffering caused by memory disorders and, hopefully, assist at least some types of amnesics in their search for lost times.

Further reading

The following selected references may be consulted for additional information on the appropriate research fields:

Molecular and cellular biology, and the mechanisms of genetic memory

Alberts, B., Bray, D., Lewis, J., Raff, M., Roberts, K., and Watson, J. D. (1989). *Molecular biology of the cell* (2nd edn). Garland Publishing, New York.

Davidson, E. H. (1986). *Gene activity in early development* (3rd edn). Academic Press, New York.

Darnell, J., Lodish, H., and Baltimore, D. (1986). *Molecular cell biology.* Scientific American Books, New York.

Lewin, B. M. (1987). *Genes* (3rd edn). Wiley, New York.

Watson, J. D., Hopkins, N. H., Roberts, J. W., Steitz, J. A., and Weiner, A. M. (1987). *Molecular biology of the gene* (4th edn). Benjamin/Cummings, Menlo Park, California.

Ethology and innate behaviours

Camhi, J. M. (1984). *Neuroethology,* Sinauer, Sunderland, Massachusetts.

Ewert, J. P. (1980). *Neuroethology.* Springer-Verlag, Berlin.

Hinde, R. A. (1970). *Animal Behavior* (2nd edn). McGraw-Hill, New York.

Hoyle, G. (1984). The scope of neuroethology. *Behav. Brain Res.* **7**, 323–54.

Lorenz, K. Z. (1981). *The foundations of ethology.* Springer-Verlag, New York.

Tinbergen, N. (1969). *The study of instinct.* Oxford University Press, New York.

General neurobiology

Kandel, E. R. and Schwartz, J. H. (ed.) (1985). *Principles of neural science* (2nd edn). Elsevier, New York.

Kuffler, S. W., Nicholls, J. G., and Martin, A. R. (1984). *From neuron to brain* (2nd edn). Sinauer, Sunderland, Massachusetts.

Shepherd, G. M. (1988). *Neurobiology* (2nd edn). Oxford University Press, New York.

Neurochemistry and neuropharmacology

Bradford, H. F. (1986). *Chemical neurobiology. An introduction to neurochemistry.* Freeman, New York.

Cooper, J. R., Bloom, F. E., and Roth, R. H. (1986). *The biochemical basis of neuropharmacology* (5th edn). Oxford University Press, New York.

Hucho, R. (1986). *Neurochemistry. Foundations and concepts.* VCH, Weinheim, W. Germany.

Siegel, G. J., Agranoff, B. W., Albers, R. W., and Molinoff, P. (ed.) (1989). *Basic neurochemistry* (4th edn). Raven Press, New York.

Phylogeny of nervous systems

Bullock, T. H., Orkland, R., and Grinnell, A. (1977). *Introduction to nervous systems.* Freeman, San Francisco.

Neuroanatomy of the mammalian nervous system

Brodal, A. (1969) *Neurological anatomy.* Oxford University Press, New York.
Carpenter, M. B. (1985). *Core text of neuroanatomy* (3rd edn). Williams and Wilkins, Baltimore.

Fundamentals of cognitive science

Glass, A. L. and Holyoak, K. J. (1986). *Cognition* (2nd edn). Random House, New York.
Stillings, N. A., Feinstein, M. H., Garfield, J. L., Rissland, E. L., Rosenbaum, D. A., Weisler, S. E., and Baker–Ward, L. (1987). *Cognitive science. An introduction.* MIT Press, Cambridge, Massachusetts.

Psychology of memory

Baddeley, A. D. (1976). *The psychology of memory.* Basic Books, New York.
Bower, G. H. and Hilgard, E. R. (1981). *Theories of learning* (5th edn). Prentice-Hall, New Jersey.
Klatzky, R. L. (1980). *Human memory: structures and processes* (2nd edn). Freeman, San Francisco.
Mackintosh, N. J. (1983). *Conditioning and associative learning.* Oxford University Press, Oxford.

Neurocomputing

Anderson, J. A. and Rosenfeld, E. (ed.) (1988). *Neurocomputing. Foundations of research.* MIT Press, Cambridge, Massachusetts.
Rumelhart, D. E. and McClelland, J. L. (ed., and the PDP research group) (1986). *Parallel distributed processing,* vol. I. MIT Press, Cambridge, Massachusetts.

Philosophy of neuroscience

Block, N. (ed.) (1980). *Readings in philosophy of psychology,* vol. 1. Harvard University Press, Cambridge, Massachusetts.
Block, N. (ed.) (1981). *Readings in philosophy of psychology,* vol. 2. Harvard University Press, Cambridge, Massachusetts.
Changeux, J.-P. (1985). *Neuronal man.* Pantheon, New York.
Churchland, P. S. (1986). *Neurophilosophy.* MIT Press, Cambridge, Massachusetts.
Cummins, R. (1989). *Meaning and mental representation.* MIT Press, Cambridge, Massachusetts.
Edelman, G. M. (1987). *Neural Darwinism. The theory of neuronal group selection.* Basic Books, New York.
Fodor, J. A. (1975). *The language of thought.* T. Y. Crowell, New York.
Fodor, J. A. (1981). *Representations.* MIT Press, Cambridge, Massachusetts.
Jackendoff, R. (1987). *Consciousness and the computational mind.* MIT Press, Cambridge, Massachusetts.
Marr, D. (1982). *Vision.* Freeman, San Francisco.

Pinker, S. (ed.) (1985). *Visual cognition.* MIT Press, Cambridge, Massachusetts.

Pylyshyn, Z. W. (1986). Computation and cognition. *Toward a foundation for cognitive science.* MIT Press, Cambridge, Massachussetts.

Roitblat, H. L. (1982). The meaning of representation in animal memory. *Behav. Brain Sci.* **5**, 353–406.

Ullman, S. (1980). Against direct perception. *Behav. Brain Sci.* **3**, 213–35.

Young, J. Z. (1978). *Programs of the brain.* Oxford University Press, Oxford.

Bibliography

Abeles, M. (1982). *Local cortical circuits. An electrophysiological study.* Springer-Verlag, Berlin.

Abeles, M. and Gerstein, G. L. (1988). Detecting spatiotemporal firing patterns among simultaneously recorded single neurons. *J. Neurophysiol.* **60**, 909–24.

Abraham, W. C., Bliss, T. V. P., and Goddard, G. V. (1985). Heterosynaptic changes accompany long-term but not short-term potentiation of the perforant path in the anaesthetized rat. *J. Physiol.* **363**, 335–49.

Abrams, T. W. (1985). Activity-dependent presynaptic facilitation: an associative mechanism in *Aplysia. Cell. Mol. Neurobiol.* **5**, 123–45.

Abrams, T. W. and Kandel, E. R. (1988). Is contiguity detection in classical conditioning a system or a cellular property? Learning in *Aplysia* suggests a possible molecular site. *Trends in Neurosci.* **11**, 128–35.

Abrams, T. W., Castellucci, V. R., Camardo, J. S., Kandel, E. R., and Lloyd, P. E. (1984). Two endogenous neuropeptides modulate the gill and siphon withdrawal reflex in *Aplysia* by presynaptic facilitation involving cAMP-dependent closure of a serotonin-sensitive potassium channel. *Proc. Natl. Acad. Sci. USA*, **81**, 7956–60.

Aceves-Pina, E. O. and Quinn, W. G. (1979). Learning in normal and mutant *Drosophila* larvae. *Science*, **206**, 93–6.

Aceves-Pina, E. O., Booker, R., Duerr, J. S., Livingstone, M. S., Quinn, W. G., Smith, R. F., Sziber, P. P., Tempel, B. L., and Tully, T. P. (1983). Learning and memory in *Drosophila*, studied with mutants. *Cold Spring Harbor Symp. Quant. Biol.* **48**, 831–40.

Acosta-Urquidi, J., Alkon, D. L., and Neary, J. T. (1984). Ca^{2+}-dependent protein kinase injection in a photoreceptor mimics biophysical effects of associative learning. *Science*, **224**, 1254–7.

Adrian, E. D. (1932). *The mechanism of nervous action: electrical studies of the neurone.* University of Pennsylvania Press, Philadelphia.

Aggleton, J. P. and Mishkin, M. (1983*a*). Memory impairments following restricted medial thalamic lesions in monkeys. *Exp. Brain Res.* **52**, 199–209.

Aggleton, J. P. and Mishkin, M. (1983*b*). Visual recognition impairment following medial thalamic lesions in monkeys. *Neuropsychologia*, **21**, 189–97.

Aggleton, J. P. and Mishkin, M. (1985). Mammillary-body lesions and visual recognition in monkeys. *Exp. Brain Res.* **58**, 190–7.

Aggleton, J. P., Friedman, D. P., and Mishkin, M. (1987). A comparison between the connections of the amygdala and hippocampus with the basal forebrain in the macaque. *Exp. Brain Res.* **67**, 556–68.

Agranoff, B. W., Davis, R. E., and Brink, J. J. (1965). Memory fixation in the goldfish. *Proc. Natl. Acad. Sci. USA*, **54**, 788–93.

Aidley, D. J. (1978). *The physiology of excitable cells,* (2nd edn). Cambridge University Press, Cambridge.

Akers, R. F. and Routtenberg, A. (1985). Protein kinase C phosphorylates a 47Mr protein (F1) directly related to synaptic plasticity. *Brain Res.* **334**, 147–51.

Akers, R. F., Lovinger, D. M., Colley, P. A., Linden, D. J., and Routtenberg, A. (1986). Translocation of protein kinase C activity may mediate hippocampal long-term potentiation. *Nature*, **231**, 587–9.

Alberts, B., Bray, D., Lewis, J., Raff, M., Roberts, K., and Watson, J. D. (1989). *Molecular biology of the cell* (2nd edn). Garland Publishing, New York.

Albus, J. S. (1971). A theory of cerebellar function. *Math. Biosci.* **10**, 25–61.

Alkon, D. L. (1983). Learning in a marine snail. *Sci. Amer.* **249**, (7), 64–74.

Alkon, D. L. (1984). Calcium-mediated reduction of ionic currents: a biophysical memory trace. *Science*, **226**, 1037–45.

Alkon, D. L., Lederhendler, I., and Shoukimas, J. L. (1982). Primary changes of membrane currents during retention of associative learning. *Science*, **215**, 693–5.

Alkon, D. L., Acosta-Urquidi, J., Olds, J., Kuzma, G., and Neary, J. T. (1983). Protein kinase injection reduces voltage-dependent potassium current. *Science*, **219**, 303–6.

Alkon, D. L., Bank, B., Chen, C. and Ram, J. (1987a). Inhibition of protein synthesis prolongs Ca^{2+}-mediated reduction of K^+ currents in molluscan neurons. *Proc. Natl. Acad. Sci. USA*, **84**, 6948–52.

Alkon, D. L., Disterhoft, J., and Coulter, D. (1987b). Conditioning-specific modification of postsynaptic membrane currents in mollusc and mammal. In *The neural and molecular bases of learning* (ed. J. P. Changeux and M. Konishi), pp. 205–38. Wiley, New York.

Altman, H. J., Stone, W. S., and Ogren, S. O. (1987). Evidence for a possible functional interaction between serotonergic and cholinergic mechanisms in memory retrieval. *Behav. Neural Biol.* **48**, 49–62.

Altman, J. and Das, G. D. (1965). Post-natal origins of microneurons in the rat brain. *Nature*, **207**, 953–6.

Alvarez-Buylla, A. and Nottebohm, F. (1988). Migration of young neurons in adult avian brain. *Nature*, **335**, 353–4.

Amaral, D. G. (1987). Memory: anatomical organization of candidate brain regions. In *Handbook of Physiology*, Section 1: The nervous system, Vol. V: Higher functions of the brain, Part 1 (ed. F. Plum), pp. 211–94. American Physiological Society, Bethesda, Md.

Andersen, P. (1987). Long-term potentiation—outstanding problems. In *Neural and molecular bases of learning* (ed. J. P. Changeux and M. Konishi), pp. 239–62. Wiley, New York.

Andersen, P., Bliss, T. V. P., and Skrede, K. (1971). Lamellar organization of hippocampal excitatory pathways. *Exp. Brain Res.* **13**, 222–38.

Andersen, P., Sundberg, S. H., Sveen, O., and Wigstrom, H. (1977). Specific long-lasting potentiation of synaptic transmission in hippocampal slices. *Nature*, **266**, 736–7.

Andersen, P., Sundberg, S. H., Sveen, O., Swann, J. W., and Wigstrom, H. (1980). Possible mechanisms for long-lasting potentiation of synaptic transmission in hippocampal slices from guinea-pig. *J. Physiol.* **302**, 463–82.

Andersen, P., Blackstad, T., Hulleberg, G., Trommald, M., and Vaaland, J. L. (1987). Dimensions of dendritic spines of rat granule cells during long-term potentiation (LTP). *J. Physiol.* **390**, 264P.

Andersen, R. A., Essick, G. A., and Sigel, R. M. (1987). Neurons of area 7 activated by both visual stimuli and oculomotor behavior. *Exp. Brain Res.* **67**, 316–22.

Anderson, J. A. and Rosenfeld, E. (ed.) (1988). *Neurocomputing. Foundations of research*. MIT Press, Cambridge, Mass.

Andrade, R., Malenka, R. C., and Nicoll, R. A. (1986). A G protein couples serotonin and $GABA_B$ receptors to the same channels in hippocampus. *Science*, **234**, 1261–5.

Andrews, E., Poser, C. M., and Kessler, M. (1982). Retrograde amnesia for forty years. *Cortex*, **18**, 441–58.

Angelergues, R. (1969). Memory disorders in neurological disease. In *Handbook of clinical neurology* (ed. P. J. Vinken and G. W. Bruyn), Vol. 3: Disorders of higher nervous activity, pp. 268–92. North-Holland, Amsterdam.

Aniksztejn, L., Ben-Ari, Y., and Gozlan, H. (1987). The lack of relation between the long-lasting enhancement of synaptic transmission produced by phorbol ester and the release of endogenous excitatory amino acids in the hippocampus of anaesthetized rats. *Neurosci.* **22** (Suppl.), S513.

Arendt, T., Allen, Y., Sinden, J., Schugens, M. M., Marchbanks, R. M., Landos, P. L., and Gray, J. A. (1988). Cholinergic-rich brain transplants reverse alcohol-induced memory deficits. *Nature* **332**, 448–50.

Arnold, A. P. (1980). Quantitative analysis of sex differences in hormone accumulation in the Zebra finch brain: methodological and theoretical issues. *J. Comp. Neurol.* **189**, 421–36.

Arnold, A. P. and Gorski, R. A. (1984). Gonadal steroid induction of structural sex differences in the central nervous system. *Ann. Rev. Neurosci.* **7**, 413–42.

Arnold, A. P. and Saltiel, A. (1979). Sexual difference in pattern of hormone accumulation in the brain of a songbird. *Science*, **205**, 702–5.

Arnold, A. P., Nottebohm, F., and Pfaff, D. W. (1976). Hormone concentrating cells in vocal control and other areas of the brain of the Zebra finch. *J. Comp. Neurol.* **165**, 487–512.

Aronson, P. S. (1985). Kinetic properties of the plasma membrane Na^+-H^+ exchanger. *Ann. Rev. Physiol.* **47**, 545–60.

Artola, A. and Singer, W. (1987). Long-term potentiation and NMDA receptors in rat visual cortex. *Nature*, **330**, 649–52.

Arvanitaki, A. and Cardot, H. (1941). Observations sur la constitution des ganglions et conducteurs nerveux et sur l'isolement du soma neuronique vivant chez les Mollusques Gastropodes. *Bull. Histol. Appl. Tech. Microscop.* **18**, 133–41.

Arvanitaki, A. and Chalazonitis, N. (1958). Configurations modales de l'activité, propres à différents neurons d'un même centre. *J. Physiol.* (Paris), **50**, 122–5.

Ascher, P. and Nowak, L. (1987). Electrophysiological studies of NMDA receptors. *Trends in Neurosci.* **10**, 284–8.

Assal, G., Favre, C., and Anderes, J. P. (1984). Non-reconnaissance d'animaux familiers chez un paysan. *Rev. Neurol.* (Paris), **140**, 580–4.

Azmitia, E. C. (1978). The serotonin-producing neurons of the midbrain median and dorsal raphe nuclei. In *Handbook of psychopharmacology*, Vol. 9: Chemical pathways in the brain (ed. L. L. Iversen, S. D. Iversen, and S. H. Snyder), pp. 223–314. Plenum Press, New York.

Bach, M. and Kruger, J. (1986). Correlated neuronal variability in monkey visual cortex revealed by microelectrode. *Exp. Brain Res.* **61**, 451–6.

Bachevalier, J. and Mishkin, M. (1984). An early and a late developing system for learning and retention in infant monkeys. *Behav. Neurosci.* **98**, 770–8.

Bachevalier, J. and Mishkin, M. (1986). Visual recognition impairment follows ventromedial but not dorsolateral prefrontal lesions in monkeys. *Behav. Brain Res.* **20**, 249–61.

Bachevalier, J., Parkinson, J. K., and Mishkin, M. (1985). Visual recognition in monkeys: effects of separate vs. combined transection of fornix and amygdalofugal pathways. *Exp. Brain Res.* **57**, 554–61.

Baddeley, A. D. (1976). *The psychology of memory.* Basic Books, New York.

Baddeley, A. D. (1986). *Working memory.* Oxford University Press, Oxford.

Bailey, C. H. and Chen, M. (1983). Morphological basis of long-term habituation and sensitization in *Aplysia. Science,* **220**, 91–3.

Bailey, C. H. and Chen, M. (1988*a*). Long-term memory in *Aplysia* modulates the total number of varicosities of single identified sensory neurons. *Proc. Natl. Acad. Sci. USA,* **85**, 2373–7.

Bailey, C. H. and Chen, M. C. (1988*b*). Morphological basis of short-term habituation in *Aplysia. J. Neurosci.* **8**, 2752–9.

Bailey, C. H. and Chen, M. (1988*c*). Long-term sensitization in *Aplysia* increases the number of presynaptic contacts onto the identified gill motor neuron L7. *Proc. Natl. Acad. Sci. USA,* **85**, 9356–9.

Baimbridge, K. G. and Miller, J. J. (1981). Calcium uptake and retention during long-term potentiation of neuronal activity in the rat hippocampal slice preparation. *Brain Res.* **221**, 299–305.

Baker, M. C. and Cunningham, M. A. (1985). The biology of bird-song dialects. *Behav. Brain Sci.* **8**, 85–133.

Baker, M. C., Spitler-Nabors, K. J., and Bradley, D. C. (1981). Early experience determines song dialect responsiveness of female sparrows. *Science,* **214**, 819–21.

Baldwin, F. M., Goldin, H. S., and Metfessel, M. (1940). Effects of testosterone propionate on female Roller Canaries under complete song isolation. *Proc. Soc. Exp. Biol. Med.* **44**, 373–5.

Ballard, D. H. (1986). Cortical connections and parallel processing: structure and function. *Behav. Brain Sci.* **9**, 67–120.

Balling, A., Technau, G. M., and Heisenberg, M. (1987). Are the structural changes in adult *Drosophila* mushroom bodies memory traces? Studies on biochemical learning mutants. *J. Neurogenet.* **4**, 65–73.

Bank, B., DeWeer, A., Kuzirian, A. M., Rasmussen, H., and Alkon, D. L. (1988). Classical conditioning induces long-term translocation of protein kinase C in rabbit hippocampal CA1 cells. *Proc. Natl. Acad. Sci. USA,* **85**, 1988–92.

Bannon, M. J., Lee, J. M., Giraud, P., Young, A., Affolter, H. U., and Bonner, T. I. (1986). Dopamine antagonist haloperidol decreases substance P, substance K, and preprotachykinin mRNAs in rat striatonigral neurons. *J. Biol. Chem.* **261**, 6640–2.

Baptista, L. F. and Petrinovich, L. (1984). Social interaction, sensitive phases and the song template hypothesis in the white-crowned sparrow. *Anim. Behav.* **32**, 172–81.

Bar, P. R., Weigant, F., Lopes da Silva, F. H., and Gispen, W. H. (1984). Tetanic stimulation affects the metabolism of phosphoinositides in hippocampal slices. *Brain Res.* **321**, 381–5.

Bargiello, T. A., Saez, L., Baylies, M. K., Gasic, G., Young, M. W., and Spray, D. C. (1987). The *Drosophila* clock gene *per* affects intercellular junctional communication. *Nature,* **328**, 686–91.

Barka, T. and Anderson, R. T. (1963). *Histochemistry: theory, practice and bibliography.* Hoeber, New York.

Barka, T., Gubits, R. M., and van der Noen, H. M. (1986). β-Adrenergic stimulation of *c-fos* gene expression in the mouse submandibular gland. *Mol. Cell. Biol.* **6**, 2984–9.

Barlow, H. B. (1972). Single units and sensation: a neuron doctrine for perceptual psychology? *Perception,* **1**, 371–94.

Barnes, C. A. (1979). Memory deficits associated with senescence: a neurophysiological and behavioral study in the rat. *J. Comp. Physiol. Psychol.* **93**, 74–104.

Barondes, S. H. and Cohen, H. D. (1966). Puromycin effect on successive phases of memory storage.

Science, **151**, 594–5.

Barondes, S. H. and Jarvik, M. E. (1964). The influence of actinomycin D on brain RNA synthesis and on memory. *J. Neurochem.* **11**, 187–95.

Barrett, P. Q., Kojima, I., Kojima, K., Zawalich, K., Isales, C. M., and Rasmussen, H. (1986). Short-term memory in the calcium messenger system. Evidence for a sustained activation of protein kinase C in adrenal glomerulosa cells. *Biochem. J.* **238**, 905–12.

Barrionuevo, G. and Brown, T. H. (1983). Associative long-term potentiation in hippocampal slices. *Proc. Natl. Acad. Sci. USA*, **80**, 7347–51.

Bateson, P. P. G. (1966). The characteristics and context of imprinting. *Biol. Rev.* **41**, 177–220.

Bateson, P. P. G. (1978). Early experience and sexual preference. In *Biological determinants of sexual behavior* (ed. J. B. Hutchison), pp. 29–53. Wiley, London.

Bateson, P. P. G. and Jaeckel, J. B. (1974). Imprinting: correlations between activities during training and testing. *Anim. Behav.* **22**, 899–906.

Bateson, P. P. G. and Reese, E. P. (1969). The reinforcing properties of conspicuous stimuli in the imprinting situation. *Anim. Behav.* **17**, 692–9.

Bateson, P. P. G. and Wainwright, A. P. (1972). The effects of prior exposure to light on the imprinting process in domestic chicks. *Behaviour*, **42**, 279–90.

Bateson, P. P. G., Horn, G., and Rose, S. P. R. (1972). Effects of early experience on regional incorporation of precursors into RNA and protein in the chick brain. *Brain Res.* **39**, 449–65.

Bateson, P. P. G., Rose, S. P. R., and Horn, G. (1973). Imprinting: lasting effects on uracil incorporation into chick brain. *Science*, **181**, 576–8.

Bateson, P. P. G., Horn, G., and Rose, S. P. R. (1975). Imprinting: correlation between behavior and incorporation of [^{14}C]uracil into chick brain. *Brain Res.* **84**, 207–20.

Baudry, M., Oliver, M., Creager, R., Wieraszko, A., and Lynch, G. (1980). Increase in glutamate receptors following repetitive electrical stimulation in hippocampal slices. *Life Sci.* **27**, 325–30.

Baum, E. B., Moody, J., and Wilczek, F. (1988). Internal representations for associative memory. *Biol. Cybernet.* **59**, 217–28.

Baxter, D. A. and Byrne, J. H. (1986). Serotonin-modulated membrane currents in *Aplysia* tail sensory neurons. *Soc. Neurosci. Abstr.* **12**, 765.

Bayer, S. A., Yackel, J. W., and Puri, P. S. (1982). Neurons in the rat dentate gyrus granular layer substantially increase during juvenile and adult life. *Science*, **216**, 890–2.

Baylis, G. C. and Rolls, E. T. (1987). Responses of neurons in the inferior temporal cortex in short term and serial recognition memory tasks. *Exp. Brain Res.* **65**, 614–22.

Bear, M. F. (Rapporteur), Cotman, C. W., Innocenti, G. M., Lomo, T., Merzenich, M. M., Meyer, R. L., Rakic, P., Seifert, W., Singer, W., Sotelo, C., and Sturmer, C. A. O. (1987). Activity-dependent modification of functional circuitry as a possible basis for learning. In *The neural and molecular bases of learning* (ed. J. P. Changeux and M. Konishi), pp. 281–300. Wiley, Chichester.

Becker, J. T. and Olton, D. S. (1982). Cognitive mapping and hippocampal function. *Neuropsychologia*, **19**, 733–41.

Belardetti, F. and Siegelbaum, S. A. (1988). Up- and down-modulation of single K^{+} channel function by distinct second messengers. *Trends in Neurosci.* **11**, 232–8.

Belardetti, F., Schacher, S., Kandel, E. R., and Siegelbaum, S. A. (1986). The growth cones of *Aplysia* sensory neurons: modulation by serotonin of action potential duration and single potassium channel currents. *Proc. Natl. Acad. Sci. USA*, **83**, 7094–8.

Bender, D. B. (1973). Visual sensitivity following inferotemporal and foveal prestriate lesions in the rhesus monkey. *J. Comp. Physiol. Psychol.* **84**, 613–21.

Benson, D. F. and Zaidel, E. (1985). *The dual brain. Hemispheric specialization in humans.* Guilford, New York.

Benzer, S. (1967). Behavioral mutants of *Drosophila* isolated by countercurrent distribution. *Proc. Natl. Acad. Sci. USA*, **58**, 1112–19.

Benzer, S. (1973). Genetic dissection of behavior. *Sci. Amer.* **229** (12), 24–37.

Berger, T. W. (1984). Long-term potentiation of hippocampal synaptic transmission affects rate of behavioral learning. *Science*, **224**, 627–30.

Berger, T. W. and Orr, W. B. (1983). Hippocampectomy selectively disrupts discrimination reversal conditioning of the rabbit nictitating membrane responses. *Behav. Brain Res.* **8**, 49–68.

Berger, T. W. and Thompson, R. F. (1978*a*). Identification of pyramidal cells as the critical elements in hippocampal neuronal plasticity during learning. *Proc. Natl. Acad. Sci. USA*, **75**, 1572–6.

Berger, T. W. and Thompson, R. F. (1978*b*). Neuronal plasticity in the limbic system during classical conditioning of the rabbit nictitating membrane response. I. The hippocampus. *Brain Res.* **145**, 323–46.

Berger, T. W., Alger, B., and Thompson, R. F. (1976). Neuronal substrate of classical conditioning in the hippocampus. *Science*, **192**, 483–5.

Berger, T. W., Laham, R. I., and Thompson, R. F. (1980). Hippocampal unit-behavior correlations during classical conditioning. *Brain Res.* **193**, 229–48.

Berger, T. W., Rinaldi, P. C., Weisz, D. J., and Thompson, R. F. (1983). Single-unit analysis of different hippocampal cell types during classical conditioning of rabbit nictitating membrane response. *J. Neurophysiol.* **50**, 1197–219.

Berk, A. J. (1986). Adenovirus promotors and E1A transactivation. *Ann. Rev. Genet.* **20**, 45–79.

Bernier, L., Castellucci, V. F., Kandel, E. R., and Schwartz, J. H. (1982). Facilitatory transmitter causes a selective and prolonged increase in adenosine 3′:5′-monophosphate in sensory neurons mediating the gill and siphon withdrawal reflex in *Aplysia. J. Neurosci.* **2**, 1682–91.

Berridge, M. (1986). Second messenger dualism in neuromodulation and memory. *Nature*, **323**, 294–5.

Berridge, M. J. (1987). Inositol triphosphate and diacylglycerol: two interacting second messengers. *Ann. Rev. Biochem.* **56**, 159–93.

Berthier, N. E., Betts, B., and Woody, C. D. (1982). Discriminative conditioning of eyeblink with aversive brain stimulation. *Soc. Neurosci. Abst.* **8**, 315.

Best, P. J. and Thompson, L. T. (1984). Hippocampal cells which have place field activity also show changes in activity during classical conditioning. *Soc. Neurosci. Abst.* **10**, 125.

Bitterman, M. E. (1975). The comparative analysis of learning. *Science*, **188**, 699–709.

Black, A. H. (1959). Heart rate change during avoidance learning in dogs. *Can. J. Psychol.* **13**, 229–42.

Black, I. B., Chikaraishi, D. M., and Lewis, E. J. (1985). Trans-synaptic increase in RNA coding for tyrosine hydroxylase in a rat sympathetic ganglion. *Brain Res.* **339**, 151–3.

Black, I. B., Adler, J. E., Dreyfus, C. F., Friedman, W. F., LaGanna, E. F., and Roach, A. H. (1987). Biochemistry of information storage in the nervous system. *Science*, **236**, 1263–8.

Blake, L., Jarvis, C. D., and Mishkin, M. (1977). Pattern discrimination thresholds after partial inferior temporal or lateral striate lesions in monkeys. *Brain Res.* **120**, 209–20.

Blasdel, G. G. and Salama, G. (1986). Voltage-sensitive dyes reveal a modular organization in monkey striate cortex. *Nature,* **321,** 579–85.

Bliss, T. V. P. and Gardner-Medwin, A. R. (1973). Long-lasting potentiation of synaptic transmission in the dentate area of the unanaesthetized rabbit following stimulation of the perforant path. *J. Physiol.* **232,** 357–74.

Bliss, T. V. P. and Lomo, T. (1973). Long-lasting potentiation of synaptic transmission in the dentate area of the anaesthetized rabbit following stimulation of the perforant path. *J. Physiol.* **232,** 331–56.

Bliss, T. V. P., Goddard, G. V., and Riives, M. (1983). Reduction of long-term potentiation in the dentate gyrus of the rat following selective depletion of monoamines. *J. Physiol.* **334,** 475–91.

Bliss, T. V. P., Douglas, R. M., Errington, M. L., and Lynch, M. A. (1986). Correlation between long-term potentiation and release of endogenous amino acids from dentate gyrus of anaesthetized rats. *J. Physiol.* **377,** 394–408.

Bliss, T. V. P., Dolphin, A. C., Errington, M. L., and Fazeli, M. S. (1987). Increase in concentration of specific extracellular proteins during long-term potentiation in the dentate gyrus of the rat. *J. Physiol.* **388,** 49P.

Block, N. (ed.) (1980). *Readings in philosophy of psychology,* vol. 1. Harvard University Press, Cambridge, Mass.

Block, N. (ed.) (1981). *Readings in philosophy of psychology,* vol. 2. Harvard University Press, Cambridge, Mass.

Bloedel, J. R. (1987). The cerebellum and memory storage. *Science,* **238,** 1728–9.

Bloom, W. and Fawcett, D. W. (1968). *A textbook of histology* (9th edn). Saunders, Philadelphia.

Boakes, R. (1984). *From Darwin to behaviorism. Psychology and the minds of animals.* Cambridge University Press, Cambridge.

Bolhuis, J. J., McCabe, B. J., and Horn, G. (1986). Androgens and imprinting. Differential effects of testosterone on filial preferences in the domestic chick. *Behav. Neurosci.* **100,** 51–6.

Bolles, R. C. (1970). Species-specific defense reactions and avoidance learning. *Psychol. Rev.* **77,** 32–48.

Booker, R. and Quinn, W. G. (1981). Conditioning of leg position in normal and mutant *Drosophila. Proc. Natl. Acad. Sci. USA,* **78,** 3940–4.

Boring, E. G. (1950). *A history of experimental psychology* (2nd edn). Prentice-Hall, New Jersey.

Bostford, J. L. (1981). Cyclic nucleotides in procaryotes. *Microbiol. Rev.* **45,** 620–42.

Botchan, M., Grodzicker, T., and Sharp, P. A. (ed.) (1986). DNA tumor viruses. Control of gene expression and replication. In *Cancer cells.* Cold Spring Harbor Laboratory, New York.

Bottjer, S. W., Miesner, E. A., and Arnold, A. P. (1984). Forebrain lesions disrupt development but not maintenance of song in passerine birds. *Science,* **224,** 901–3.

Bottjer, S. W., Glaessner, S. L., and Arnold, A. P. (1985). Ontogeny of brain nuclei controlling song learning and behavior in Zebra finches. *J. Neurosci.* **5,** 1556–62.

Bourne, H. R. (1986). One molecular machine can transduce diverse signals. *Nature,* **321,** 814–16.

Bower, G. H. and Hilgard, E. R. (1981). *Theories of learning* (5th edn). Prentice-Hall, New Jersey.

Boyd, I. A. and Martin, A. R. (1956). The end-plate potential in mammalian muscle. *J. Physiol.* **132,** 74–91.

Boyle, M. B., Klein, M., Smith, S. J., and Kandel, E. R. (1984). Serotonin increases intracellular Ca^{2+} transients in voltage-clamped sensory neurons of *Aplysia californica. Proc. Natl. Acad. Sci. USA,* **81,** 7642–6.

Bradford, H. F. (1986). *Chemical neurobiology. An introduction to neurochemistry.* Freeman, New York.

Bradshaw, J. L. and Nettleton, N. C. (1983). *Human cerebral asymmetry.* Prentice-Hall, New Jersey.

Brandon, J. G. and Coss, R. G. (1982). Rapid dendritic spine stem shortening during one-trial learning: the honeybee's first orientation flight. *Brain Res.* **252**, 51–61.

Brazier, M. A. B. (1988). *A history of neuropsychology in the 19th century.* Raven Press, New York.

Brenowitz, E. A. and Arnold, A. P. (1986). Interspecific comparisons of the size of neural song control regions and song complexity in duetting birds: evolutionary implications. *J. Neurosci.* **6**, 2875–9.

Bridge, M. S. and Crow, T. (1986). Neural correlates of conditioning in lateral and medial B-photo-receptors of *Hermissenda*. *Soc. Neurosci. Abst.* **12**, 861.

Broca, P. (1878). Anatomie compare des circonvolutions cérébrales. Le grand lobe limbique et la scissure limbique dans la série des mammifières. *Rev. Anthrop.* **1**, 385–498.

Brodal, A. (1969). *Neurological anatomy.* Oxford University Press, New York.

Brody, B. A. and Pribram, K. H. (1978). The role of frontal and parietal cortex in cognitive processing. *Brain*, **101**, 607–33.

Brons, J. F. and Woody, C. D. (1980). Long-term changes in excitability of cortical neurons after Pavlovian conditioning and extinction. *J. Neurophysiol.* **44**, 605–15.

Brons, J. F., Woody, C. D., and Alon, N. (1982). Changes in the excitability to weak intensity stimulation of units of the pericruciate cortex in cats. *J. Neurophysiol.* **47**, 377–88.

Brooks, D. N. and Baddeley, A. D. (1976). What can amnesic patients learn? *Neuropsychologia*, **14**, 111–22.

Brown, J. S., Kalish, H. I., and Farber, I. E. (1951). Conditional fear as revealed by magnitude of startle response to an auditory stimulus. *J. Exp. Psychol.* **41**, 317–27.

Brown, P. L. and Jenkins, H. M. (1968). Auto-shaping of the pigeon's key-peck. *J. Exp. Anal. Behav.* **11**, 1–8.

Brown, S. and Schafer, E. A. (1888). An investigation into the functions of the occipital and temporal lobes of the monkey's brain. *Phil. Trans. R. Soc. Lond. B*, **179**, 303–29.

Brownell, G. L., Budinger, T. F., Lauterbur, P. C., and McGeer, P. L. (1982). Positron tomography and nuclear magnetic resonance imaging. *Science*, **215**, 619–26.

Brunelli, M., Castellucci, V., and Kandel, E. R. (1976). Synaptic facilitation and behavioral sensitization in *Aplysia*: possible role of serotonin and cyclic AMP. *Science*, **194**, 1178–81.

Bruyer, R., Laterre, C., Seron, X., Feyereisen, P., Strypstein, E., Pierrard, E., and Rectem, D. (1983). A case of prosopagnosia with some preserved covert remembrance of familiar faces. *Brain and Cognition*, **2**, 257–84.

Bullock, T. H., Orkland, R., and Grinnell, A. (1977). *Introduction to nervous systems.* Freeman, San Francisco.

Burd, G. D. and Nottebohm, F. (1985). Ultrastructural characterization of synaptic terminals formed on newly generated neurons in a song control nucleus of the adult sensory forebrain. *J. Comp. Neurol.* **240**, 143–52.

Burke, R. E. (1987). Synaptic efficacy and the control of neuronal input–output relations. *Trends in Neurosci.* **10**, 42–5.

Butters, N. (1984). Alcoholic Korsakoff's syndrome: an update. *Semin. Neurol.* **4**, 226–44.

Buxbaum, J. and Dudai, Y. (1988). A microtiter based assay for protein kinase activity, suitable for a large number of assays, and its application to the analysis of *Drosophila* learning mutants. *Anal. Biochem.* **169**, 209–15.

Buxbaum, J. and Dudai, Y. (1989). A quantitative model for the kinetics of cAMP-dependent protein kinase (type II) activity: long-term activation of the kinase and its possible relevance to learning and memory. *J. Biol. Chem.* (in press).

Byers, D. (1980). Studies on learning and cyclic AMP phosphodiesterase of the *dunce* mutant of *Drosophila melanogaster*. Ph.D. thesis, California Institute of Technology, Pasadena.

Byers, D., Davis, R. L., and Kiger J. A. (1981). Defect in cyclic AMP phosphodiesterase due to the *dunce* mutation of learning in *Drosophila melanogaster*. *Nature*, **289**, 79–81.

Byrne, J. H. (1985). Neural and molecular mechanisms underlying information storage in *Aplysia*: implications for learning and memory. *Trends in Neurosci.* **8**, 478–82.

Byrne, J. H. (1987). Cellular analysis of associative learning. *Physiol. Rev.* **67**, 329–439.

Byrne, J., Castellucci, V., and Kandel, E. R. (1974). Receptive fields and response properties of mechanoreceptor neurons innervating siphon skin and mantle shelf in *Aplysia*. *J. Neurophysiol.* **37**, 1041–64.

Byrne, J. H., Castellucci, V. F., and Kandel, E. R. (1978). Contribution of individual mechanoreceptor sensory neurons to defensive gill-withdrawal reflex in *Aplysia*. *J. Neurophysiol.* **41**, 418–31.

Byrne, W. L., Samuel, D., Bennett, E. L., Rosenzweig, M. R., Wasserman, E., Wagner, A. R., Gardner, F., Galambos, R., Berger, B. D., Margules, D. L., Fenichel, R. L., Stein, L., Corson, J. A., Enesco, H. E., Chorover, S. L., Holt, C. E., Schiller, P. H., Chippetta, L., Jarvik, M. E., Leaf, R. C., Dutcher, J. D., Horovitz, Z. P., and Carlson, P. L. (1966). Memory transfer. *Science*, **153**, 658–9.

Cairns, R. B. (1966). Attachment behavior of mammals. *Psychol. Rev.* **73**, 409–26.

Cajal, S. R. (1911). *Histologie du système nerveux de l'homme et des vertébrés*. Malonie, Paris.

Calford, M. B. and Tweedale, R. (1988). Immediate and chronic changes in responses of somatosensory cortex in adult flying-fox after digit amputation. *Nature*, **332**, 446–8.

Camhi, J. M. (1984). *Neuroethology*. Sinauer, Sunderland, Mass.

Camhi, J. M. (1988). Escape behavior in the cockroach: distributed neural processing. *Experientia*, **44**, 401–8.

Canady, R. A., Kroodsma, D. E., and Nottebohm, F. (1984). Population differences in complexity of a learned skill are correlated with the brain space involved. *Proc. Natl. Acad. Sci. USA*, **81**, 6232–4.

Candland, D. K. (1969). Discriminability of facial regions used by the domestic chicken in maintaining the social dominance order. *J. Comp. Physiol. Psychol.* **69**, 281–5.

Carew, T. J. and Kandel, E. R. (1973). Acquisition and retention of long-term habituation in *Aplysia*: correlation of behavioural and cellular processes. *Science*, **182**, 1158–60.

Carew, T. J. and Sahley, C. L. (1986). Invertebrate learning and memory: from behavior to molecules. *Ann. Rev. Neurosci.* **9**, 435–87.

Carew, T. J., Castellucci, V., and Kandel, E. R. (1971). An analysis of dishabituation and sensitization of the gill-withdrawal reflex in *Aplysia*. *Int. J. Neurosci.* **2**, 79–98.

Carew, T. J., Pinsker, H., and Kandel, E. R. (1972). Long-term habituation of a defensive withdrawal reflex in *Aplysia*. *Science*, **175**, 451–4.

Carew, T. J., Walters, E. T., and Kandel, E. R. (1981). Classical conditioning in a simple withdrawal reflex in *Aplysia californica*. *J. Neurosci.* **1**, 1426–37.

Carew, T. J., Hawkins, R. D., and Kandel, E. R. (1983). Differential classical conditioning of a defensive withdrawal reflex in *Aplysia californica*. *Science*, **219**, 397–400.

Carew, T. J., Hawkins, R. D., Abrams, T. W., and Kandel, E. R. (1984). A test of Hebb's postulate at identified synapses which mediate classical conditioning in *Aplysia*. *J. Neurosci.* **4**, 1217–24.

Carey, S. and Diamond, R. (1980). Maturational determination of the developmental course of face encoding. In *Biological studies of mental processes* (ed. D. Kaplan), pp. 60–93. MIT Press, Cambridge, Mass.

Carpenter, M. B. (1985). *Core text of neuranatomy* (3rd edn). Williams and Williams, Baltimore.

Castellucci, V. F. and Kandel, E. R. (1974). A quantal analysis of the synaptic depression underlying habituation of the gill-withdrawal reflex in *Aplysia. Proc. Natl. Acad. Sci. USA*, **71**, 5004–8.

Castellucci, V. F. and Kandel, E. R. (1976). Presynaptic facilitation as a mechanism for behavioural sensitization in *Aplysia. Science*, **194**, 1176–8.

Castellucci, V. F., Pinsker, H., Kupfermann, I., and Kandel, E. R. (1970). Neuronal mechanisms of habituation and dishabituation of the gill-withdrawal reflex in *Aplysia. Science*, **167**, 1745–8.

Castellucci, V. F., Kandel, E. R., Schwartz, J. H., Wilson, F. D., Nairn, A. C., and Greengard, P. (1980). Intracellular injection of the catalytic subunit of cyclic AMP-dependent protein kinase simulates facilitation of transmitter release underlying behavioral sensitization in *Aplysia. Proc. Natl. Acad. Sci. USA*, **77**, 7492–6.

Castellucci, V. F., Nairn, A., Greengard, P., Schwartz, J. H., and Kandel, E. R. (1982). Inhibitor of adenosine 3′:5′-monophosphate dependent protein kinase blocks presynaptic facilitation in *Aplysia. J. Neurosci.* **2**, 1673–81.

Castellucci, V. F., Kennedy, T. E., Kandel, E. R., and Goelet, P. (1988). A quantitative analysis of 2-D gels identifies proteins in which labelling is increased following long-term sensitization in *Aplysia. Neuron*, **1**, 321–8.

Cedar, H., Kandel, E. R., and Schwartz, J. H. (1972). Cyclic adenosine monophosphate in the nervous system of *Aplysia california*. I. Increased synthesis in response to synaptic stimulation. *J. Gen. Physiol.* **60**, 558–69.

Chang, F. L. F. and Greenough, W. T. (1984). Transient and enduring morphological correlates of synaptic activity and efficacy change in the rat hippocampal slice. *Brain Res.* **309**, 35–46.

Chang, J. J. and Gelperin, A. (1980). Rapid taste-aversion learning by an isolated molluscan central nervous system. *Proc. Natl. Acad. Sci. USA*, **77**, 6204–6.

Changeux, J. P. (1985). *Neuronal man*. Pantheon, New York.

Changeux, J. P. (1986). Coexistence of neuronal messengers and molecular selection. *Prog. Brain Res.* **68**, 373–403.

Changeux, J. P. and Danchin, A. (1976). Selective stabilisation of developing synapses as a mechanism for the specification of neuronal networks. *Nature*, **264**, 705–12.

Changeux, J. P., Heidmann, T., and Patte, P. (1984). Learning by selection. In *The biology of learning* (ed. P. Marler and H. S. Terrace), pp. 115–33. Springer-Verlag, Berlin.

Changeux, J. P., Klarsfeld, A., and Heidmann, T. (1987). The acetylcholine receptor and molecular models for short- and long-term learning. In *The neural and molecular bases of learning* (J. P. Changeux and M. Konishi), pp. 31–84. Wiley, New York.

Chapouthier, G. (1983). Protein synthesis and memory. In *The physiological basis of memory* (2nd edn), (ed. J. A. Deutsch), pp. 1–47. Academic Press, New York.

Charniak, E. and McDermott, D. (1985). *Introduction to artificial intelligence*. Addison Wesley, Reading, Mass.

Chen, C. N., Denome, S., and Davis, R. L. (1986). Molecular analysis of cDNA clones and the corresponding genomic coding sequences of the *Drosophila dunce*[+] gene, the structural gene for cAMP phosphodiesterase. *Proc. Natl. Acad. Sci. USA*, **83**, 9313–17.

Chen, C. N., Malone, T., Beckendorf, S. K., and Davis, R. L. (1987). At least two genes reside within the *dunce* gene of *Drosophila. Nature*, **329**, 721–4.

Cheney, D., Seyfarth, R., and Smuts, B. (1986). Social relationships and social cognition in nonhuman primates. *Science*, **234**, 1361–66.

Cherubini, E., Ben-Ari, Y., Gho, M., Bidard, J. N., and Lazdunski, M. (1987). Long-term potentiation of synaptic transmission in the hippocampus induced by a bee venom peptide. *Nature*, **328**, 70–3.

Chien, S. (1985). Cerebral blood flow and metabolism. In *Principles of neural science* (2nd edn) (ed. E. R. Kandel, and J. H. Schwartz), pp. 845–52. Elseiver, New York.

Chomsky, N. (1959). Review of *Verbal Behaviour* by B. F. Skinner. *Language*, **35**, 26–58.

Chow, K. L. (1951). Effects of partial extirpation of posterior association cortex on visually mediated behavior in monkeys. *Comp. Physiol. Monog.* **20**, 187–217.

Christian, E. P. and Deadwyler, S. A. (1986). Behavioural functions and hippocampal cell types: evidence for two nonoverlapping populations in the rat. *J. Neurophysiol.* **55**, 331–48.

Churchland, P. S. (1986). *Neurophilosophy.* MIT Press, Cambridge, Mass.

Cipolla-Neto, J., Horn, G., and McCabe, B. J. (1982). Hemispheric asymmetry and imprinting: the effect of sequential lesions to the hyperstriatum ventrale. *Exp. Brain. Res.* **48**, 22–7.

Clark, G. A., McCormick, D. A., Lavond, D. G., and Thompson, R. F. (1984). Effects of lesions of cerebellar nuclei on conditioned behavioral and hippocampal neuronal responses. *Brain Res.* **291**, 125–36.

Clark, R. B. (1960). Habituation of the polychaete *Nereis* to sudden stimuli. 1. General properties of the habituation process. *Anim. Behav.* **8**, 82–91.

Clark, S. A., Allard, T., Jenkins, W. M., and Merzenich, H. M. (1988). Receptive fields in the body-surface map in adult cortex defined by temporally correlated inputs. *Nature*, **332**, 444–5.

Cleary, L. J. and Byrne, J. H. (1986). Associative learning of the gill and siphon withdrawal reflex in *Aplysia*: interneurons mediating the unconditioned response. *Soc. Neurosci. Abst.* **12**, 397.

Cline, H. T., Debski, E. A., and Constantine-Paton, M. (1987). *N*-methyl-D-aspartate receptor antagonist desegregates eye-specific stripes. *Proc. Natl. Acad. Sci. USA*, **84**, 4342–5.

Coan, E. J. and Collingridge, G. L. (1985). Magnesium ions block an *N*-methyl-D-aspartate receptor-mediated component of synaptic transmission in rat hippocampus. *Neurosci. Lett.* **53**, 21–6.

Cochran, B. H., Reffel, A. C., and Stiles, C. D. (1983). Molecular cloning of gene sequences regulated by platelet-derived growth factor. *Cell*, **33**, 939–47.

Cohen, D. H. (1984). Identification of vertebrate neurons modified during learning: analysis of sensory pathways. In *Primary neural substrates of learning and behavioural change* (ed. D. L. Alkon and J. Farley), pp. 129–54. Cambridge University Press.

Cohen, D. H. and Goff, D. G. (1978). Conditioned heart rate change in the pigeon: analysis and prediction of acquisition patterns. *Physiol. Psychol.* **6**, 127–41.

Cohen, D. H. and Pitts, L. H. (1968). Vagal and sympathetic components of conditioned cardio-acceleration in the pigeon. *Brain Res.* **9**, 15–31.

Cohen, D. H., Ervin, F., and Barondes, S. H. (1966). Puromycin and cycloheximide: different effects on hippocampal electrical activity. *Science*, **154**, 1557–8.

Cohen, D. H., Gibbs, C. M., Siegelman, P., Gamlin, P., and Broyles, J. (1982). Is locus coeruleus involved in plasticity of lateral geniculate neurons during learning? *Soc. Neurosci. Abst.* **8**, 666.

Cohen, L. B., Salzberg, B. M., and Grinvald, A. (1978). Optical methods for monitoring neuron activity. *Ann. Rev. Neurosci.* **1**, 171–82.

Cohen, N. J. (1984). Preserved learning capacity in amnesia: evidence for multiple memory systems.

In *Neuropsychology of memory* (ed. L. R. Squire and N. Butters), pp. 83–103. Guilford, New York.

Cohen, N. J. and Squire, L. R. (1980). Preserved learning and retention of pattern-analyzing skill in amnesia: dissociation of knowing how and knowing that. *Science*, **210**, 207–9.

Collingridge, G. L. (1985). Long term potentiation in the hippocampus: mechanisms of initiation and modulation by neurotransmitters. *Trends in Pharmacol. Sci.* **6**, 407–11.

Collingridge, G. L. (1987). The role of NMDA receptors in learning and memory. *Nature*, **330**, 604–5.

Collingridge, G. L. and Bliss, T. V. P. (1987). NMDA receptors—their role in long-term potentiation. *Trends in Neurosci.* **10**, 288–93.

Collingridge, G. L., Kehl, S. J., and McLennan, H. (1983). Excitatory amino acids in synaptic transmission in the schaffer collateral-commissural pathway of the rat hippocampus. *J. Physiol.* **334**, 33–46.

Comb, M., Hayman, S. E., and Goodman, H. M. (1987). Mechanism of trans-synaptic regulation of gene expression. *Trends in Neurosci.* **10**, 473–8.

Connor, J. A. and Alkon, D. L. (1984). Light- and voltage-dependent increases of calcium ion concentration in molluscan photoreceptors. *J. Neurophysiol.* **51**, 745–52.

Coombs, C. H. (1938). Adaptation of the galvanic response to auditory stimuli. *J. Exp. Psychol.* **22**, 244–68.

Cooper, J. R., Bloom, F. E., and Roth, R. H. (1986). *The biochemical basis of neuropharmacology* (5th edn). Oxford University Press, New York.

Cooper, L. N. (1973). A possible organization of animal memory and learning. In *Proceedings of the Nobel Symposium on collective properties of physical systems* (ed. B. Lundquist and S. Lundquist), pp. 252–64. Academic Press, New York.

Cooperman, S. S., Grubman, S. A., Barchi, R., Goodman, R. H., and Mandel, G. (1987). Modulation of sodium-channel mRNA levels in rat skeletal muscle. *Proc. Natl. Acad. Sci. USA*, **84**, 8721–5.

Coopersmith, R. and Leon, M. (1986). Enhanced neural response by adult rats to odors experienced early in life. *Brain Res.* **371**, 400–3.

Coopersmith, R., Henderson, S. R., and Leon, M. (1986). Odor specificity of the enhanced neural response following early odor experience in rats. *Dev. Brain. Res.* **27**, 191–7.

Corfas, G. and Dudai, Y. (1989). Habituation and dishabituation of a cleaning reflex in normal and mutant *Drosophila. J. Neurosci.* **9**, 56–62.

Corkin, S. (1984). Lasting consequences of bilateral medial temporal lobectomy: clinical course and experimental findings in H.M. *Semin. Neurol.* **4**, 249–59.

Correll, R. and Scoville, W. B. (1965). Performance on delayed match following lesions of medial temporal structures. *J. Comp. Physiol. Psychol.* **60**, 360–7.

Cotman, C. W. (ed.) (1985). *Synaptic plasticity.* Guilford, New York.

Cotman, C. W. and Nieto-Sampedro, M. (1984). Cell biology of synaptic plasticity. *Science*, **225**, 1287–94.

Cowan, T. M. and Siegel, R. W. (1986). *Drosophila* mutations that alter ionic conduction disrupt acquisition and retention of a conditioned odor avoidance response. *J. Neurogenet.* **3**, 187–201.

Cowey, A. (1982). Sensory and non-sensory visual disorders in man and monkey. *Phil. Trans. R. Soc. Lond. B,* **298**, 3–13.

Cowey, A. and Gross, C. G. (1970). Effects of foveal prestriate and inferotemporal lesions on visual discrimination by rhesus monkeys. *Exp. Brain Res.* **11**, 128–44.

Coyle, J. T., Price, D. L., and DeLong, M. R. (1983). Alzheimer's disease: a disorder of cortical cholinergic innervation. *Science*, **219**, 1184–90.

Crepel, F. and Krupa, M. (1988). Activation of protein kinase C induces a long-term depression of glutamate sensitivity of cerebellar Purkinje cells. An *in vitro* study. *Brain Res.* **458**, 397–401.

Crick, F. (1982). Do dendritic spines twitch? *Trends in Neurosci.* **5**, 44–6.

Crick, F. (1984*a*). Memory and molecular turnover. *Nature*, **312**, 101.

Crick, F. (1984*b*). Function of the thalamic reticular complex: the searchlight hypothesis. *Proc. Natl. Acad. Sci. USA*, **81**, 4586–90.

Cross, H. A., Halcomb, C. G., and Matter, W. W. (1967). Imprinting or exposure learning in rats given early auditory stimulation. *Psychon. Sci.* **7**, 233–4.

Crow, T. (1983). Conditioned modification of locomotion in *Hermissenda crassicornis*: analysis of time-dependent associative and nonassociative components. *J. Neurosci.* **3**, 2621–8.

Crow, T. (1984). A behavioural and cellular neurophysiological analysis of associative learning in *Hermissenda*. In *Primary neural substrates of learning and behavioural change* (ed. D. L. Alkon and J. Farley), pp. 205–27. Cambridge University Press, Cambridge.

Crow, T. (1985*a*). Conditioned modification of phototactic behavior in *Hermissenda*. I. Analysis of light intensity. *J. Neurosci.* **5**, 209–14.

Crow, T. (1985*b*). Conditioned modification of phototactic behavior in *Hermissenda*. II. Differential adaptation of B-photoreceptors. *J. Neurosci.* **5**, 215–23.

Crow, T. (1985*c*). Cellular mechanisms of associative learning in *Hermissenda*: contribution of light-activated conductances. *Soc. Neurosci. Abst.* **11**, 794.

Crow, T. J. and Alkon, D. L. (1978). Retention of an associative behavioral change in *Hermissenda*. *Science*, **201**, 1239–41.

Crow, T. J. and Alkon, D. L. (1980). Associative behavioral modification in *Hermissenda*: cellular correlates. *Science*, **209**, 412–14

Crow, T. and Forrester, J. (1986). Light paired with serotonin mimics the effect of conditioning on phototactic behavior in *Hermissenda. Proc. Natl. Acad. Sci. USA*, **83**, 7975–8.

Crow, T. and Offenbach, N. (1983). Modification of the initiation of locomotion in *Hermissenda*: behavioral analysis. *Brain Res.* **271**, 301–10.

Crow, T. J., Heldman, E., Hacopian, V., Enos, R., and Alkon, D. L. (1979). Ultrastructure of photo-receptors in the eye of *Hermissenda* labelled with intracellular injection of horseradish peroxidase. *J. Neurocytol.* **8**, 181–95.

Cull-Candy, S. G. and Usowicz, M. M. (1987). Multiple-conductance channels activated by excitatory amino acids in cerebellar neurons. *Nature*, **325**, 525–8.

Cummings, J. L. (1985). Hemispheric asymmetries in visual-perceptual and visual-spatial function. In *The dual brain* (ed. D. F. Benson and E. Zaidel), pp. 233–46. Guilford, New York.

Dagan, D. and Levitan, I. B. (1981). Isolated identified *Aplysia* neurons in cell culture. *J. Neurosci.* **1**, 736–40.

Dale, H. H. (1953). *Adventures in physiology.* Pergamon Press, London.

Dale, N., Kandel, E. R. and Schacher, S. (1987). Serotonin produces long-term changes in the excitability of *Aplysia* sensory neurons in culture that depend on new protein synthesis. *J. Neurosci.* **7**, 2232–8.

Damasio, A. R. (1985). Prosopagnosia. *Trends in Neurosci.* **8**, 132–5.

Damasio, A. R. and Geschwind, N. (1984). The neural basis of language. *Ann. Rev. Neurosci.* **7**, 127–47.

Darnell, J., Lodish, H., and Baltimore, D. (1986). *Molecular cell biology.* Scientific American Books, New York.

Darwin, C. (1871). *The descent of man, and selection in relation to sex.* J. Murray, London. (Reprinted 1981, Princeton University Press, Princeton.)

Darwin, C. (1872). *The expression of the emotions in man and animals.* Revised and abridged by C. M. Beadnell (1934). Watts, London.

Davidson, E. H. (1986). *Gene activity in early development* (3rd edn). Academic Press, New York.

Davies, D. C., Horn, G., and McCabe, B. J. (1985). Noradrenaline and learning: effects of the noradrenergic neurotoxin DSP4 on imprinting in the domestic chick. *Behav. Neurosci.* **99**, 652–60.

Davies, J. D. (1971). *Phrenology, fad and science.* Archon Books, Shoestring Publishing, New Haven, CT.

Davis, H. P. and Squire, L. R. (1984). Protein synthesis and memory: a review. *Psychol. Bull.* **96**, 518–59.

Davis, L., Banker, G. A., and Steward, O. (1987). Selective dendritic transport of RNA in hippocampal neurons in culture. *Nature*, **330**, 477–9.

Davis, R. L. and Davidson, N. (1986). The memory gene *dunce*[+] encodes a remarkable set of RNAs with internal heterogeneity. *Mol. Cell. Biol.* **6**, 1464–70.

Davis, R. L. and Kauvar, L. M. (1984). *Drosophila* cyclic nucleotide phosphodiesterase. *Adv. Cyc. Nucl. Prot. Phosphoryl. Res.* **16**, 393–402.

Davis, R. L. and Kiger, J. A. (1981). *dunce* mutants of *Drosophila melanogaster*: mutants defective in the cyclic AMP phosphodiesterase enzyme system. *J. Cell Biol.* **90**, 101–7.

Davis, W. J., Villet, J., Lee, D., Rigler, M., Gillette, R., and Prince, E. (1980). Selective and differential avoidance learning in the feeding and withdrawal behavior of *Pleurobranchaea californica. J. Comp. Physiol.* **138**, 157–65.

De Graan, P. N. E., Oestreicher, A. B., Schrama, L. H., and Gispen, W. H. (1986). Phosphoprotein B-50: localization and function. *Prog. Brain Res.* **69**, 37–50.

Dehaene, S., Changeux, J. P., and Nadal, J. P. (1987). Neural networks that learn temporal sequences by selection. *Proc. Natl. Acad. Sci. USA*, **84**, 2727–31.

de Jonge, M. and Racine, R. J. (1985). The effects of repeated induction of long-term potentiation in the dentate gyrus. *Brain Res.* **328**, 181–5.

Delacour, J. (1977). Cortex inferotemporal et mémoire visuelle a court terme chez le singe. Nouvelles données. *Exp. Brain Res.* **28**, 301–10.

del Castillo, J. and Katz, B. (1954). Quantal components of the end-plate potential. *J. Physiol.* **124**, 560–73.

De Renzi, E. (1982). Memory disorders following focal neocortical damage. *Phil. Trans. R. Soc. Lond. B*, **298**, 73–83.

Descartes, R. (1649). Les passions de l'âme (trans. R. Stoothoff). In *The philosophical writings of Descartes*, Vol. I. Cambridge University Press, Cambridge (1985).

Descartes, R. (1664). L'homme (trans. R. Stoothoff). In *The philosophical writings of Descartes*, Vol. I. Cambridge University Press, Cambridge (1985).

Desimone, R. and Gross, C. G. (1979). Visual areas in the temporal cortex of the macaque. *Brain Res.* **178**, 363–80.

Desmond, J. E. and Moore, J. W. (1986). Dorsolateral pontine tegmentum and the classically conditioned nictitating membrane response: analysis of CR-related single-unit activity. *Exp. Brain Res.* **65**, 59–74.

Desmond, N. L. and Levy, W. B. (1983). Synaptic correlates of associative potentiation/depression: an ultrastructural study in the hippocampus. *Brain Res.* **265**, 21–30.

De Wied, D. and Gispen, W. H. (1977). Behavioral effects of peptides. In *Peptides in neurobiology* (ed. H. Gainer), pp. 397–448. Plenum Press, New York.

DeYoe, E. A. and Van Essen, D. C. (1988). Concurrent processing streams in monkey visual cortex. *Trends in Neurosci.* **11**, 219–26.

Dingledine, R. (1983). *N*-methyl aspartate activates voltage-dependent calcium conductance in rat hippocampal pyramidal cells. *J. Physiol.* **343**, 385–405.

Dingman, W. and Sporn, M. B. (1961). The incorporation of 8-azaguanine into rat brain RNA and its effects on maze learning by the rat: an inquiry into the chemical basis of memory. *J. Psychiat. Res.* **1**, 1–11.

Disterhoft, J. F. and Olds, J. (1972). Differential development of conditioned unit changes in thalamus and cortex of rat. *J. Neurophysiol.* **35**, 665–79.

Disterhoft, J. F. and Segal, M. (1978). Neuron activity in rat hippocampus and motor cortex during discrimination reversal. *Brain Res. Bull.* **3**, 583–8.

Disterhoft, J. F. and Stuart, D. K. (1976). Trial sequence of changed unit activity in auditory system of alert rat during conditioned response acquisition and extinction. *J. Neurophysiol.* **39**, 266–81.

Disterhoft, J. F., Kwan, H. H., and Lo, W. D. (1977). Nictitating membrane conditioning to tone in the immobilized albino rabbit. *Brain Res.* **137**, 127–43.

Disterhoft, J. F., Quinn, K. J., Weiss, C., and Shipley, M. T. (1985). Accessory abducens nucleus and conditioned eye retraction/nictitating membrane extension in rabbit. *J. Neurosci.* **5**, 941–50.

Disterhoft, J. F., Coulter, D. A., and Alkon, D. L. (1986). Conditioning-specific membrane changes of rabbit hippocampal neurons measured *in vitro*. *Proc. Natl. Acad. Sci. USA*, **83**, 2733–7.

Dolphin, A. C. (1985). Long-term potentiation at peripheral synapses. *Trends in Neurosci.* **8**, 376–8.

Dolphin, A. C., Errington, M. L., and Bliss, T. V. P. (1982). Long-term potentiation of the perforant path *in vivo* is associated with increased glutamate release. *Nature*, **297**, 496–8.

Domjan, M. (1983). Biological constraints on instrumental and classical conditioning: implications for general process theory. In *The psychology of learning and motivation*, Vol. 17 (ed. G. H. Bower), pp. 215–77. Academic Press, New York.

Donato, R. (1986). S-100 proteins. *Cell Calc.* **7**, 123–45.

Donegan, N. H., Gluck, M. A., and Thompson, R. F. (1988). Integrating behavioural and biological models of classical conditioning. In: *Computational models of learning in simple neural systems* (ed. R. D. Hawkins and G. H. Bower), Academic Press, New York (in press).

Dragunow, M. and Robertson, H. A. (1987). Kindling stimulation induces *c-fos* protein(s) in granule cells of the rat dentate gyrus. *Nature*, **329**, 441–2.

Dretske, F. I. (1983). Precis of knowledge and the flow of information. (With peer review comments.) *Behav. Brain Sci.* **6**, 55–90.

Dudai, Y. (1977). Properties of learning and memory in *Drosophila melanogaster*. *J. Comp. Physiol.* **114**, 69–89.

Dudai, Y. (1979). Behavioral plasticity in a *Drosophila* mutant, *dnc*[DB276]. *J. Comp. Physiol.* **114**, 69–89.

Dudai, Y. (1983). Mutations affect storage and use of memory differentially in *Drosophila*. *Proc. Natl. Acad. Sci. USA*, **80**, 5445–8.

Dudai, Y. (1985). Some properties of adenylate cyclase which might be important for learning. *FEBS Lett.* **191**, 165–70.

Dudai, Y. (1987). The cAMP cascade in the nervous system: molecular sites of action and possible relevance to neuronal plasticity. *CRC Crit. Rev. Biochem.* **22**, 221–81.

Dudai, Y. (1988*a*). Genetic dissection of learning and short-term memory in *Drosophila. Ann. Rev. Neurosci.* **11**, 537–63.

Dudai, Y. (1988*b*). Molecular dissection of complex behaviors: elementary mechanistic rules in search of content. In *From neuron to reading* (ed. A. Galaburda). MIT Press, Cambridge, Mass. (in press).

Dudai, Y. and Zvi, S. (1984). Adenylate cyclase in the *Drosophila* memory mutant *rutabaga* displays an altered Ca^{2+} sensitivity. *Neurosci. Lett.* **47**, 119–24.

Dudai, Y., Jan, Y.-N., Byers, D., Quinn, W. G., and Benzer, S. (1976). *dunce*, a mutant of *Drosophila* deficient in learning. *Proc. Natl. Acad. Sci. USA*, **73**, 1684–8.

Dudai, Y., Uzzan, A., and Zvi, S. (1983). Abnormal activity of adenylate cyclase in the *Drosophila* memory mutant *rutabaga. Neurosci. Lett.* **42**, 207–12.

Dudai, Y., Zvi, S., and Segel, S. (1984). A defective conditioned behavior and a defective adenylate cyclase in the *Drosophila* mutant *rutabaga. J. Comp. Physiol.* **155**, 569–76.

Dudai, Y., Sher, B., Segal, D., and Yovell, Y. (1985). Defective responsiveness of adenylate cyclase to forskolin in the *Drosophila* mutant *rut. J. Neurogenet.* **2**, 365–80.

Dudai, Y., Buxbaum, J., Corfas, G., Orgad, S., Segal, D., Sher, B., Uzzan, A., and Zvi, S. (1986). Defective cAMP metabolism and defective memory in *Drosophila. Acta Biochim. Biophys. Hung.* **21**, 177–92.

Dudai, Y. (Rapporteur), Amari, S. I., Bienestock, E., Dehaene, S., Fuster, J., Goddard, G. V., Konishi, M., Menzel, R., Mishkin, M., Muller, C. M., Rolls, E. T., Schwegler, H. H., and von der Malsburg, C. (1987). On neuronal assemblies and memories. In *The neuronal and molecular bases of learning* (ed. J.-P. Changeux and M. Konishi), pp. 399–410. Wiley, New York.

Dudai, Y., Corfas, G., and Hazvi, S. (1988). What is the possible contribution of Ca^{2+}-calmodulin stimulated adenylate cyclase to acquisition, consolidation and retention of an associative olfactory memory in *Drosophila. J. Comp. Physiol.* **162**, 101–9.

Duerr, J. S. and Quinn, W. G. (1982). Three *Drosophila* mutations that block associative learning also affect habituation and sensitization. *Proc. Natl. Acad. Sci. USA*, **79**, 3646–50.

Duffy, C. J., Teyler, T. J., and Shashoua, V. E. (1981). Long-term potentiation in the hippocampal slice: evidence for stimulated secretion of newly synthesized proteins. *Science*, **212**, 1148–51.

Duffy, F. H. (1987). Brain electrical activity mapping. In *Encyclopedia of Neuroscience*, Vol. I (ed. G. Adelman), pp. 162–5. Birkhauser, Boston.

Dumuis, A., Sebben, M., Haynes, L., Pin, J.-P., and Bockaert, J. (1988). NMDA receptors activate the arachidonic acid cascade system in striatal neurons. *Nature*, **336**, 68–70.

Dunlap, K., Holz, G. G., and Rane, S. G. (1987). G proteins as regulators of ion channel function. *Trends in Neurosci.* **10**, 241–4.

Dunwiddie, T. and Lynch, G. (1978). Long-term potentiation and depression of synaptic responses in the rat hippocampus: localization and frequency dependency. *J. Physiol.* **276**, 353–67.

Dunwiddie, T. V. and Lynch, G. (1979). The relationship between extracellular calcium concentrations and the induction of hippocampal long-term potentiation. *Brain Res.* **169**, 103–10.

Dunwiddie, T., Madison, D., and Lynch, G. (1978). Synaptic transmission is required for initiation of long-term potentiation. *Brain Res.* **150**, 413–17.

Ebbinghaus, H. (1885). *Ueber das Gedachtnis.* Translated into English as *Memory, a contribution to experimental psychology* (1913). (Republished (1964), Dover, New York.)

Eccles, J. C. (1977). An instruction-selection theory of learning in the cerebellar cortex. *Brain Res.* **127**, 327–52.

Eccles, J. C., Ito, M., and Szentagothai, J. (1967). *The cerebellum as a neuronal machine.* Springer-Verlag, Berlin.

Edelman, G. M. (1984). Expression of cell adhesion molecules during embryogenesis and regeneration. *Exp. Cell Res.* **161**, 1–16.

Edelman, G. M. (1987). *Neural Darwinism. The theory of neuronal group selection.* Basic Books, New York.

Eichenbaum, H., Fagan, A., and Cohen, N. J. (1986). Normal olfactory discrimination learning set and facilitation of reversal learning after medial-temporal damage in rats: implications for an account of preserved learning abilities in amnesia. *J. Neurosci.* **6**, 1876–84.

Eichenbaum, H., Kuperstein, M., Fagan, A., and Nagode, J. (1987). Cue-sampling and goal-approach correlates of hippocampal unit activity in rats performing an odor-discrimination task. *J. Neurosci.* **7**, 716–32.

Eliot, L., Dudai, Y., Kandel, E. R., and Abrams, T. W. (1986). Activation of adenylate cyclase in *Aplysia* by Ca^{2+}/calmodulin: a possible molecular site of stimulus convergence in associative conditioning. *Soc. Neurosci. Abst.* **12**, 400.

Emson, P. C. and Lindvall, O. (1986). Neuroanatomical aspects of neurotransmitters affected in Alzheimer's disease. *Brit. Med. Bull.* **42**, 57–62.

Erickson, R. C. and Scott, M. L. (1977). Clinical memory testing: a review. *Psychol. Bull.* **84**, 1130–49.

Errington, M. L., Lynch, M. A., and Bliss, T. V. P. (1987). Long-term potentiation in the dentate gyrus: induction and increased glutamate release are blocked by D(−)aminophosphonovalerate. *Neurosci.* **20**, 279–84.

Estes, W. K. and Skinner, B. F. (1941). Some quantitative properties of anxiety. *J. Exp. Psychol.* **29**, 390–400.

Ewert, J. P. (1980). *Neuroethology.* Springer-Verlag, Berlin.

Fagan, J. F. (1985). A new look at infant intelligence. In *Current topics in human intelligence*, Vol. 1, Research methods (ed. D. K. Detterman), pp. 223–46. Ablex, Norwood, NJ.

Fantz, R. L. (1964). Visual experience in infants: decreased attention to familiar patterns relative to novel ones. *Science*, **146**, 668–70.

Farley, J. (1986). Cellular mechanisms of causal detection in a mollusc. In *Neural mechanisms of conditioning* (ed. D. L. Alkon and C. D. Woody), pp. 19–54. Plenum Press, New York.

Farley, J. and Alkon, D. L. (1980). Neural organization predicts stimulus specificity for a retained associative behavioral change. *Science*, **210**, 1373–5.

Farley, J. and Alkon, D. L. (1982). Associative neural and behavioural change in *Hermissenda*: consequences of nervous system orientation for light and pairing specificity. *J. Neurophysiol.* **48**, 785–807.

Farley, J. and Auerbach, S. (1986). Protein kinase C activation induces conductance changes in *Hermissenda* photoreceptors like those seen in associative learning. *Nature*, **319**, 220–3.

Farley, J., Richards, W. G., Ling, L. J., Liman, E., and Alkon, D. L. (1983). Membrane changes in a single photoreceptor cause associative learning in *Hermissenda*. *Science*, **221**, 1201–3.

Farley, J. M., Sakakibara, M., and Alkon, D. (1984). Associative-training correlated changes in I_{Ca-K} in *Hermissenda* type B photoreceptors. *Soc. Neurosci. Abst.* **10**, 270.

Feldberg, W. (1977). The early history of synaptic and neuromuscular transmission by acetylcholine: reminiscences of an eye witness. In *The pursuit of Nature. Informal essays on the history of physiology,* pp. 65–83. Cambridge University Press.

Fibiger, H. C. (1982). The organization and some projections of cholinergic neurons of the mammalian forebrain. *Brain Res. Rev.* **4**, 327–88.

Fifkova, E. (1985). A possible mechanism of morphometric changes in dendritic spines induced by stimulation. *Cell. Mol. Neurobiol.* **5**, 47–63.

Fifkova, E. and Anderson, C. L. (1981). Stimulation-induced changes in dimensions of stalks of dendritic spines in the dentate molecular layer. *Exp. Neurol.* **74**, 621–7.

Fifkova, E. and Van Harreveld, A. (1977). Long-lasting morphological changes in dendritic spines of dentate granular cells following stimulation of the entorhinal area. *J. Neurocytol.* **6**, 211–30.

Flaherty, C. F. (1985). *Animal learning and cognition.* Knopf, New York.

Flexner, J. B., Flexner, L. B., and Stellar, E. (1963). Memory in mice as affected by intracerebral puromycin. *Science*, **141**, 57–9.

Flexner, L. B. and Goodman, R. H. (1975). Studies on memory: inhibitors of protein synthesis also inhibit catecholamine synthesis. *Proc. Natl. Acad. Sci. USA*, **72**, 4660–3.

Flood, J. F., Smith, G. E., and Cherkin, A. (1985). Memory enhancement: supra-additive effect of subcutaneous cholinergic drug combinations in mice. *Psychopharmacol.* **86**, 61–7.

Fodor, J. A. (1975). *The language of thought.* T. Y. Crowell, New York.

Fodor, J. A. (1981). *Representations.* MIT Press, Cambridge, Mass.

Fodor, J. A. (1983). *The modularity of mind.* MIT Press, Cambridge, Mass.

Folkers, E. (1982). Visual learning and memory of *Drosophila melanogaster* wild type C-S and the mutants *dunce*[1], *amnesiac*, *turnip* and *rutabaga*. *J. Insect Physiol.* **28**, 535–9.

Folkers, E. and Spatz, H. Ch. (1984). Visual learning performance of *Drosophila melanogaster* is altered by neuropharmaca affecting phosphodiesterase activity and acetylcholine transmission. *J. Insect Physiol.* **30**, 957–65.

Forman, R. R. (1984). Leg position learning by an insect. I. A heat avoidance learning paradigm. *J. Neurobiol.* **15**, 127–40.

Forrester, J. and Crow, T. (1987). Short-term and long-term changes in identified B-photoreceptors of *Hermissenda* produced by light and serotonin. *Soc. Neurosci. Abst.* **13**, 618.

Foster, A. C. and Fagg, G. E. (1984). Acidic amino acid binding sites in mammalian neuronal membranes: their characteristics and relationship to synaptic receptors. *Brain Research Rev.* **7**, 103–64.

Foster, A. C. and Fagg, G. E. (1987). Taking apart NMDA receptors. *Nature*, **329**, 395–6.

Franz, S. I. (1912). New phrenology. *Science*, **35**, 321–8.

Frazier, W. T., Kandel, E. R., Kupfermann, I., Waziri, R., and Coggeshall, R. E. (1967). Morphological and functional properties of identified neurons in the abdominal ganglion of *Aplysia californica*. *J. Neurophysiol.* **30**, 1288–351.

Fregnac, Y. and Imbert, M. (1984). Development of neuronal selectivity in primary visual cortex of cat. *Physiol. Rev.* **64**, 325–434.

Fregnac, Y., Shulz, D., Thorpe, S., and Bienenstock, E. (1988). A cellular analogue of visual cortical plasticity. *Nature*, **333**, 367–70.

Frieder, B. and Allweis, C. (1982). Memory consolidation: further evidence for the four-phase model from the time-courses of diethyldithiocarbamate and ethacrinic acid amnesias. *Physiol. Behav.* **29**, 1071–5.

Friedman, H. R. and Goldman-Rakic, P. S. (1988). Activation of the hippocampus and dentate gyrus by working memory: a 2-deoxyglucose study of behaving rhesus monkeys. *J. Neurosci.* **8**, 4693–706.

Frost, W. N. and Kandel, E. R. (1984). Sensitizing stimuli reduce the effectiveness of the L30 inhibitory interneurons in the siphon withdrawal reflex circuit of *Aplysia. Soc. Neurosci. Abst.* **10**, 510.

Frost, W. N., Castellucci, V. F., Hawkins, R. D., and Kandel, E. R. (1985*a*). Monosynaptic connections made by the sensory neurons of the gill- and siphon-withdrawal reflex in *Aplysia* participate in the storage of long-term memory for sensitization. *Proc. Natl. Acad. Sci. USA*, **82**, 8266–9.

Frost, W. N., Clark, G. A., and Kandel, E. R. (1985*b*). Changes in cellular excitability in a new class of siphon motor neurons during sensitization in *Aplysia. Soc. Neurosci. Abst.* **11**, 643.

Frost, W. N., Clark, G. A., and Kandel, E. R. (1988). Parallel processing of short-term memory for sensitization in *Aplysia. J. Neurobiol.* **19**, 297–334.

Fuster, J. M. (1973). Unit activity in prefrontal cortex during delayed-response performance: neuronal correlates of transient memory. *J. Neurophysiol.* **36**, 61–78.

Fuster, J. M. (1980). *The prefrontal cortex. Anatomy, physiology and neuropsychology of the frontal lobe.* Raven Press, New York.

Fuster, J. M. and Alexander, G. E. (1971). Neuron activity related to short-term memory. *Science*, **173**, 652–4.

Fuster, J. M. and Jervey, J. P. (1981). Inferotemporal neurons distinguish and retain behaviorally relevant features of visual stimuli. *Science*, **212**, 952–5.

Fuster, J. M. and Jervey, J. P. (1982). Neuronal firing in the inferotemporal cortex of the monkey in a visual memory task. *J. Neurosci.* **2**, 361–75.

Fuster, J. M., Bauer, R. H., and Jervey, J. P. (1981). Effects of cooling inferotemporal cortex on performance of visual memory tasks. *Exp. Neurol.* **71**, 398–409.

Fuster, J. M., Bauer, R. H., and Jervey, J. P. (1985). Functional interactions between inferotemporal and prefrontal cortex in a cognitive task. *Brain Res.* **330**, 299–307.

Gaffan, D. (1974). Recognition impaired and association intact in the memory of monkeys after transection of the fornix. *J. Comp. Physiol. Psychol.* **86**, 1100–9.

Gaffan, D. and Harrison, S. (1987). Amygdalectomy and disconnection in visual learning for auditory secondary reinforcement by monkeys. *J. Neurosci.* **7**, 2285–92.

Gaffan, E. A., Gaffan, D., and Harrison, S. (1988). Disconnection of the amygdala from visual association cortex impairs visual reward-association learning in monkeys. *J. Neurosci.* **8**, 3144–50.

Gailey, D. A., Jackson, F. R., and Siegel, R. W. (1982). Male courtship in *Drosophila*: the conditioned response to immature males and its genetic control. *Genetics*, **102**, 771–82.

Garcia, J. and Koelling, R. A. (1966). Relation of cue to consequence in avoidance learning. *Psychon. Sci.* **4**, 123–4.

Garcia, J., Ervin, F. R., and Koelling, R. A. (1966). Learning with prolonged delay of reinforcement. *Psychon. Sci.* **5**, 121–2.

Gardner, H. (1983). *Frames of mind: the theory of multiple intelligences.* Basic Books, New York.

Garthwaite, J., Charles, S. L., and Chess-Williams, R. (1988). Endothelium-derived relaxing factor release on activation of NMDA receptors suggests role as intercellular messenger in the brain. *Nature*, **336**, 385–8.

Gellman, R. S. and Miles, F. A. (1985). A new role for the cerebellum in conditioning? *Trends in Neurosci.* **8**, 181–2.

Gelperin, A. (1975). Rapid food-aversion learning by a terrestrial mollusc. *Science*, **189**, 567–70.

Gelperin, A., Hopfield, J. J., and Tank, D. W. (1985). The logic of *Limax* learning. In *Model neural networks and behavior* (ed. A. I. Selverston), pp. 237–61. Plenum Press, New York.

Georgopoulos, A. P., Kettner, R. E., and Schwartz, A. B. (1988). Primate motor cortex and free arm movements to visual targets in three-dimensional space. II. Coding of the direction of movement by a neuronal population. *J. Neurosci.* **8**, 2928–37.

Geschwind, N. (1965*a*). Disconnexion syndromes in animals and man. Part I. *Brain*, **88**, 237–94.

Geschwind, N. (1965*b*). Disconnexion syndromes in animals and man. Part II. *Brain*, **88**, 585–644.

Geschwind, N. and Galaburda, A. M. (ed.) (1984). *Cerebral dominance. The biological foundation.* Harvard University Press, Cambridge, Mass.

Geschwind, N. and Galaburda, A. (1985). Cerebral lateralization. Biological mechanisms, associations and pathology: I. A hypothesis and a program for research. *Arch. Neurol.* **42**, 428–59.

Ghez, C. and Fahn, S. (1985). The cerebellum. In *Principles of neural science* (2nd edn) (ed. E. R. Kandel and J. H. Schwartz), pp. 502–22. Elsevier, New York.

Gibbs, C. M., Cohen, D. H., and Broyles, J. L. (1986). Modification of the discharge of lateral geniculate neurons during visual conditioning. *J. Neurosci.* **6**, 627–36.

Gibbs, M. E. and Ng, K. T. (1977). Psychobiology of memory: towards a model of memory formation. *Biobehav. Rev.* **1**, 113–36.

Gilbert, P. F. C. and Thach, W. T. (1977). Purkinje cell activity during motor learning. *Brain Res.* **128**, 309–28.

Gingrich, K. J. and Byrne, J. H. (1985). Simulation of synaptic depression, posttetanic potentiation, and presynaptic facilitation of synaptic potentials from sensory neurons mediating gill-withdrawal reflex in *Aplysia. J. Neurophysiol.* **53**, 652–69.

Glass, A. L. and Holyoak, K. J. (1986). *Cognition* (2nd edn). Random House, New York.

Glaser, E. M. (1966). *The physiological basis of habituation.* Oxford University Press, London.

Glisky, E. L., Schacter, D. L., and Tulving, E. (1986). Computer learning by memory-impaired patients: acquisition and retention of complex knowledge. *Neuropsychologia*, **24**, 313–28.

Gloor, P., Oliver, A., Quesney, L. F., Andermann, F., and Horowitz, S. (1982). The role of the limbic system in experiential phenomena of temporal lobe epilepsy. *Ann. Neurol.* **12**, 129–44.

Goelet, P., Castellucci, V. F., Schacher, S., and Kandel, E. R. (1986). The long and the short of long-term memory—a molecular framework. *Nature*, **322**, 419–22.

Goh, Y. and Alkon, D. L. (1984). Sensory, interneuronal, and motor interactions within *Hermissenda* visual pathway. *J. Neurophysiol.* **52**, 156–69.

Gold, M. R. and Cohen, D. H. (1981). Modification of the discharge of vagal cardiac neurons during learned heart rate change. *Science*, **214**, 345–7.

Gold, M. R. and Cohen, D. H. (1984). The discharge characteristics of vagal cardiac neurons during classically conditioned heart rate change. *J. Neurosci.* **4**, 2963–71.

Goldenring, J. R., Wasterlain, C. G., Oestreicher, A. B., de Graan, P. N. E., Farber, D. B., Glaser, G., and DeLorenzo, R. J. (1986). Kindling induces a long-lasting change in the activity of a hippocampal membrane calmodulin-dependent protein kinase system. *Brain Res.* **377**, 47–53.

Goldman, S. A. and Nottebohm, F. (1983). Neuronal production, migration, and differentiation in a vocal control nucleus of the adult female canary brain. *Proc. Natl. Acad. Sci. USA*, **80**, 2390–4.

Goldman-Rakic, P. S. (1987). Circuitry of primate prefrontal cortex and regulation of behavior by representational knowledge. In *Handbook of physiology*, Section 1: The nervous system, Vol. V: Higher cortical function, Part 1 (ed. F. Plum), pp. 373–417. American Physiological Society,

Bethesda, Md.

Goldman-Rakic, P. S. (1988). Topography of cognition: parallel distributed networks in primate association cortex. *Ann. Rev. Neurosci.* **11**, 137–56.

Goldman-Rakic, P. S., Selemon, L. D., and Schwartz, M. L. (1984). Dual pathways connecting the dorsolateral prefrontal cortex with the hippocampal formation and parahippocampal cortex in the rhesus monkey. *Neurosci.* **12**, 719–43.

Gonshor, A. and Melvill Jones, G. (1976). Extreme vestibulo–ocular adaptation induced by prolonged optical reversal of vision. *J. Physiol.* **256**, 381–414.

Gormezano, I., Schneiderman, N., Deaux, E., and Fuentes, I. (1962). Nictitating membrane: classical conditioning and extinction in the albino rabbit. *Science*, **138**, 33–4.

Gould, J. L. (1984). Natural history of honey bee learning. In *The biology of learning* (ed. P. Marler and H. S. Terrace), pp. 149–80. Springer-Verlag, Berlin.

Graf, P., Squire, L. R., and Mandler, G. (1984). The information that amnesic patients do not forget. *J. Exp. Psychol. LMC*, **10**, 164–78.

Gray, R. and Johnston, D. (1987). Noradrenaline and β-receptor agonists increase activity of voltage dependent calcium channels in hippocampal neurons. *Nature*, **327**, 620–2.

Graziadei, P. P. C. and Monti Graziadei, G. A. (1979). Neurogenesis and neuron regeneration in th olfactory system of mammals. I. Morphological aspects of differentiation and structural organiza tion of the olfactory sensory neurons. *J. Neurocytol.* **8**, 1–18.

Greenberg, M. E., Greene, L. A., and Ziff, E. B. (1985). Nerve growth factor and epidermal growtl factor induce rapid transient changes in proto-oncogene transcription in PC12 cells. *J. Biol. Chem* **260**, 14101–10.

Greenberg, M. E., Hermanowski, A. L., and Ziff, E. B. (1986*a*). Effect of protein synthesis inhibitor on growth factor activation of *c-fos*, *c-myc*, and actin gene transcription. *Mol. Cell. Biol.* **6**, 1050–7.

Greenberg, M. E., Ziff, E. B., and Greene, L. A. (1986*b*). Stimulation of neuronal acetylcholine receptors induces rapid gene transcription. *Science*, **234**, 80–3.

Greenberg, S. M., Castellucci, V. F., Bayley, H., and Schwartz, J. H. (1987). A molecular mechanism for long-term sensitization in *Aplysia. Nature*, **329**, 62–5.

Greenough, W. T. (1984). Structural correlates of information storage in the mammalian brain: a review and hypothesis. *Trends in Neurosci.* **7**, 229–33.

Grinvald, A. (1985). Real-time optical mapping of neuronal activity: from single growth cones to the intact mammalian brain. *Ann. Rev. Neurosci.* **8**, 263–305.

Grinvald, A., Lieke, E., Frostig, R. D., Gilbert, C. D., and Wiesel, T. N. (1986). Functional architectur of cortex revealed by optical imaging of intrinsic signals. *Nature*, **324**, 361–4.

Gross, C. G., Bender, D. B., and Rocha-Miranda, C. E. (1969). Visual receptive fields of neurons ii inferotemporal cortex of the monkey. *Science*, **166**, 1303–6.

Gross, C. G., Rocha-Miranda, C. E., and Bender, D. B. (1972). Visual properties of neurons ii inferotemporal cortex of the macaque. *J. Neurophysiol.* **35**, 96–111.

Gross, C. G., Bender, D. B., and Gerstein, G. L. (1979). Activity of inferior temporal neurons ir behaving monkeys. *Neuropsychologia*, **17**, 215–29.

Groves, P. M. and Thompson, R. F. (1970). Habituation: a dual-process theory. *Psychol. Rev.* **77**, 419–50.

Gurney, M. E. (1981). Hormonal control of cell form and number in the Zebra finch song system. *J. Neurosci.* **1**, 658–73.

Gurney, M. E. (1982). Behavioral correlates of sexual differentiation in the Zebra finch song system. *Brain Res.* **231**, 153–72.

Gurney, M. E. and Konishi, M. (1980). Hormone-induced sexual differentiation of brain and behavior in Zebra finches. *Science*, **208**, 1380–3.

Gustafsson, B. and Wigstrom, H. (1986). Hippocampal long-lasting potentiation produced by pairing single volleys and brief conditioning tetani evoked in separate afferents. *J. Neurosci.* **6**, 1575–82.

Gustafsson, B., Wigstrom, H., Abraham, W. C., and Huang, Y. Y. (1987). Long-term potentiation in the hippocampus using depolarizing current pulses as the conditioning stimulus to single volley synaptic potentials. *J. Neurosci.* **7**, 774–80.

Hagan, J. J. and Morris, R. G. M. (1988). The cholinergic hypothesis of memory: a review of animal experiments. In *Psychopharmacology of the aging nervous system*, Vol. 20 (ed. L. L. Iversen, S. D. Iversen, and S. H. Snyder), pp. 237–323. Plenum Press, New York.

Halazonetis, T. D., Georgopoulos, K., Greenberg, N. E., and Leder, P. (1988). c-Jun dimerizes with itself and with c-fos, forming complexes of different DNA binding affinities. *Cell*, **55**, 917–24.

Halgren, E., Walter, R. D., Cherlow, D. G., and Crandall, P. H. (1978). Mental phenomena evoked by electrical stimulation of the human hippocampal formation and amygdala. *Brain*, **101**, 83–117.

Hall, J. (1986). Learning and rhythms in courting, mutant *Drosophila. Trends in Neurosci.* **9**, 414–18.

Harlow, H. F. (1949). The formation of learning sets. *Psychol. Rev.* **56**, 51–65.

Harlow, H. F. and Bromer, J. A. (1938). A test apparatus for monkeys. *Psychol. Rec.* **2**, 434–6.

Harris, D. A., Falls, D. L., Dill-Devor, R. M., and Fischbach, G. D. (1988). Acetylcholine receptor-inducing factor from chicken brain increases the level of mRNA encoding the receptor α subunit. *Proc. Natl. Acad. Sci. USA*, **85**, 1983–7.

Harris, E. W. and Cotman, C. W. (1986). Long-term potentiation of guinea pig mossy-fiber responses is not blocked by N-methyl-D-aspartate antagonists. *Neurosci. Lett.* **70**, 132–7.

Harris, E. W., Ganong, A. H., and Cotman, C. W. (1984). Long-term potentiation in the hippocampus involves activation of N-methyl-D-aspartate receptors. *Brain Res.* **323**, 132–7.

Harris, J. D. (1943). Habituatory response decrement in the intact organism. *Psychol. Bull.* **40**, 385–422.

Hashimoto, Y., Schworer, C. M., Colbran, R. J., and Soderling, T. R. (1987). Autophosphorylation of Ca^{2+}/calmodulin-dependent protein kinase. Effects on total and Ca^{2+}-independent activities and kinetic parameters. *J. Biol. Chem.* **262**, 8051–5.

Hawkins, R. D. and Abrams, T. W. (1984). Evidence that activity-dependent facilitation underlying classical conditioning in *Aplysia* involves modulation of the same ionic current as normal presynaptic facilitation. *Soc. Neurosci. Abst.* **10**, 268.

Hawkins, R. D., Castellucci, V. F., and Kandel, E. R. (1981). Interneurons involved in mediation and modulation of gill-withdrawal reflex in *Aplysia*. I. Identification and characterization. *J. Neurophysiol.* **45**, 304–14.

Hawkins, R. D., Carew, T. J., and Kandel, E. R. (1983*a*). Effects of interstimulus interval and contingency on classical conditioning in *Aplysia. Soc. Neurosci. Abst.* **9**, 168.

Hawkins, R. D., Abrams, T. W., Carew, T. J., and Kandel, E. R. (1983*b*). A cellular mechanism of classical conditioning in *Aplysia*: activity-dependent amplification of presynaptic facilitation. *Science*, **219**, 400–5.

Heathcote, R. D. and Sargent, P. B. (1985). Loss of supernumerary axons during neuronal morphogenesis. *J. Neurosci.* **5**, 1940–6.

Hebb, D. O. (1949). *The organization of behavior: a neuropsychological theory.* Wiley, New York.

Heidmann, A., Heidmann, T. M., and Changeux, J. P. (1984). Stabilisation sélective de représentations neuronals par résonance entre 'preprésentations' spontanées du reseau cérébral et 'percepts' evoqués par interaction avec le monde extérieur. *C.R. Acad. Sci. III,* **299**, 839–44.

Heidmann, T. and Changeux, J. P. (1982). Un modèle moléculaire de régulation d'efficacité au niveau postsynaptique d'un synapse chimique. *C.R. Acad. Sci. Paris,* **295**, 665–70.

Heiligenberg, W. and Bastian, J. (1984). The electric sense of weakly electric fish. *Ann. Rev. Physiol.* **46**, 561–83.

Heilman, K. M. and Valenstein, E. (1979). *Clinical neuropsychology.* Oxford University Press, New York.

Heisenberg, M., Borst, A., Wagner, S., and Byers, D. (1985). *Drosophila* mushroom body mutants are deficient in olfactory learning. *J. Neurogenet.* **2**, 1–30.

Heit, G., Smith, M. E., and Halgren, E. (1988). Neuronal encoding of individual words and faces by the human hippocampus and amygdala. *Nature,* **333**, 773–5.

Henderson, C. E. (1987). Activity and the regulation of neuronal growth factor metabolism. In *The neuronal and molecular bases of learning* (ed. J.-P. Changeux and M. Konishi), pp. 99–118. Wiley, New York.

Hess, E. H. (1959). Imprinting. *Science,* **130**, 133–41.

Hewitt, J. K., Fulker, D. W., and Hewitt, C. A. (1983). Genetic architecture of olfactory discriminative avoidance conditioning in *Drosophila melanogaster. J. Comp. Psychol.* **97**, 52–8.

Higashima, M. and Yamamoto, C. (1985). Two components of long-term potentiation in mossy fiber-induced excitation in hippocampus. *Exp. Neurol.* **90**, 529–39.

Hilgard, E. R. and Marquis, D. G. (1940). *Conditioning and learning.* Appleton-Century-Crofts, New York.

Hille, B. (1984). *Ionic channels of excitable membranes.* Sinauer, Sunderland, Mass.

Hillyard, S. A. (1985). Electrophysiology of human selective attention. *Trends in Neurosci.* **8**, 400–5.

Hinde, R. A. (1970). *Animal behavior* (2nd edn). McGraw-Hill, New York.

Hinton, G. E. and Anderson, J. A. (ed.) (1981). *Parallel models of associative memory.* Erlbaum, New Jersey.

Hirano, T., Woody, C., Birt, D., Aou, S., Miyake, J., and Nenov, V. (1987). Pavlovian conditioning of discriminatively elicited eyeblink responses with short onset latency attributable to lengthened interstimulus intervals. *Brain Res.* **400**, 171–5.

Hirsh, R. (1974). The hippocampus and contextual retrieval of information from memory: a theory. *Behav. Biol.* **12**, 421–44.

Hochner, B., Klein, M., Schacher, S., and Kandel, E. R. (1986*a*). Action-potential duration and modulation of transmitter release from the sensory neurons of *Aplysia* in presynaptic facilitation and behavioral sensitization. *Proc. Natl. Acad. Sci. USA,* **83**, 8410–4.

Hochner, B., Klein, M., Schacher, S., and Kandel, E. R. (1986*b*). Additional component in the cellular mechanism of presynaptic facilitation contributes to behavioral dishabituation in *Aplysia. Proc. Natl. Acad. Sci. USA,* **83**, 8794–8.

Hochner, B., Braha, O., Klein, M., and Kandel, E. R. (1986*c*). Distinct processes in presynaptic facilitation contribute to sensitization and dishabituation in *Aplysia*: possible involvement of C kinase in dishabituation. *Soc. Neurosci. Abst.* **12**, 1340.

Hodgkin, A. L. (1977). Chance and design in electrophysiology: an informal account of certain experiments on nerve carried out between 1934 and 1952. In *The pursuit of Nature. Informal essays on the history of physiology*, pp. 1–21, Cambridge University Press, Cambridge.

Hodgson, T. M. and Crow, T. (1987). Characterization of a subset of putative motor neurons in the pedal ganglia of *Hermissenda* whose light responses are due to synaptic input from photo-receptors. *Soc. Neurosci. Abst.* **13**, 618.

Hopkins, W. F. and Johnston, D. (1984). Frequency-dependent noradrenergic modulation of long-term potentiation in the hippocampus. *Science*, **226**, 350–2.

Horel, J. A. (1978). The neuroanatomy of amnesia. A critique of the hippocampal memory hypothesis. *Brain*, **101**, 403–45.

Horn, G. (1985). *Memory, imprinting, and the brain.* Oxford University Press, Oxford.

Horn, G. and McCabe, B. J. (1984). Predispositions and preferences. Effects on imprinting of lesions to the chick brain. *Anim. Behav.* **32**, 288–92.

Horn, G., Horn, A. L. D., Bateson, P. P. G., and Rose, S. P. R. (1971). Effects of imprinting on uracil incorporation into brain RNA in the 'split brain' chick. *Nature*, **229**, 131–2.

Horn, G., Rose, S. P. R., and Bateson, P. P. G. (1973). Monocular imprinting and regional incorporation of tritiated uracil into the brain of intact and 'split brain' chicks. *Brain Res.* **56**, 227–37.

Horn, G., McCabe, B. J., and Bateson, P. P. G. (1979). An autoradiographic study of the chick brain after imprinting. *Brain Res.* **168**, 361–73.

Horn, G., McCabe, B. J., and Cipolla-Neto, J. (1983). Imprinting in the domestic chick: the role of each side of the hyperstriatum ventrale in acquisition and retention. *Exp. Brain Res.* **53**, 91–8.

Horn, G., Bradley, P., and McCabe, B. J. (1985). Changes in the structure of synapses associated with learning. *J. Neurosci.* **5**, 3161–8.

Horridge, G. A. (1962). Learning of leg position by the ventral nerve cord in headless insects. *Proc. R. Soc. Lond. B. Biol. Sci.* **157**, 33–52.

Hoyle, G. (1980). Learning, using natural reinforcements, in insect preparations that permit cellular neuronal analysis. *J. Neurobiol.* **11**, 323–54.

Hoyle, G. (1984). The scope of neuroethology. *Behav. Brain Res.* **7**, 367–412.

Hu, G.-Y., Hvalby, O., Walaas, S. I., Albert, K. A., Skjeflo, P., Andersen, P., and Greengard, P. (1987). Protein kinase C injection into hippocampal pyramidal cells elicits features of long term potentiation. *Nature*, **328**, 426–9.

Huang, Y.-Y., Wigstrom, H., and Gustafsson, B. (1987). Facilitated induction of hippocampal long-term potentiation in slices perfused with low concentrations of magnesium. *Neurosci.* **22**, 9–16.

Hubel, D. H. and Livingstone, M. S. (1987). Segregation of form, color, and stereopsis in primate area 18. *J. Neurosci.* **7**, 3378–415.

Hubel, D. H., Wiesel, T. N., and LeVay, S. (1977). Plasticity of ocular dominance columns in monkey striate cortex. *Phil. Trans. R. Soc. Lond. B*, **278**, 377–409.

Hucho, F. (1986). *Neurochemistry. Foundations and concepts.* VCH, Weinheim, W. Germany.

Hunt, S. P., Pini, A., and Evan, G. (1987). Induction of *c-fos*-like protein in spinal cord neurons following sensory stimulation. *Nature*, **328**, 632–4.

Hunter, T. (1987). A thousand and one protein kinases. *Cell*, **50**, 823–9.

Hunter, W. S. (1913). The delayed reaction in animals and children. *Behav. Monog.* **2**(6), 1–86.

Hunter, W. S. (1930). A consideration of Lashley's theory of the equipotentiality of cerebral action. *J. Gen. Psychol.* **3**, 455–68.

Huppert, F. A. and Piercy, M. (1978). Dissociation between learning and remembering in organic amnesia. *Nature*, **275**, 317–18.

Huppert, F. A. and Piercy, M. (1979). The role of trace strength in recency and frequency judgments by amnesic and control subjects. *Q. J. Exp. Psychol.* **30**, 347–54.

Hyden, H. and Egyhazi, E. (1962). Nuclear RNA changes in nerve cells during a learning experiment in rats. *Proc. Natl. Acad. Sci. USA*, **48**, 1366–72.

Hyden, H. and Lange, P. (1965). A differentiation in RNA response in neurons early and late in learning. *Proc. Natl. Acad. Sci. USA*, **53**, 946–52.

Hyvärinen, J. (1982). Posterior parietal lobe of the primate brain. *Physiol. Rev.* **62**, 1060–129.

Imagawa, M., Chiu, R., and Karin, M. (1987). Transcription factor AP-2 mediates induction by two different signal-transduction pathways: protein kinase C and cAMP. *Cell*, **51**, 251–60.

Ito, M. (1972). Neural design of the cerebellar motor control system. *Brain Res.* **40**, 81–4.

Ito, M. (1984). *The cerebellum and neural control*. Raven Press, New York.

Ito, M. (1987). Characterization of synaptic plasticity in the cerebellar and cerebral neocortex. In *The neuronal and molecular bases of learning* (ed. J.-P. Changeux and M. Konishi), pp. 263–80. Wiley, New York.

Ito, M., Sakurai, M., and Tongroach, P. (1982). Climbing fibre induced depression of both mossy fibre responsiveness and glutamate sensitivity of cerebellar Purkinje cells. *J. Physiol.* **324**, 113–34.

Iversen, S. D. (1976). Do hippocampal lesions produce amnesia in animals? *Int. Rev. Neurobiol.* **19**, 1–49.

Iwai, E. (1985). Neuropsychological basis of pattern vision in macaque monkeys. *Vision Res.* **25**, 425–39.

Jackendoff, R. (1987). *Consciousness and the computational mind*. MIT Press, Cambridge, Mass.

Jacklet, J. W. and Rine, J. (1977). Facilitation at neuromuscular junctions: contribution to habituation and dishabituation of the *Aplysia* gill withdrawal reflex. *Proc. Natl. Acad. Sci. USA*, **74**, 1267–71.

Jacklet, J. W., Peretz, B., and Lukowiak, K. (1975). Habituation of the gill reflex in *Aplysia*: sites in the peripheral nervous system. *Fed. Proc.* **34**, 359.

Jackson, J. H. (1888). On a particular variety of epilepsy ('intelectual aura'), one case with symptoms of organic brain disease. *Brain*, **11**, 179–207.

Jacobsen, C. F. (1936). Studies of cerebral functions in primates. I. The function of the frontal association areas in monkeys. *Comp. Psychol. Monogr.* **13**, 1–60.

Jacobsen, C. F. and Nissen, H. W. (1937). Studies of cerebral function in primates. IV. The effects of frontal lobe lesions on the delayed alternation habit in monkeys. *J. Comp. Psychol.* **23**, 101–12.

Jacobson, R. D., Virag, I., and Skene, J. H. P. (1986). A protein associated with axon growth, GAP-43, is widely distributed and developmentally regulated in rat CNS. *J. Neurosci.* **6**, 1843–55.

Jahr, C. E. and Stevens, C. F. (1987). Glutamate activates multiple single channel conductances in hippocampal neurons. *Nature*, **325**, 522–5.

James, W. (1890). *Principles of psychology*. Published in an abridged version as *Psychology* (1892). (Republished (1948), World Publishing, Cleveland.)

Jarvik, M. E. (1953). Discrimination of colored food and food signs by primates. *J. Comp. Physiol. Psychol.* **46**, 390–2.

Jenkins, H. M. and Moore, B. R. (1973). The form of autoshaped response with food or water reinforcers. *J. Exp. Anal. Behav.* **20**, 163–81.

Jenkins, W. M. and Merzenich, M. M. (1987). Reorganization of neocortical representations after brain injury: a neurophysiological model of the bases of recovery from stroke. *Prog. Brain Res.* **71**,

249–66.

Jennings, H. S. (1906). *The behavior of lower organisms.* Columbia University Press, New York.

John, E. R. (1967). *Mechanisms of memory.* Academic Press, New York.

John, E. R. (1972). Switchboard versus statistical theories of learning and memory. *Science*, **177**, 850–64.

John, E. R., Tang, Y., Brill, A. B., Young, R., and Ono, K. (1986). Double-labeled metabolic maps of memory. *Science*, **233**, 1167–75.

Johns, P. R. (1977). Growth of the adult goldfish eye. III. Source of the new retinal cells. *J. Comp. Neurol.* **176**, 343–58.

Johnson, M. H. and Horn, G. (1986). Dissociation of recognition memory and associative learning by a restricted lesion of the chick forebrain. *Neuropsychologia*, **24**, 329–40.

Johnson, M. H. and Horn, G. (1987). The role of a restricted region of the chick forebrain in the recognition of individual conspecifics. *Behav. Brain Res.* **23**, 269–75.

Johnson, M. H., Bolhuis, J. J., and Horn, G. (1985). Interaction between acquired preferences and developing predispositions during imprinting. *Anim. Behav.* **33**, 1000–6.

Johnston, T. D. (1981). Contrasting approaches to a theory of learning. *Behav. Brain Sci.* **4**, 125–73.

Jones, B. and Mishkin, M. (1972). Limbic lesions and the problem of stimulus-reinforcement associations. *Exp. Neurol.* **36**, 362–77.

Jones, E. G. and Powell, T. P. S. (1970). An anatomical study of converging sensory pathways within the cerebral cortex of the monkey. *Brain*, **93**, 793–820.

Joseph, J. P. and Barone, P. (1987). Prefrontal unit activity during a delayed oculomotor task in the monkey. *Exp. Brain Res.* **67**, 460–8.

Kaczmarek, L. K. (1987). The role of protein kinase C in the regulation of ion channels and neurotransmitter release. *Trends in Neurosci.* **10**, 30–4.

Kalil, K. and Skene, J. H. P. (1986). Elevated synthesis of an axonally transported protein correlates with axon outgrowth in normal and injured pyramidal tracts. *J. Neurosci.* **6**, 2563–70.

Kanamatsu, T., Unsworth, C. D., Diliberto, E. J., Viveros, O. H., and Hong, J. S. (1986). Reflex splanchnic nerve stimulation increases levels of proenkephalin A mRNA and proenkephalin A-related peptides in the rat adrenal medulla. *Proc. Natl. Acad. Sci. USA*, **83**, 9245–9.

Kandel, E. R. (1976). *Cellular basis of behavior.* Freeman, San Francisco.

Kandel, E. R. (1979). *Behavioral biology of* Aplysia. *A contribution to the comparative study of opisthobranch molluscs.* Freeman, San Francisco.

Kandel, E. R. and Schwartz, J. H. (1982). Molecular biology of learning: modulation of transmitter release. *Science*, **218**, 433–43.

Kandel, E. R. and Schwartz, J. H. (ed.) (1985). *Principles of neural science* (2nd edn). Elsevier, New York.

Kandel, E. R. and Spencer, W. A. (1968). Cellular neurophysiological approaches in the study of learning. *Psychol. Rev.* **48**, 65–134.

Kandel, E. R., Brunelli, M., Byrne, J., and Castellucci, V. (1983). A common presynaptic locus for the synaptic changes underlying short-term habituation and sensitization of the gill-withdrawal reflexes in *Aplysia. Cold Spring Harbor Symp. Quant. Biol.* **48**, 465–82.

Kano, M. and Kato, M. (1985). Specific glutamate sensitivity involved in the long-term depression of parallel fiber-Purkinje cell transmission in rabbit cerebellar cortex. *Neurosci. Lett. Supp.* **22**, S26.

Kaplan, M. S. and Hinds, J. W. (1977). Neurogenesis in the adult rat: electron microscopic analysis of light radioautographs. *Science*, **197**, 1092–4.

Katz, B. (1966). *Nerve, muscle, and synapse.* McGraw-Hill, New York.

Katz, B. and Miledi, R. (1968). The role of calcium in neuromuscular facilitation. *J. Physiol.* **195**, 481–92.

Katz, L. C. and Gurney, M. E. (1981). Auditory responses in the Zebra finch's motor system for song. *Brain Res.* **211**, 192–7.

Kauer, J. A., Malenka, R. C., and Nicoll, R. A. (1988*a*). NMDA application potentiates synaptic transmission in the hippocampus. *Nature*, **334**, 250–2.

Kauer, J. A., Malenka, R. C., and Nicoll, R. A. (1988*b*). A persistent postsynaptic modification mediates long-term potentiation in the hippocampus. *Neuron*, **1**, 911–17.

Kelley, D. B. (1986). The genesis of male and female brains. *Trends in Neurosci.* **9**, 499–502.

Kelso, S. R. and Brown, T. H. (1986). Differential conditioning of associative synaptic enhancement in hippocampal brain slices. *Science*, **232**, 85–7.

Kelso, S. R., Ganong, A. H., and Brown, T. H. (1986). Hebbian synapses in hippocampus. *Proc. Natl. Acad. Sci. USA*, **83**, 5326–30.

Kendrick, K. M. and Baldwin, B. A. (1987). Cells in the temporal cortex of conscious sheep can respond preferentially to the sight of faces. *Science*, **236**, 448–50.

Kim, E. H.-J., Woody, C. D., and Berthier, N. E. (1983). Rapid acquisition of conditioned eye blink responses in cats following pairing of an auditory CS with glabella tap US and hypothalamic stimulation. *J. Neurophysiol.* **49**, 767–79.

King, A. P. and West, M. J. (1983). Epigenesis of cowbird song—a joint endeavour of males and females. *Nature*, **305**, 704–6.

Kishimoto, A., Kajikawa, N., Shiota, M., and Nishizuka, Y. (1983). Proteolytic activation of calcium-activated, phospholipid-dependent protein kinase by calcium-dependent neutral protease. *J. Biol. Chem.* **258**, 1156–64.

Kistler, H. B., Jr, Hawkins, R. D., Koester, J., Steinbusch, H. W. M., Kandel, E. R., and Schwartz, J. H. (1985). Distribution of serotonin-immunoreactive cell bodies and processes in the abdominal ganglion of mature *Aplysia. J. Neurosci.* **5**, 72–80.

Klarsfeld, A. and Changeux, J. P. (1985). Activity regulates the levels of acetylcholine receptor α-subunit mRNA in cultured chicken myotubes. *Proc. Natl. Acad. Sci. USA*, **82**, 4558–62.

Klatzky, R. L. (1980). *Human memory: structures and processes* (2nd edn). Freeman, San Francisco.

Klein, D. F. (ed.) (1987). *Anxiety.* Karger, Basel.

Klein, M. and Kandel, E. R. (1978). Presynaptic modulation of voltage-dependent Ca^{2+} current: mechanism for behavioral sensitization in *Aplysia californica. Proc. Natl. Acad. Sci. USA*, **75**, 3512–16.

Klein, M. and Kandel, E. R. (1980). Mechanism of calcium current modulation underlying presynaptic facilitation and behavioral sensitization in *Aplysia. Proc. Natl. Acad. Sci. USA*, **77**, 6912–16.

Klein, M., Shapiro, E., and Kandel, E. R. (1980). Synaptic plasticity and the modulation of the Ca^{2+} current. *J. Exp. Biol.* **89**, 117–57.

Klein, M., Camardo, J., and Kandel, E. R. (1982). Serotonin modulates a specific potassium current in the sensory neurons that show presynaptic facilitation in *Aplysia. Proc. Natl. Acad. Sci. USA*, **79**, 5713–17.

Klein, M., Hochner, B., and Kandel, E. R. (1986). Facilitatory transmitters and cAMP can modulate accommodation as well as transmitter release in *Aplysia* sensory neurons: evidence for parallel processing in a single cell. *Proc. Natl. Acad. Sci. USA*, **83**, 7994–8.

Kleinschmidt, A., Bear, M. F., and Singer, W. (1987). Blockade of 'NMDA' receptors disrupts experience-dependent plasticity of kitten striate cortex. *Science*, **238**, 355–7.

Kley, N., Loeffler, J. P., Pittius, C. W., and Hollt, V. (1987). Involvement of ion channels in the induction of proenkephalin A gene expression by nicotine and cAMP in bovine chromaffin cells. *J. Biol. Chem.* **262**, 4083–9.

Kligman, D. and Marshak, D. R. (1985). Purification and characterization of a neurite extension factor from bovine brain. *Proc. Natl. Acad. Sci. USA*, **82**, 7136–9.

Klüver, H. and Bucy, P. C. (1938). An analysis of certain effects of bilateral temporal lobectomy in the rhesus monkey, with special reference to 'psychic blindness'. *J. Psychyol.* **5**, 33–54.

Knudsen, E. I. (1985). Experience alters the spatial tuning of auditory units in the optic tectum during a sensitive period in the barn owl. *J. Neurosci.* **5**, 3094–109.

Knudsen, E. I., du Lac, S., and Esterly, S. D. (1987). Computational maps in the brain. *Ann. Rev. Neurosci.* **10**, 41–65.

Koch, C. and Poggio, T. (1983). A theoretical analysis of electrical properties of spines. *Proc. R. Soc. Lond. B*, **218**, 455–77.

Koch, C., Poggio, T., and Torre, V. (1986). Computations in the vertebrate retina: gain enhancement, differentiation and motion discrimination. *Trends in Neurosci.* **9**, 204–11.

Koehler, O. (1951). Der Vogelsang als Vorstufe von Musik und Sprache. *J. Ornithol.* **93**, 3–20.

Koffka, K. (1935). *Principles of gestalt psychology.* Harcourt, Brace and World, New York.

Kohler, W. (1925). *The mentality of apes.* Harcourt, Brace and World, New York.

Kojima, S. and Goldman-Rakic, P. S. (1984). Functional analysis of spatially discriminative neurons in prefrontal cortex of rhesus monkey. *Brain Res.* **291**, 229–40.

Koller, K. J., Wolff, R. S., Warden, M. K., and Zoeller, R. T. (1987). Thyroid hormones regulate levels of thyrotropin-releasing-hormone mRNA in paraventricular nucleus. *Proc. Natl. Acad. Sci. USA*, **84**, 7329–33.

Konishi, M. (1965). The role of auditory feedback in the control of vocalization in the white-crowned sparrow. *Z. Tierpsychol.* **22**, 770–83.

Konishi, M. (1985). Birdsong: from behavior to neuron. *Ann. Rev. Neurosci.* **8**, 125–70.

Konishi, M. (1986). Centrally synthesized maps of sensory space. *Trends in Neurosci.* **9**, 163–8.

Konishi, M. (1989). From stimulus to map. (In preparation).

Konishi, M. and Akutagawa, E. (1981). Androgen increases protein synthesis within the avian brain vocal control system. *Brain Res.* **222**, 442–6.

Konishi, M. and Akutagawa, E. (1988). A critical period for estrogen action on neurons of the song control system in the zebra finch. *Proc. Natl. Acad. Sci. USA*, **85**, 7006–7.

Konorski, J. (1948). *Conditioned reflexes and neuron organization.* Cambridge University Press, Cambridge.

Konorski, J. (1967). *Integrative activity of the brain. An interdisciplinary approach.* University of Chicago Press, Chicago.

Kornblith, C. and Olds, J. (1973). Unit activity in brain stem reticular formation of the rat during learning. *J. Neurophysiol.* **36**, 489–501.

Korsakoff, S. S. (1887). Disturbance of psychic function in alcoholic paralysis and its relation to the disturbance of the psychic sphere in multiple neuritis of non-alcoholic origin. *Vestn. Psychiatrii.* **4**, (2).

Koshaka, S. I., Takamatsu, K., Aoki, E., and Tsukada, Y. (1979). Metabolic mapping of chick brain after imprinting using [^{14}C]2-deoxyglucose technique. *Brain Res.* **172**, 539–44.

Koshland, D. (1980). Bacterial chemotaxis in relation to neurobiology. *Ann. Rev. Neurosci.* **3**, 43–75.

Kosslyn, S. M. (1980). *Image and mind.* Harvard University Press, Cambridge, Mass.

Koutalidis, O., Foster, A., and Weisz, D. J. (1988). Parallel pathways can conduct visual CS information during classical conditioning of the NM response. *J. Neurosci.* **8**, 417–27.

Krebs, J. R. (1977). The significance of song repertoires: the Beau Geste hypothesis. *Anim. Behav.* **25**, 475–8.

Kroodsma, D. E. (1977). Correlates of song organization among North American wrens. *Am. Nat.* **111**, 995–1008.

Kroodsma, D. E. and Miller, E. H. (ed.) (1982a). *Acoustic communication in birds*, Vol. 1: Production, perception and design features of sounds. Academic Press, New York.

Kroodsma, D. E. and Miller, E. H. (1982b). *Acoustic communication in birds*, Vol. 2: Song learning and its consequence. Academic Press, New York.

Kroodsma, D. E. and Pickert, R. (1980). Environmentally dependent sensitive periods for avian vocal learning. *Nature*, **288**, 477–9.

Kruger, J. (1983). Simultaneous individual recordings from many cerebral neurons: techniques and results. *Rev. Physiol. Biochem. Pharmacol.* **98**, 177–232.

Kruijer, W., Schubert, D., and Verma, I. M. (1985). Induction of the proto-oncogene *fos* by nerve growth factor. *Proc. Natl. Acad. Sci. USA*, **82**, 7330–4.

Kubota, K. and Funahashi, S. (1982). Direction-specific activities of dorsolateral prefrontal and motor cortex pyramidal tract neurons during visual tracking. *J. Neurophysiol.* **47**, 362–76.

Kucharski, D. and Hall, W. G. (1987). New routes to old memories. *Science*, **238**, 786–8.

Kuffler, S. W., Nicholls, J. G., and Martin, A. R. (1984). *From neuron to brain* (2nd edn). Sinauer, Sunderland, Mass.

Kuhnt, U., Mihaly, A., and Joo, F. (1985). Increased binding of calcium in the hippocampal slice during long-term potentiation. *Neurosci. Lett.* **53**, 149–54.

Kupfermann, I., Castellucci, V., Pinsker, H., and Kandel, E. R. (1970). Neuronal correlates of habituation and dishabituation of the gill-withdrawal reflex in *Aplysia. Science*, **167**, 1743–5.

Kupfermann, I., Pinsker, H., Castellucci, V., and Kandel, E. R. (1971). Central and peripheral control of gill movements in *Aplysia. Science*, **174**, 1252–6.

Kushner, M. J., Rosenquist, A., Alavi, A., Rosen, M., Dann, R., Fazekas, F., Bosley, T., Greenberg, J. and Reivich, M. (1988). Cerebral metabolism and patterned visual stimulation: a positron emission tomography study of the human visual cortex. *Neurology*, **38**, 89–95.

Kyriacou, C. P. and Hall, J. C. (1984). Learning and memory mutations impair acoustic priming of mating behaviour in *Drosophila. Nature*, **308**, 62–5.

Lai, Y., Nairn, A. C., and Greengard, P. (1986). Autophosphorylation reversibly regulates the Ca^{2+}/calmodulin-dependence of Ca^{2+}/calmodulin-dependent protein kinase II. *Proc. Natl. Acad. Sci. USA*, **83**, 4253–7.

Laroche, S., Errington, M. L., Lynch, M. A., and Bliss, T. V. P. (1987). Increase in [^{3}H]glutamate release from slices of dentate gyrus and hippocampus following classical conditioning in the rat. *Behav. Brain Res.* **25**, 23–9.

Larrabee, M. G. and Bronk, D. W. (1947). Prolonged facilitation of synaptic excitation in sympathetic ganglia. *J. Neurophysiol.* **10**, 139–54.

Lashley, K. S. (1929). *Brain mechanisms and intelligence.* University of Chicago Press, Chicago.

Lashley, K. S. (1949). Persistent problems in the evolution of mind. *Q. Rev. Biol.* **24**, 28–42.

Lashley, K. S. (1950). In search of the engram. *Symp. Soc. Exp. Biol.* **4**, 454–82.

Lashley, K. S., Chow, K. L., and Semms, J. (1951). An examination of the electrical field theory of cerebral integration. *Psychol. Rev.* **58**, 123–36.

Lau, L. F. and Nathans, D. (1987). Expression of a set of growth-related immediate early genes in BALB/c 3T3 cells: coordinate regulation with *c-fos* or *c-myc*. *Proc. Natl. Acad. Sci. USA*, **84**, 1182–6.

Lavond, D. G., Hembree, T. L., and Thompson, R. F. (1985). Effect of kainic acid lesions of the cerebellar interpositus nucleus on eyelid conditioning in the rabbit. *Brain Res.* **326**, 179–82.

Lavond, D. G., Knowlton, B. J., Steinmetz, J. E., and Thompson, R. F. (1987*a*). Classical conditioning of the rabbit eyelid response with a mossy-fiber stimulation CS: II. Lateral reticular nucleus stimulation. *Behav. Neurosci.* **101**, 676–82.

Lavond, D. G., Steinmetz, J. E., Yokaitis, M. H., and Thompson, R. F. (1987*b*). Reacquisition of classical conditioning after removal of cerebellar cortex. *Exp. Brain Res.* **67**, 569–93.

Lawler, K. A. and Cowey, A. (1987). On the role of posterior parietal and prefrontal cortex in visuo-spatial perception and attention. *Exp. Brain Res.* **65**, 695–8.

Lederhendler, I. I., Gart, S., and Alkon, D. L. (1986). Classical conditioning of *Hermissenda*: origin of a new response. *J. Neurosci.* **6**, 1325–31.

LeDoux, J. E. (1987). Emotion. In *Handbook of physiology*, Section 1: The nervous system, Vol. V: Higher functions of the brain, Part 1 (ed. F. Plum), pp. 419–59. American Physiological Society, Bethesda, Md.

Lee, C., Rohrer, W. H., and Sparks, D. L. (1988). Population coding of saccadic eye movements by neurons in the superior colliculus. *Nature*, **332**, 357–60.

Lee, K. S. (1983). Cooperativity among afferents for the induction of long-term potentiation in the CA1 region of the hippocampus. *J. Neurosci.* **3**, 1369–72.

Lee, K. S., Schottler, F., Oliver, M., and Lynch, G. (1980). Brief bursts of high-frequency stimulation produce two types of structural change in rat hippocampus. *J. Neurophysiol.* **44**, 247–58.

Lee, W., Mitchell, P., and Tjian, R. (1987). Purified transcription factor AP-1 interacts with TPA-inducible enhancer elements. *Cell*, **49**, 741–52.

Leon, M. (1987). Plasticity of olfactory output circuits related to early olfactory learning. *Trends in Neurosci.* **10**, 434–8.

Leonard, S. L. (1939). Induction of singing in female canaries by injection of male hormone. *Proc. Soc. Exp. Biol. Med.* **41**, 229–30.

Levinthal, C. F., Tartell, R. H., Margolin, C. M., and Fishman, H. (1985). The CS–US interval (ISI) function in rabbit nictitating membrane response conditioning with very long intertrial intervals. *Anim. Learn. Behav.* **13**, 228–32.

Levitan, I. B. (1985). Phosphorylation of ionic channels. *J. Memb. Biol.* **87**, 177–90.

Lev-Tov, A. and Rahamimoff, R. (1980). A study of tetanic and post-tetanic potentiation of minia-ture end-plate potentials at the frog neuromuscular junction. *J. Physiol.* **309**, 247–73.

Levy, W. B. and Steward, O. (1979). Synapses as associative memory elements in the hippocampal formation. *Brain Res.* **175**, 233–45.

Levy, W. B. and Steward, O. (1983). Temporal contiguity requirements for long-term associative potentiation/depression in the hippocampus. *Neurosci.* **8**, 791–7.

Lewin, B. M. (1987). *Genes* (3rd edn). Wiley, New York.

Lewis, E. J., Harrington, C. A., and Chikaraishi, S. M. (1987). Transcriptional regulation of the tyrosine hydroxylase gene by glucocorticoid and cyclic AMP. *Proc. Natl. Acad. Sci. USA*, **84**, 3550–

4.

Lewis, H. R. and Papadimitriou, C. H. (1981). *Elements of the theory of computation*. Prentice-Hall, New Jersey.

Liberman, A. M., Cooper, F. S., Shankweiler, D. P., Studder-Kennedy, M. (1967). Perception of the speech code. *Psychol. Rev.* **74**, 431–61.

Lincoln, J. S., McCormick, D. A., and Thompson, R. F. (1982). Ipsilateral cerebellar lesions prevent learning of the classically conditioned nictitating membrane/eyelid response. *Brain Res.* **242**, 190–3.

Linden, D. J., Murakami, K., and Routtenberg, A. (1986). A newly discovered protein kinase C activator (oleic acid) enhances long-term potentiation in the intact hippocampus. *Brain Res.* **379**, 358–63.

Linden, D. J., Shew, F. S., Murakami, K., and Routtenberg, A. (1987). Enhancement of long-term potentiation by *cis*-unsaturated fatty acid: relation to protein kinase C and phospholipase A_2. *J. Neurosci.* **7**, 3783–92.

Lisman, J. E. (1985). A mechanism for memory storage insensitive to molecular turnover: a bistable autophosphorylating kinase. *Proc. Natl. Acad. Sci. USA*, **82**, 3055–7.

Livingston, K. E. and Escobar, A. (1971). Anatomical bias of the limbic system concept. A proposed reorientation. *Arch. Neurol.* **24**, 17–21.

Livingstone, M. S. (1985). Genetic dissection of *Drosophila* adenylate cyclase. *Proc. Natl. Acad. Sci. USA*, **82**, 5992–6.

Livingstone, M. S., Sziber, P. P., and Quinn, W. G. (1984). Loss of calcium/calmodulin responsiveness in adenylate cyclase of *rutabaga*, a *Drosophila* learning mutant. *Cell*, **37**, 205–15.

Llinas, R. R. (1988). The intrinsic electrophysiological properties of mammalian neurons: insights into central nervous system function. *Science*, **242**, 1654–64.

Llinas, R., Walton, K., Hillman, D. E., and Sotelo, C. (1975). Inferior olive: its role in motor learning. *Science*, **190**, 1230–1.

Loeb, J. (1900). *Comparative physiology of the brain and comparative psychology*. Putnam, New York.

Lomo, T. (1966). Frequency potentiation of excitatory synaptic activity in the dentate area of the hippocampal formation. *Acta Physiol. Scand.* **68**, suppl. 277, 128.

London, J. A. and Gillette, R. (1986). Mechanism for food avoidance learning in the central pattern generator of feeding behavior of *Pleurobrancheae californica*. *Proc. Natl. Acad. Sci. USA*, **83**, 4058–62.

Lorenz, K. (1937). The companion in the bird's world. *Auk*, **54**, 245–73.

Lorenz, K. Z. (1981). *The foundations of ethology*. Springer-Verlag, New York.

LoTurco, J. J., Coulter, D. A., and Alkon, D. L. (1988). Enhancement of synaptic potentials in rabbit CA1 pyramidal neurons following classical conditioning. *Proc. Natl. Acad. Sci. USA*, **85**, 1672–76.

Lou, L. L., Lloyd, S. J., and Schulman, H. (1986). Activation of the multifunctional Ca^{2+}/calmodulin-dependent protein kinase by autophosphorylation: ATP modulates production of an autonomous enzyme. *Proc. Natl. Acad. Sci. USA*, **83**, 9497–501.

Lovinger, D. M., Colley, P. A., Akers, R. F., Nelson, R. B., and Routtenberg, A. (1986). Direct relation of long-term synaptic potentiation to phosphorylation of membrane protein F1, a substrate for membrane protein kinase C. *Brain Res.* **399**, 205–11.

Lovinger, D. M., Wong, K. L., Murakami, K., and Routtenberg, A. (1987). Protein kinase C inhibitors eliminate hippocampal long-term potentiation. *Brain Res.* **436**, 177–83.

Lukowiak, K. (1979). The development of central nervous system control of the gill withdrawal reflex by siphon stimulation in *Aplysia. Can. J. Physiol. Pharmacol.* **57**, 987–97.

Lukowiak, K. and Jacklet, J. W. (1972). Habituation and dishabituation: interactions between peripheral and central nervous systems in *Aplysia. Science,* **178**, 1306–8.

Lukowiak, K. and Sahley, C. (1981). The *in vitro* classical conditioning of the gill withdrawal reflex of *Aplysia californica. Science,* **212**, 1516–18.

Lund, R. D. (1978). *Development and plasticity of the brain.* Oxford University Press, New York.

Luria, A. R. (1966). *Higher cortical function in man* (2nd edn). Basic Books, New York.

Lynch, G. and Baudry, M. (1984). The biochemistry of memory: a new specific hypothesis. *Science,* **224**, 1057–63.

Lynch, G. S., Dunwiddie, T., and Gribkoff, V. (1977). Heterosynaptic depression: a postsynaptic correlate of long-term potentiation. *Nature,* **266**, 737–9.

Lynch, G., Halpain, S., and Baudry, M. (1982). Effects of high-frequency synaptic stimulation on glutamate receptor binding studied with a modified *in vitro* hippocampal slice preparation. *Brain Res.* **244**, 101–11.

Lynch, G., Larson, J., Kelso, S., Barrionuevo, G., and Schottler, F. (1983). Intracellular injections of EGTA block induction of hippocampal long-term potentiation. *Nature,* **305**, 719–21.

Lynch, M. A., Errington, M. L., and Bliss, T. V. P. (1985). Long-term potentiation of synaptic transmission in the dentate gyrus: increased release of [¹⁴C]glutamate without increase in receptor binding. *Neurosci. Lett.* **62**, 123–9.

McCabe, B. J. and Horn, G. (1988). Learning and memory: regional changes in N-methyl-D-aspartate receptors in the chick brain after imprinting. *Proc. Natl. Acad. Sci. USA,* **85**, 2849–53.

McCabe, B. J., Horn, G., and Bateson, P. P. G. (1981). Effects of restricted lesions of the chick fore-brain on the acquisition of filial preferences during imprinting. *Brain Res.* **205**, 29–37.

McCabe, B. J., Cipolla-Neto, J., Horn, G., and Bateson, P. (1982). Amnesic effects of bilateral lesions placed in the hyperstriatum ventrale of the chick after imprinting. *Exp. Brain Res.* **48**, 13–21.

McCarthy, R. A. and Warrington, E. K. (1988). Evidence for modality-specific meaning systems in the brain. *Nature,* **334**, 428–30.

McCasland, J. S. (1987). Neuronal control of bird song production. *J. Neurosci.* **7**, 23–39.

McCasland, J. S. and Konishi, M. (1981). Interaction between auditory and motor activities in an avian song control nucleus. *Proc. Natl. Acad. Sci. USA,* **78**, 7815–9.

McCormick, D. A. and Thompson, R. F. (1984*a*). Cerebellum: essential involvement in the classically conditioned eyelid response. *Science,* **223**, 296–9.

McCormick, D. A. and Thompson, R. F. (1984*b*). Neuronal responses of the rabbit cerebellum during acquisition and performance of a classically conditioned nictitating membrane-eyelid response. *J. Neurosci.* **4**, 2811–22.

McCormick, D. A., Clark, G. A., Lavond, D. G., and Thompson, R. F. (1982*a*). Initial localization of the memory trace for a basic form of learning. *Proc. Natl. Acad. Sci. USA,* **79**, 2731–5.

McCormick, D. A., Guyer, P. E., and Thompson, R. F. (1982*b*). Superior cerebellar peduncle lesions selectively abolish the ipsilateral classically conditioned nictitating membrane/eyelid response of the rabbit. *Brain Res.* **244**, 347–50.

McCormick, D. A., Lavond, D. G., and Thompson, R. F. (1982*c*). Concomitant classical conditioning of the rabbit nictitating membrane and eyelid response: correlations and implications. *Physiol. Behav.* **28**, 769–75.

McCormick, D. A., Steinmetz, J. E., and Thompson, R. F. (1985). Lesions of the inferior olivary complex cause extinction of the classically conditioned eyeblink response. *Brain Res.* **359**, 120–30.

MacDermott, A. B., Mayer, M. L., Westbrook, G. L., Smith, S. J., and Barker, J. L. (1986). NMDA-receptor activation increases cytoplasmic calcium concentration in cultured spinal cord neurons. *Nature*, **321**, 519–22.

McGaugh, J. L. (1966). Time-dependent processes in memory storage. *Science*, **153**, 1351–8.

McGaugh, J. L. (1983). Hormonal influences on memory. *Ann. Rev. Psychol.* **34**, 297–323.

Mackey, S. L., Glanzman, D. L., Small, S. A., Dyke, A. M., Kandel, E. R., and Hawkins, R. D. (1987). Tail shock produces inhibition as well as sensitization of the siphon–withdrawal reflex of *Aplysia*: possible behavioural role for presynaptic inhibition mediated by the peptide Phe-Met-Arg-Phe-NH$_2$. *Proc. Natl. Acad. Sci. USA*, **84**, 8730–4.

Mackey, S. L., Lalevic, N., Hawkins, R. D., and Kandel, E. R. (1988). Comparison of dishabituation and sensitization of the gill-withdrawal reflex in *Aplysia*. *Soc. Neurosci. Abst.* **14**, 842.

Mackintosh, N. J. (1983). *Conditioning and associative learning.* Oxford University Press, Oxford.

Macko, K. A., Jarvis, C., Kennedy, C., Miyaoka, M., Shinohara, M., Sokoloff, L., and Mishkin, M. (1982). Mapping the primate visual system with [2-^{14}C]deoxyglucose. *Science*, **218**, 394–7.

MacLean, P. D. (1970). The tribune brain, emotion, and scientific bias. In *The Neurosciences*, Second Study Program (ed. F. O. Schmitt), pp. 336–49. The Rockefeller University Press, New York.

McNaughton, B. L. (1982). Long-term synaptic enhancement and short-term potentiation in rat fascia dentata act through different mechanisms. *J. Physiol.* **324**, 249–62.

McNaughton, B. L., Douglas, R. M., and Goddard, G. W. (1978). Synaptic enhancement in fascia dentata: cooperativity among coactive afferents. *Brain Res.* **157**, 277–93.

McNaughton, B. L., Barnes, C. A., and O'Keefe, J. (1983). The contributions of position, direction, and velocity to single unit activity in the hippocampus of freely-moving rats. *Exp. Brain Res.* **52**, 41–9.

McNaughton, B. L., Barnes, C. A., Rao, G., Baldwin, J., and Rasmussen, M. (1986). Long-term enhancement of hippocampal synaptic transmission and the acquisition of spatial information. *J. Neurosci.* **6**, 563–71.

Macphail, E. (1987). The comparative psychology of intelligence. *Behav. Brain Sci.* **10**, 645–95.

Madison, D. V., Malinow, R., and Tsien, R. W. (1988). Induction, maintenance and expression of long-term potentiation (LTP) are distinguished by blockers and stimulators of protein kinase C. *Soc. Neurosci. Abst.* **14**, 18.

Magleby, K. L. and Zengel, J. E. (1982). A quantitative description of stimulation-induced changes in transmitter release in the frog neuromuscular junction. *J. Gen. Physiol.* **80**, 613–38.

Mahut, H. (1971). Spatial and object reversal learning in monkeys with partial temporal lobe ablations. *Neuropsychologia*, **9**, 409–24.

Mahut, H. (1972). A selective spatial deficit in monkeys after transection of the fornix. *Neuropsychologia*, **10**, 65–74.

Mahut, H., Zola-Morgan, S., and Moss, M. (1982). Hippocampal resections impair associative learning and recognition memory in the monkey. *J. Neurosci.* **2**, 1214–29.

Maier, N. R. F. and Schneirla, T. C. (1964). *Principles of animal psychology.* Dover, New York.

Maier, V. and Scheich, H. (1983). Acoustic imprinting leads to differential 2-deoxy-D-glucose uptake in the chick forebrain. *Proc. Natl. Acad. Sci. USA*, **80**, 3860–4.

Maioli, M. G., Squatrito, S., Galletti, C., Battaglini, P. P., and Sanseveriono, E. R. (1983). Cortico-cortical connections from the visual region of the superior temporal sulcus to frontal eye field in the macaque. *Brain Res.* **265**, 294–9.

Malenka, R. C., Madison, D. V., and Nicoll, R. A. (1986). Potentiation of synaptic transmission in the hippocampus by phorbol esters. *Nature,* **321**, 175–7.

Malinow, R. and Miller, J. P. (1986). Postsynaptic hyperpolarization during conditioning reversibly blocks induction of long-term potentiation. *Nature,* **320**, 529–31.

Malinow, R., Madison, D. V. and Tsien, R. W. (1988). Persistent protein kinase activity underlying long-term potentiation. *Nature,* **335**, 820–4.

Mallet, J., Faucon Biguet, N., Buda, M., Lamouroux, A., and Samolyk, D. (1983). Detection and regulation of the tyrosine hydroxylase mRNA levels in rat adrenal medulla and brain tissues. *Cold Spring Harbor Symp. Quant. Biol.* **83**, 305–8.

Mamounas, L. A., Thompson, R. F., Lynch, G., and Baudry, M. (1984). Classical conditioning of the rabbit eyelid response increases glutamate receptor binding in hippocampal synaptic membranes. *Proc. Natl. Acad. Sci. USA,* **81**, 2548–52.

Marcus, E. A., Nolen, T. G., Rankin, C. H., and Carew, T. J. (1988). Behavioral dissociation of dishabituation, sensitization, and inhibition in *Aplysia. Science,* **241**, 210–13.

Margoliash, D. (1983). Acoustic parameters underlying the responses of song-specific neurons in the white-crowned sparrow. *J. Neurosci.* **3**, 1039–57.

Margoliash, D. (1986). Preference for autogenous song by auditory neurons in a song system nucleus of the white-crowned sparrow. *J. Neurosci.* **6**, 1643–61.

Margoliash, D. and Konishi, M. (1985). Auditory representation of autogenous song in the song system of the white-crowned sparrows. *Proc. Natl. Acad. Sci. USA,* **82**, 5997–6000.

Mariani, J. (1983). Elimination of synapses during the development of the central nervous system. *Prog. Brain Res.* **58**, 383–92.

Mariath, H. A. (1985). Operant conditioning in *Drosophila melanogaster* wild-type and learning mutants with defects in the cyclic AMP metabolism. *J. Insect Physiol.* **31**. 779–87.

Marks, I. M. (1987). *Fears, phobias, and rituals. Panic, anxiety, and their disorders.* Oxford University Press, New York.

Marler, P. (1984). Song learning: innate species differences in the learning process. In *The biology of learning* (ed. P. Marler and H. S. Terrace), pp. 289–309. Springer-Verlag, Berlin.

Marler, P. and Peters, S. (1977). Selective vocal learning in a sparrow. *Science,* **198**, 519–21.

Marler, P. and Peters, S. (1981). Sparrows learn adult songs and more from memory. *Science,* **213**, 780–2.

Marler, P. and Sherman, V. (1983). Song structure without auditory feedback: emendation of the auditory template hypothesis. *J. Neurosci.* **3**, 517–31.

Marler, P. and Tamura, M. (1964). Culturally transmitted patterns of vocal behavior in sparrows. *Science,* **146**, 1483–6.

Marler, P., Peters, S., Ball, G. F., Duffy, A. M. Jr., and Wingfield, J. C. (1988). The role of sex steroids in the acquisition and production of birdsong. *Nature,* **336**, 770–2.

Marr, D. (1969). A theory of cerebellar cortex. *J. Physiol.* **202**, 437–70.

Marr, D. (1982). *Vision,* Freeman, San Francisco.

Marshall, J. (1980). The new organology. *Behav. Brain Sci.* **3**, 23–5.

Marslen-Wilson, W. D. and Teuber, H. L. (1975). Memory for remote events in anterograde amnesia: recognition of public figures from newsphotographs. *Neuropsychologia,* **13**, 353–64.

Martinez, J. L., Jensen, R. A., and McGaugh, J. L. (1981). Attenuation of experimentally-induced amnesia. *Prog. Neurobiol.* **16**, 155–86.

Martino, G. (1939). The conditioned reflex of blinking. *J. Neurophysiol.* **2**, 173–7.

Mason, S. T. (1984). *Catecholamines and behavior.* Cambridge University Press, Cambridge.

Mastropaolo, J., Nadi, N. S., Ostrowski, N. L., and Crawley, J. N. (1988). Galanin antagonizes acetylcholine on a memory task in a basal forebrain-lesioned rats. *Proc. Natl. Acad. Sci. USA*, **85**, 9841–5.

Matsumura, M. and Woody, C. D. (1982). Excitability changes of facial motoneurons of cats related to conditioned and unconditioned facial motor responses. In *Conditioning: representation of involved neural functions* (ed. C. D. Woody), pp. 451–8. Plenum Press, New York.

Matthies, H. (1980). Pharmacology of learning and memory. *Trends in Pharmacol. Sci.* **1**, 333–6.

Mauk, M. D. and Thompson, R. F. (1987). Retention of classically conditioned eyelid responses following acute decerebration. *Brain Res.* **403**, 89–95.

Mauk, M. D., Steinmetz, J. E., and Thompson, R. F. (1986). Classical conditioning using stimulation of the inferior olive as the unconditioned stimulus. *Proc. Natl. Acad. Sci. USA*, **83**, 5349–53.

Maunsell, J. H. R. and Newsome, W. T. (1987). Visual processing in monkey extrastriate cortex. *Ann. Rev. Neurosci.* **10**, 363–401.

Mayer, M. L., Westbrook, G. L., and Guthrie, P. B. (1984). Voltage-dependent block by Mg^{2+} of NMDA responses in spinal cord neurons. *Nature*, **309**, 261–3.

Mayr, E. (1982). *The growth of biological thought.* Harvard University Press, Cambridge, Mass.

Mazziotta, J. C., Phelps, M. E., Carson, R. E., and Kuhl, D. E. (1982). Tomographic mapping of human cerebral metabolism: auditory stimulation. *Neurology*, **32**, 921–37.

Mazziotta, J. C., Phelps, M. E., and Carson, R. E. (1984). Tomographic mapping of human cerebral metabolism: subcortical responses to auditory and visual stimulation. *Neurology*, **34**, 825–8.

Melloni, E., Pontremoli, S., Michetti, M., Sacco, O., Sparatore, B., Salamino, F., and Horecker, B. L. (1985). Binding of protein kinase C to neutrophil membranes in the presence of Ca^{2+} and its activation by a Ca^{2+}-requiring proteinase. *Proc. Natl. Acad. Sci. USA*, **82**, 6435–9.

Melloni, E., Pontremoli, S., Michetti, M., Sacco, O., Sparatore, B., and Horecker, B. L. (1986). The involvement of calpain in the activation of protein kinase C in neutrophils stimulated by phorbol myristic acid. *J. Biol. Chem.* **261**, 4101–5.

Menne, D. and Spatz, H.Ch. (1977). Colour vision in *Drosophila melanogaster*. *J. Comp. Physiol.* **114**, 301–12.

Menzel, R. (1983). Neurobiology of learning and memory: the honey bee as a model system. *Naturwissen.* **70**, 504–11.

Menzel, R. and Erber, J. (1978). Learning and memory in bees. *Sci. Amer.* **239**, (7), 80–7.

Menzel, R. and Mercer, A. (ed.) (1987). *Neurobiology and behavior of honeybees.* Springer-Verlag, Berlin.

Merlie, J. P., Isenberg, K. E., Russell, S. D., and Sanes, J. R. (1984). Denervation supersensitivity in skeletal muscle: analysis with a cloned cDNA probe. *J. Cell Biol.* **99**, 332–5.

Merzenich, M. M. (1987). Dynamic neocortical processes and the origins of higher brain functions. In *The neuronal and molecular bases of learning*, (ed. J.-P. Changeux and M. Konishi), pp. 337–58. Wiley, New York.

Merzenich, M. M., Nelson, R. J., Kaas, J. H., Stryker, M. P., Jenkins, W. M., Zook, J. M., Cynader, M. S., and Schoppmann, A. (1987). Variability in hand surface representations in areas 3b and 1 in adult owl and squirrel monkeys. *J. Comp. Neurol.* **258**, 281–96.

Mesulam, M. M., Mufson, E. J., Levey, A. I., and Wainer, B. H. (1983). Cholinergic innervation of cortex by the basal forebrain: cytochemistry and cortical connections of the septal area, diagonal band nuclei, nucleus basalis (substantia innominata), and hypothalamus in the rhesus monkey. *J. Comp. Neurol.* **214**, 170–97.

Mesulam, M. M., Van Hoesen, G. W., Pandya, D. N., and Geschwind, N. (1977). Limbic and sensory connections of the inferior parietal lobule (area PG) in the rhesus monkey: a study with a new method for horseradish peroxidase histochemistry. *Brain Res.* **136**, 393–414.

Metfessel, M. (1935). Roller canary song produced without learning from external sources. *Science*, **81**, 470.

Michell, B. (1988). Transmembrane signalling. *Trends in Pharmacol. Sci.* **9** (4), vi–vii.

Mikami, A. and Kubota, K. (1980). Inferotemporal neuron activities and color discrimination with delay. *Brain Res.* **182**, 65–78.

Miles, F. A. and Fuller, J. H. (1974). Adaptive plasticity in the vestibulo-ocular responses of the rhesus monkey. *Brain Res.* **80**, 512–16.

Miles, F. A. and Lisberger, S. G. (1981). Plasticity in the vestibular-ocular reflex: a new hypothesis. *Ann. Rev. Neurosci.* **4**, 273–99.

Miller, G. A. (1956). The magical number seven, plus or minus two: some limits on our capacity for processing information. *Psychol. Rev.* **63**, 81–97.

Miller, M. W., Lee, S. C., and Krasne, F. B. (1987). Cooperativity-dependent long-lasting potentiation in the crayfish lateral giant escape reaction circuit. *J. Neurosci.* **7**, 1081–92.

Miller, R. J. (1986). Protein kinase C: a key regulator of neuronal excitability? *Trends in Neurosci.* **9**, 538–41.

Miller, S. G. and Kennedy, M. B. (1986). Regulation of brain type II Ca^{2+}/calmodulin-dependent protein kinase by autophosphorylation: a Ca^{2+}-triggered molecular switch. *Cell*, **44**, 861–70.

Miller, S. and Konorski, J. (1928). Sur une forme particulière des réflexes conditionnels. *C. R. Sean. Soc. Biol.* **99**, 1155–7.

Miller, S. G., Patton, B. L., and Kennedy, M. B. (1988). Sequences of autophosphorylation sites in neuronal type II CaM kinase that control Ca^{2+}-independent activity. *Neuron*, **1**, 593–604.

Milner, A. D., Ockleford, E. M., and Dewar, W. (1977). Visuo-spatial performance following posterior parietal and lateral posterior frontal lesions in stumptail macaques. *Cortex*, **13**, 350–60.

Milner, B. (1963). Effects of different brain lesions on card sorting. *Arch. Neurol.* **9**, 100–10.

Milner, B. (1968). Visual recognition and recall after right temporal-lobe excision in man. *Neuropsychologia*, **6**, 191–209.

Milner, B. (1972). Disorders of learning and memory after temporal lobe lesions in man. *Clin. Neurosurg.* **19**, 421–46.

Milner, B. and Petrides, M. (1984). Behavioral effects of frontal-lobe lesions in man. *Trends in Neurosci.* **7**, 403–7.

Milner, B., Corkin, S., and Teuber, H. L. (1968). Further analysis of the hippocampal amnesic syndrome: 14-year follow-up study of H.M. *Neuropsychologia*, **6**, 215–34.

Mishkin, M. (1954). Visual discrimination performance following partial ablations of the temporal lobe: II. Ventral surface vs. hippocampus. *J. Comp. Physiol. Psychol.* **47**, 187–93.

Mishkin, M. (1972). Cortical visual areas and their interactions. In *Brain and human behavior* (ed. A. G. Karczmar and J. C. Eccles) pp. 187–208. Springer-Verlag, Berlin.

Mishkin, M. (1978). Memory in monkeys severely impaired by combined but not by separate removal of amygdala and hippocampus. *Nature*, **273**, 297–8.

Mishkin, M. (1982). A memory system in the monkey. *Phil. Trans. R. Soc. Lond. B*, **298**, 85–95.

Mishkin, M. and Appenzeller, T. (1987). The anatomy of memory. *Sci. Amer.* **256** (6), 62–71.

Mishkin, M. and Delacour, J. (1975). An analysis of short-term visual memory in the monkey. *J. Exp. Psychol. Anim. Behav. Process.* **1**, 326–34.

Mishkin, M. and Manning, F. J. (1978). Non-spatial memory after selective prefrontal lesions in monkeys. *Brain Res.* **143**, 313–23.

Mishkin, M. and Petri, H. L. (1984). Memories and habits: some implications for the analysis of learning and retention. In *Neuropsychology of memory* (ed. L. R. Squire and N. Butters) pp. 287–96. Guilford, New York.

Mishkin, M. and Pribram, K. H. (1954). Visual discrimination performance following partial ablations of the temporal lobe: I. Ventral vs. lateral. *J. Comp. Physiol. Psychol.* **47**, 14–20.

Mishkin, M. and Ungerleider, L. G. (1982). Contribution of striate inputs to the visuospatial functions of parieto-preoccipital cortex in monkeys. *Behav. Brain Res.* **6**, 57–77.

Mishkin, M., Lewis, M. E., and Ungerleider, L. G. (1982). Equivalence of parieto-preoccipital subareas for visuospatial ability in monkeys. *Behav. Brain Res.* **6**, 41–55.

Mishkin, M., Malamut, B., and Bachevalier, J. (1984). Memories and habits: two neural systems. In *Neurobiology of learning and memory* (ed. G. Lynch, J. L. McGaugh, and N. M. Weinberger), pp. 65–77. Guilford, New York.

Miyashita, Y. and Chang, H. S. (1988). Neuronal correlate of pictorial short-term memory in the primate temporal cortex. *Nature*, **331**, 68–70.

Mizumori, S. J. Y., Sakai, D. H., Rosenzweig, M. R., Bennett, E. L., and Wittreich, P. (1987). Investigations into the neuropharmacological basis of temporal stages of memory formation in mice trained in an active avoidance task. *Behav. Brain Res.* **23**, 239–50.

Monaghan, D. T., Olverman, H. J., Nguyen, L., Watkins, J. C., and Cotman, C. W. (1988). Two classes of N-methyl-D-aspartate recognition sites: Differential distribution and differential regulation by glycine. *Proc. Natl. Acad. Sci. USA*, **85**, 9836–40.

Montarolo, P. G., Goelet, P., Castellucci, V. F., Morgan, J., Kandel, E. R., and Schacher, S. (1986). A critical period for macromolecular synthesis in long-term heterosynaptic facilitation in *Aplysia*. *Science*, **234**, 1249–54.

Montarolo, P. G., Kandel, E. R., and Schacher, S. (1988). Long-term heterosynaptic inhibition in *Aplysia*. *Nature*, **333**, 171–4.

Montminy, M. R. and Bilezikjian, L. M. (1987). Binding of a nuclear protein to the cyclic-AMP response element of the somatostatin gene. *Nature*, **328**, 175–8.

Montminy, M. R., Low, M. J., Tapia-Arancibia, L., Reichlin, S., Mandel, G. and Goodman, R. H. (1986). Cyclic AMP regulates somatostatin mRNA accumulation in primary diencephalic cultures and in transfected fibroblast cells. *J. Neurosci.* **6**, 1171–6.

Moore, B. R. and Stuttard, S. (1979). Dr. Guthrie and *Felix domesticus* or: tripping over the cat. *Science*, **205**, 1031–3.

Moore, R. Y. and Bloom, F. E. (1978). Central catecholamine neuron systems: anatomy and physiology of the dopamine system. *Ann. Rev. Neurosci.* **1**, 129–69.

Moore, R. Y. and Bloom, F. E. (1979). Central catecholamine neuron systems: anatomy and physiology of the norepinephrine and epinephrine system. *Ann. Rev. Neurosci.* **2**, 113–68.

Morgan, C. L. (1894). *An introduction to comparative psychology*. Scott, London.

Morgan, C. L. (1900). *Animal behavior*. Edward Arnold, London.

Morgan, J. I. and Curran, T. (1986). Role of ion flux in the control of *c-fos* expression. *Nature*, **322**, 552–5.

Morgan, J. I., Cohen, D. R., Hempstead, J. L., and Curran, T. (1987). Mapping patterns of *c-fos* expression in the central nervous system after seizure. *Science*, **237**, 192–97.

Morielli, A. D., Matera, E. M., Kovac, M. P., Shrum, R. G., McCormac, K. J., and Davis, W. J. (1986). Cholinergic suppression: a postsynaptic mechanism of long-term associative learning. *Proc. Natl. Acad. Sci. USA,* **83**, 4556–60.

Morris, R. G. M., Anderson, E., Lynch, G. S., and Baudry, M. (1986). Selective impairment of learning and blockade of long-term potentiation by an *N*-methyl-D-aspartate receptor antagonist, AP5. *Nature*, **319**, 774–6.

Moss, M., Mahut, H., and Zola-Morgan, S. (1981). Concurrent discrimination learning of monkeys after hippocampal, entorhinal or fornix lesions. *J. Neurosci.* **1**, 227–40.

Mountcastle, V. B., Motter, B. C., Steinmetz, M. A., and Sestokas, A. K. (1987). Common and differential effects of attentive fixation on the excitability of parietal and prestriate (V4) cortical visual neurons in the macaque monkey. *J. Neurosci.* **7**, 2239–55.

Mowrer, O. H. (1939). A stimulus-response analysis of anxiety and its role as a reinforcing agent. *Psychol. Rev.* **46**, 553–65.

Mowrer, O. H. (1947). On the dual nature of learning—a re-interpretation of 'conditioning' and 'problem-solving'. *Harvard Educ. Rev.* **17**, 102–48.

Mpitsos, G. J. and Davis, W. J. (1973). Learning: classical and avoidance conditioning in the mollusc *Pleurobranchaea*. *Science*, **180**, 317–20.

Muller, C. M. and Leppelsack, H.-J. (1985). Feature extraction and itonotopic organization in the avian auditory forebrain. *Exp. Brain Res.* **59**, 587–99.

Muller, D., Joly, M., and Lynch, G. (1988*a*). Contributions of quisqualate and NMDA receptors to the induction and expression of LTP. *Science*, **242**, 1694–7.

Muller, D., Turnbull, J., Baudry, M., and Lynch, G. (1988*b*). Phorbol ester-induced synaptic facilitation is different than long-term potentiation. *Proc. Natl. Acad. Sci. USA,* **85**, 6997–7000.

Muller, R. U. and Kubie, J. L. (1987). The effects of changes in the environment on the spatial firing of hippocampal complex-spike cells. *J. Neurosci.* **7**, 1951–68.

Muller, R. U., Kubie, J. L., and Ranck, J. B. (1987). Spatial firing patterns of hippocampal complex-spike cells in a fixed environment. *J. Neurosci.* **7**, 1935–50.

Munn, N. L. (1950). *Handbook of psychological research on the rat. An introduction to animal psychology.* Houghton Mifflin, Boston.

Murakami, F., Oda, Y., and Tsukahara, N. C. (1988). Synaptic plasticity in the red nucleus and learning. *Behav. Brain Res.* **28**, 175–9.

Murphy, G. (1949). *Historical introduction to modern psychology* (5th edn). Routledge and Kegan Paul, London.

Murray, E. A. and Mishkin, M. (1985). Amygdalectomy impairs crossmodal association in monkeys. *Science*, **228**, 604–6.

Nachmansohn, D. (1959). *Chemical and molecular basis of nerve activity.* Academic Press, New York.

Nagamine, Y. and Reich, E. (1985). Gene expression and cAMP. *Proc. Natl. Acad. Sci. USA*, **82**, 4606–10.

Nagel, E. (1961). *The structure of science: problems in the logic of scientific explanation.* Harcourt, Brace and World, New York.

Neer, E. J. and Clapham, D. E. (1988). Roles of G protein subunits in transmembrane signalling.

Nature, **333**, 129–34.

Nelson, R. J., Sur, M., Felleman, D. J., and Kaas, J. H. (1981). Representations of the body surface in postcentral parietal cortex of *Macaca fascicularis. J. Comp. Neurol.* **192**, 611–44.

Nestler, E. J. and Greengard, P. (1984). *Protein phosphorylation in the nervous system.* Wiley, New York.

Neuman, R., Cherubini, E., and Ben-Ari, Y. (1987). Is activation of *N*-methyl-D-aspartate receptor channels sufficient to induce long term potentiation? *Neurosci. Lett.* **80**, 283–8.

Niedermeyer, E. and Lopes da Silva, F. (1982). *Electroencephalography: basic principles, clinical applications and related fields.* Urban and Schwarzenberg, Baltimore, Md.

Niki, H. and Watanabe, M. (1976). Prefrontal unit activity and delayed response: relation to cue location versus direction of response. *Brain Res.* **105**, 79–88.

Nishijo, H., Ono, T., and Nishino, H. (1988). Single neuron responses in amygdala of alert monkey during complex sensory stimulation with affective significance. *J. Neurosci.* **8**, 3570–83.

Nishizuka, Y. (1988). The molecular heterogeneity of protein kinase C and its implication for cellular regulation. *Nature*, **334**, 661–5.

Nordeen, K. W. and Nordeen, E. J. (1988*a*). Projection neurons within a vocal motor pathway are born during song learning in zebra finches. *Nature*, **334**, 149–51.

Nordeen, E. J. and Nordeen, K. W. (1988*b*). Sex and regional differences in the incorporation of neurons born during song learning in zebra finches. *J. Neurosci.* **8**, 2869–74.

Norman, R. J., Buchwald, J. S., and Villablanca, J. R. (1977). Classical conditioning with auditory discrimination of the eyeblink in decerebrate cats. *Science*, **196**, 551–3.

Nottebohm, F. (1969). The 'critical period' for song learning. *Ibis*, **111**, 386–7.

Nottebohm, F. (1970). Ontogeny of bird song. *Science*, **167**, 950–6.

Nottebohm, F. (1971). Neural lateralization of vocal control in a passerine bird. I. Song. *J. Exp. Zool.* **177**, 229–62.

Nottebohm, F. (1980*a*). Brain pathways for vocal learning in birds: a review of the first 10 years. *Prog. Psychobiol. Physiol. Psychol.* **9**, 85–124.

Nottebohm, F. (1980*b*). Testosterone triggers growth of brain vocal control nuclei in adult female canaries. *Brain Res.* **189**, 429–36.

Nottebohm, F. (1981). A brain for all seasons: cyclical anatomical changes in song control nuclei in the canary brain. *Science*, **214**, 1368–70.

Nottebohm, F. (1985). Neuronal replacement in adulthood. *Ann. NY Acad. Sci.* **457**, 143–61.

Nottebohm, F. and Arnold, A. P. (1976). Sexual dimorphism in vocal control areas of the songbird brain. *Science*, **194**, 211–13.

Nottebohm, F. and Nottebohm, M. E. (1976). Left hypoglossal dominance in the control of canary and white-crowned sparrow song. *J. Comp. Physiol.* **108**, 171–92.

Nottebohm, F., Stokes, T. M., and Leonard, C. M. (1976). Central control of song in the canary, *Serinus canarius. J. Comp. Neurol.* **165**, 457–86.

Nottebohm, F., Kasparian, S. and Pandazias, C. (1981). Brain space for a learned task. *Brain Res.* **213**, 99–109.

Nottebohm, F., Kelley, D. B., and Paton, J. A. (1982). Connections of vocal control nuclei in the canary telencephalon. *J. Comp. Neurol.* **207**, 344–57.

Nowak, L., Bregestovski, P., Ascher, P., Herbert, A., and Prochiantz, A. (1984). Magnesium gates glutamate-activated channels in mouse central neurons. *Nature*, **307**, 462–5.

Oakley, D. A. and Russell, I. S. (1977). Subcortical storage of pavlovian conditioning in the rabbit. *Physiol. and Behav.* **18**, 931–7.

Ocorr, K. A. and Byrne, J. H. (1985). Membrane responses and changes in cAMP levels in *Aplysia* sensory neurons produced by serotonin, tryptamine, FMRFamide and small cardioactive peptide B (SCPB). *Neurosci. Lett.* **55**, 113–18.

Ocorr, K. A., Walters, E. T., and Byrne, J. H. (1985). Associative conditioning analog selectively increases cAMP levels of tail sensory neurons in *Aplysia. Proc. Natl. Acad. Sci. USA*, **82**, 2548–52.

Ogren, S. O. (1985). Evidence for a role of brain serotonergic neurotransmission in avoidance learning. *Acta. Physiol. Scand.* **125**, supp. 544.

O'Keefe, J. (1979). A review of the hippocampal place cells. *Prog. Neurobiol.* **13**, 419–39.

O'Keefe, J. and Conway, D. H. (1978). Hippocampal place units in the freely moving rat: why they fire where they fire. *Exp. Brain Res.* **31**, 573–90.

O'Keefe, J. and Conway, D. H. (1980). On the trail of the hippocampal engram. *Physiol. Psychol.* **8**, 229–38.

O'Keefe, J. and Nadel, L. (1978). *The hippocampus as a cognitive map.* Oxford University Press, Oxford.

O'Keefe, J. and Speakman, A. (1987). Single unit activity in the rat hippocampus during a spatial memory task. *Exp. Brain Res.* **68**, 1–27.

Olds, J., Disterhoft, J. F., Segal, M., Kornblith, C. L., and Hirsh, R. (1972). Learning centers of rat brain mapped by measuring latencies of conditioned unit responses. *J. Neurophysiol.* **35**, 202–19.

Olds, J., Nienhuis, R., and Olds, M. E. (1978). Patterns of conditioned unit responses in the auditory system of the rat. *Exp. Neurol.* **59**, 209–28.

Olton, D. S., Becker, J. T., and Handelmann, G. E. (1979). Hippocampus, space and memory. *Behav. Brain Sci.* **2**, 313–65.

Olton, D. S., Becker, J. T., and Handelmann, G. E. (1980). Hippocampal function: working memory or cognitive mapping? *Physiol. Psychol.* **8**, 239–46.

Optican, L. M. and Richmond, B. J. (1987). Temporal encoding of two-dimensional patterns by single units in primate inferior temporal cortex. III. Information theoretic analysis. *J. Neurophysiol.* **57**, 162–78.

Orbach, J., Milner, B., and Rasmussen, T. (1960). Learning and retention in monkeys after amygdala-hippocampus resection. *Arch. Neurol.* **3**, 230–51.

Osofsky, J. D. (ed.) (1979). *Handbook of infant development.* Wiley, New York.

Overstreet, D. H. (1984). Behavioural plasticity and the cholinergic system. *Prog. Neuropsychol. Biol. Psychiat.* **8**, 133–51.

Paivio, A. (1986). *Mental representations. A dual coding approach.* Oxford University Press, New York.

Palm, G. (1982). *Neural assemblies. An alternative approach to artificial intelligence.* Springer-Verlag, Berlin.

Palm, G. and Aertsen, A. (eds.) (1986). *Brain Theory.* Springer-Verlag, Berlin.

Papez, J. W. (1937). A proposed mechanism of emotion. *Arch. Neurol. Psychiat.* **38**, 725–43.

Parkin, A. J. (1982). Residual learning capability in organic amnesia. *Cortex*, **18**, 417–40.

Parkin, A. J. (1984). Amnesic syndrome: a lesion-specific disorder? *Cortex*, **20**, 479–508.

Parkinson, J. K., Murray, E. A., and Mishkin, M. (1988). A selective mnemonic role for the hippocampus in monkeys: memory for the location of objects. *J. Neurosci.* **8**, 4159–67.

Parsons, B., Rainbow, T. C., Pfaff, D. W., and McEwan, B. S. (1982). Hypothalamic protein synthesis essential for the activation of the lordosis reflex in the female rat. *Endocrinol.* **110**, 620–4.

Passingham, R. (1975). Delayed matching after selective prefrontal lesions in monkeys (*Macaca mulatta*). *Brain Res.* **92**, 89–102.

Passingham, R. E. (1985). Memory of monkeys (*Macaca mulatta*) with lesions in prefrontal cortex.

Behav. Neurosci. **99**, 3–21.

Paterson, A. and Zangwill, O. L. (1945). A case of topographical disorientation associated with a unilateral cerebral lesion. *Brain* **68**, 188–212.

Paton, J. A. and Nottebohm, F. N. (1984). Neurons generated in the adult brain are recruited into functional circuits. *Science*, **225**, 1046–8.

Paton, J. A., O'Loughlin, B. E., and Nottebohm, F. (1985). Cells born in adult canary forebrain are local interneurons. *J. Neurosci.* **5**, 3088–93.

Pavlov, I. P. (1906). The scientific investigation of the physical faculties or processes in the higher animals. *Science*, **24**, 613–19.

Pavlov, I. P. (1927). *Conditioned reflexes. An investigation of the physiological activity of the cerebral cortex.* Oxford University Press, London.

Payne, J. K. and Horn, G. (1984). Long-term consequences of exposure to an imprinting stimulus on 'spontaneous' impulse activity in the chick brain. *Behav. Brain. Res.* **13**, 155–62.

Payne, R. B., Thompson, W. L., Fiala, K. L., and Sweany, L. L. (1981). Local song traditions in indigo buntings: cultural transmission of behavior patterns across generations. *Behavior*, **77**, 199–221.

Peckham, G. W. and Peckham, E. G. (1887). Some observations on the mental power of spiders. *J. Morphol.* **1**, 383–419.

Pellionisz, A. and Llinas, R. (1985). Tensor network theory of the metaorganization of functional geometries in the central nervous system. *Neurosci.* **16**, 245–73.

Penfield, W. and Milner, B. (1958). Memory deficit produced by bilateral lesions in the hippocampal zone. *Arch. Neurol. Psychiat.* **79**, 475–97.

Penfield, W. and Perot, P. (1963). The brain's record of auditory and visual experience. *Brain*, **86**, 595–696.

Peretz, B. and Lukowiak, K. D. (1975). Age-dependent CNS control of the habituating gill withdrawal reflex and of correlated activity in identified neurons in *Aplysia. J. Comp. Physiol.* **103**, 1–17.

Perkins, C. C., Jr (1968). An analysis of the concept of reinforcement. *Psychol. Rev.* **75**, 155–72.

Perrett, D. I., Rolls, E. T., and Caan, W. (1982). Visual neurons responsive to faces in the monkey temporal cortex. *Exp. Brain Res.* **47**, 329–42.

Perrett, D. I., Mistlin, A. J., and Chitty, A. J. (1987). Visual neurons responsive to faces. *Trends in Neurosci.* **10**, 358–64.

Peters, A. and Kaiserman-Abramof, I. R. (1970). The small pyramidal neurons of the cerebral cortex. The perikaryon, dendrites and spines. *Am. J. Anat.* **127**, 321–56.

Peterson, M. R. and Jusczyk, P. W. (1984). On perceptual predispositions for human speech and monkey vocalizations. In *The biology of learning* (ed. P. Marler and H. S. Terrace), pp. 585–616. Springer-Verlag, Berlin.

Petrides, M. and Iversen, S. D. (1979). Restricted posterior parietal lesions in the rhesus monkey and performance on visuospatial tasks. *Brain Res.* **161**, 63–77.

Pfaffenberger, C. J. and Scott, J. P. (1959). The relationship between socialization and trainability in guide dogs. *J. Genet. Psychol.* **95**, 145–55.

Pfenninger, K. H. (1986). Of nerve growth cones, leukocytes and memory: second messenger systems and growth-regulated proteins. *Trends in Neurosci.* **9**, 562–5.

Phelps, M. E. and Mazziotta, J. C. (1985). Positron emission tomography: human brain function and biochemistry. *Science*, **228**, 799–809.

Phillips, A. G. and Carr, G. D. (1987). Cognition and basal ganglia: a possible substrate for procedural knowledge. *Can. J. Neurol. Sci.* **14**, 381–5.

Pin, J. P., Bockaert, J., and Recasesn, M. (1984). The Ca^{2+}/Cl^- dependent L-[³H]glutamate binding: a new receptor or a particular transport process? *FEBS Lett.* **175**, 31–6.

Pinker, S. (ed.) (1985). *Visual cognition.* MIT Press, Cambridge, Mass.

Pinsker, H., Kupfermann, I., Castellucci, V., and Kandel, E. R. (1970). Habituation and dishabituation of the gill-withdrawal reflex in *Aplysia. Science,* **167**, 1740–42.

Piomelli, D., Volterra, A., Dale, N., Siegelbaum, S. A., Kandel, E. R., Schwartz, J. H., and Belardetti, F. (1987). Lipoxygenase metabolites of arachidonic acid as second messengers for presynaptic inhibition of *Aplysia* sensory cells. *Nature,* **328**, 38–43.

Pohl, W. (1973). Dissociation of spatial discrimination deficits following frontal and parietal lesions in monkeys. *J. Comp. Physiol. Psychol.* **82**, 227–39.

Pons, T. P., Garraghty, P. E., Friedman, D. P., and Mishkin, M. (1987). Physiological evidence for serial processing in somatosensory cortex. *Science,* **237**, 417–20.

Pons, T. P., Garraghty, P. E., and Mishkin, M. (1988). Lesion-induced plasticity in the second somatosensory cortex of adult macaques. *Proc. Natl. Acad. Sci. USA,* **85**, 5279–81.

Popper, K. R. and Eccles, J. C. (1981). *The self and its brain.* Springer-Verlag, Berlin.

Powell, D. A. and Buchanan, S. (1980). Autonomic-somatic relationships in the rabbit (*Oryctolagus cuniculus*): effects of hippocampal lesions. *Physiol. Psychol.* **8**, 455–62.

Prince, A. and Pinker, S. (1988). Rules and connections in human language. *Trends in Neurosci.* **11**, 195–202.

Prove, E. (1974). Der Einfluss von Kastration und Testosteronsubstitution auf das Sexualverhalten mannlicher Zebrafinken. *J. Ornithol.* **115**, 338–47.

Purves, D. and Voyvodic, J. T. (1987). Imaging mammalian nerve cells and their connections over time in living animals. *Trends in Neurosci.* **10**, 398–404.

Pylyshyn, Z. W. (1980). Computation and cognition: issues in the foundations of cognitive science. *Behav. Brain Sci.* **3**, 111–69.

Quach, T. T., Tang, F., Kageyama, H., Mocchetti, I., Guidotti, A., Meak, J. L., Costa, E., and Schwartz, J. P. (1984). Enkephalin biosynthesis in adrenal medulla. Modulation of proenkephalin mRNA content of cultured chromaffin cells by 8-bromo-adenosine 3'-5'-monophosphate. *Mol. Pharmacol.* **26**, 255–60.

Quinn, W. G. and Dudai, Y. (1976). Memory phases in *Drosophila. Nature* **262**, 576–7.

Quinn, W. G., Harris, W. A., and Benzer, S. (1974). Conditioned behavior in *Drosophila melanogaster. Proc. Natl. Acad. Sci. USA,* **71**, 708–12.

Quinn, W. G., Sziber, P. P., and Booker, R. (1979). The *Drosophila* memory mutant *amnesiac. Nature,* **277**, 212–14.

Racine, R. J., Milgram, N. W., and Hafner, S. (1983). Long-term potentiation phenomena in the rat limbic forebrain. *Brain Res.* **260**, 217–31.

Rafal, R. D. and Posner, M. I. (1987). Deficits in human spatial attention following thalamic lesions. *Proc. Natl. Acad. Sci. USA,* **84**, 7349–53.

Rainbow, T. C. (1979). Role of RNA and protein synthesis in memory formation. *Neurochem. Res.* **4**, 297–312.

Rakic, P. (1988). Specialization of cerebral cortical areas. *Science,* **241**, 170–6.

Rall, W. (1978). Dendritic spines and synaptic potency. In *Studies in neurophysiology* (ed. R. Porter), pp.

203–9. Cambridge University Press.

Ranck, J. B. (1973). Studies on single neurons in dorsal hippocampal formation and septum in unrestrained rats. Part I. Behavioral correlates and firing repertoires. *Exp. Neurol.* **41**, 461–531.

Rankin, C. H. and Carew, T. J. (1987). Development of learning and memory in *Aplysia*. II. Habituation and dishabituation. *J. Neurosci.* **7**, 133–43.

Rankin, C. H. and Carew, T. J. (1988). Dishabituation and sensitization emerge as separate processes during development in *Aplysia*. *J. Neurosci.* **8**, 197–211.

Rauschecker, J. P. and Singer, W. (1981). The effects of early visual experience on the cat's visual cortex and their possible explanations by Hebb synapses. *J. Physiol.* **310**, 215–39.

Rayport, S. G. and Schacher, S. (1986). Synaptic plasticity *in vitro*: cell culture of identified *Aplysia* neurons mediating short-term habituation and sensitization. *J. Neurosci.* **6**, 759–63.

Rescorla, R. A. (1967). Pavlovian conditioning and its proper control procedures. *Psychol. Rev.* **74**, 71–80.

Rescorla, R. A. (1988). Behavioral studies of Pavlovian conditioning. *Ann. Rev. Neurosci.* **11**, 329–52.

Rescorla, R. A. and Solomon, R. L. (1967). Two-process learning theory: relationships between Pavlovian conditioning and instrumental learning. *Psychol. Rev.* **74**, 151–82.

Reymann, K. G., Frey, U., Jork, R., and Matthies, H. (1988). Polymyxin B, an inhibitor of protein kinase C, prevents the maintenance of synaptic long-term potentiation in hippocampal CA_1 neurons. *Brain Res.* **440**, 305–14.

Riabowol, K. T., Fink, J. S., Gilman, M. Z., Walsh, D. A., Goodman, R. H., and Feramisco, J. R. (1988). The catalytic subunit of cAMP-dependent protein kinase induces expression of genes containing cAMP-responsive enhancer elements. *Nature*, **336**, 83–6.

Ribot, T. A. (1882). *Diseases of memory.* Appleton–Century–Crofts, New York.

Richards, G. and Ashbruner, M. (1984). Insect hormones and the regulation of genetic activity. In *Biological regulation and development*, Vol. 3B (ed. R. F. Goldberger and K. R. Yamamoto), pp. 213–53. Plenum Press, New York.

Richards, W. and Farley, J. (1984). Associative learning changes intrinsic to *Hermissenda* Type A photoreceptors. *Soc. Neurosci. Abst.* **10**, 623.

Richmond, B. J., Optican, L. M., Podell, M., and Spitzer, H. (1987). Temporal encoding of two-dimensional patterns by single units in primate inferior temporal cortex. I. Response characteristics. *J. Neurophysiol.* **57**, 132–46.

Roach, A., Adler, J. E., and Black, I. B. (1987). Depolarizing influences regulate preprotachykinin mRNA in sympathetic neurons. *Proc. Natl. Acad. Sci. USA*, **84**, 5078–81.

Robinson, D. A. (1976). Adaptive gain control of vestibuloocular reflex by the cerebellum. *J. Neurophysiol.* **39**, 954–69.

Robinson, G. B. and Racine, R. J. (1985). Long-term potentiation in the dentate gyrus: effects of noradrenaline depletion in the awake rat. *Brain Res.* **325**, 71–8.

Rogers, L. J. (1982). Light experience and asymmetry of brain function in chickens. *Nature*, **297**, 223–5.

Roitblat, H. L. (1982). The meaning of representation in animal memory. *Behav. Brain Sci.* **5**, 353–406.

Roland, P. E. and Friberg, L. (1985). Localization of cortical areas activated by thinking. *J. Neurophysiol.* **53**, 1219–43.

Romanes, G. J. (1882). *Animal intelligence.* Kegan, Paul and Trensh, London.

Rose, S. P. R. (1981). What should a biochemistry of learning and memory be about? *Neurosci.* **6**, 811–21.

Rose, S. P. R. and Csillag, A. (1985). Passive avoidance training results in lasting changes in deoxy-glucose metabolism in left hemisphere regions of chick brain. *Behav. Neural Biol.* **44**, 315–24.

Rose, S. P. R. and Harding, S. (1984). Training increases [³H]fucose incorporation in chick brain only if followed by memory storage. *Neurosci.* **12**, 663–7.

Rosenfield, M. E. and Moore, J. W. (1983). Red nucleus lesions disrupt the classically conditioned nictitating membrane response in rabbits. *Behav. Brain Res.* **10**, 393–8.

Rosenfield, M. E., Dovydaitis, A., and Moore, J. W. (1985). Brachium conjunctivum and rubrobulbar tract: brain stem projections of red nucleus essential for the conditioned nictitating membrane response. *Physiol. Behav.* **34**, 751–9.

Routtenberg, A. and Lovinger, D. M. (1985). Selective increase in phosphorylation of a 47-kDa protein (F1) directly related to long-term potentiation. *Behav. Neural Biol.* **43**, 3–11.

Rowe, C. J. (1980). *An outline of psychiatry* (7th edn). W. C. Brown, Iowa.

Rozengurt, E. (1986). Early signals in the mitogenic response. *Science*, **234**, 161–6.

Rubens, A. (1979). Agnosia. In *Clinical neuropsychology* (ed. K. M. Heilman and E. Valenstein), pp. 233–67. Oxford University Press, New York.

Rumelhart, D. E. and McClelland, J. L. (ed., with the PDP research group) (1986a). *Parallel distributed processing*, vol. 1. MIT Press, Cambridge, Mass.

Rumelhart, D. E. and McClelland, J. L. (1986b). On learning the past tenses of English verbs. In *Parallel distributed processing*, Vol. 2 (ed. J. L. McClelland and D. E. Rumelhart), pp. 217–71. MIT Press, Cambridge, Mass.

Russell, B. (1961). *History of western philosophy* (2nd edn). George Allen and Unwin, London.

Sackmann, B. and Neher, E. (ed.) (1983). *Single-channel recording.* Plenum Press, New York.

Sacks, O. (1985). *The man who mistook his wife for a hat, and other clinical stories.* Summit, New York.

Sacktor, T. C., O'Brien, C. A., Weinstein, I. B., and Schwartz, J. H. (1986). Translocation from cytosol to membrane of protein kinase C after stimulation of *Aplysia* neurons with serotonin. *Soc. Neurosci. Abst.* **12**, 1340.

Sahgal, A. and Iversen, S. D. (1978). Categorization and retrieval after selective inferotemporal lesions in monkeys. *Brain Res.* **146**, 341–50.

Sahley, C. (1985). Co-activation, cell assemblies and learning. *Trends in Neurosci.* **8**, 423–4.

Sahley, C. L., Galperin, A., and Rudy, J. W. (1981a). One-trial learning modifies food odor preferences of a terrestrial mollusc. *Proc. Natl. Acad. Sci. USA*, **78**, 640–2.

Sahley, C. L., Rudy, J. W., and Galperin, A. (1981b). An analysis of associative learning in a terrestrial mollusc. I. Higher-order conditioning, blocking, and a transient US pre-exposure effect. *J. Comp. Physiol.* **144**, 1–8.

Saitoh, T. and Schwartz, J. H. (1985). Phosphorylation-dependent subcellular translocation of a Ca^{2+}/calmodulin-dependent protein kinase produces an autonomous enzyme in *Aplysia. J. Cell Biol.* **100**, 835–42.

Sakai, M. (1974). Prefrontal unit activity during visually guided lever pressing reaction in the monkey. *Brain Res.* **81**, 297–309.

Sakakibara, M. D., Alkon, D. L., Neary, J. T., DeLorenzo, R., Gould, R., and Heldman, E. (1985). Ca^{2+}-mediated reduction of Ca^{2+}-currents is enhanced by injection of IP3 or neuronal Ca^{2+}/calmodulin kinase type II. *Soc. Neurosci. Abst.* **11**, 956.

Sakakibara, M., Collin, C., Kuzirian, A., Alkon, D. L., Heldman, E., Naito, S., and Lederhendler, I.

(1987). Effects of α_2-adrenergic agonists and antagonists on photoreceptor membrane currents. *J. Neurochem.* **48**, 405–16.

Sakurai, M. (1987). Synaptic modification of parallel fibre-Purkinje cell transmission in in vitro guinea-pig cerebellar slices. *J. Physiol.* **394**, 463–80.

Salmon, D. P., Zola-Morgan, S., and Squire, L. R. (1987). Retrograde amnesia following combined hippocampus-amygdala lesions in monkeys. *Psychobiology*, **15**, 37–47.

Salzen, E. A., Parker, D. M., and Williamson, A. J. (1975). A forebrain lesion preventing imprinting in domestic chicks. *Exp. Brain Res.* **24**, 145–57.

Salzen, E. A., Parker, D. M., and Williamson, A. J. (1978). Forebrain lesions and retention of imprinting in domestic chicks. *Exp. Brain Res.* **31**, 107–16.

Sanders, H. I. and Warrington, E. K. (1971). Memory for remote events in amnesic patients. *Brain*, **94**, 661–8.

Sanes, J. N., Suner, S., Lando, J. F., and Donoghue, J. P. (1988). Rapid reorganization of adult motor cortex somatic representation patterns after motor nerve injury. *Proc. Natl. Acad. Sci. USA*, **85**, 2003–7.

Sassone-Corsi, P., Sisson, J. C., and Verma, I. N. (1988). Transcriptional autoregulation of the proto-oncogene *fos*. *Nature*, **334**, 314–19.

Sastry, B. R. and Goh, J. W. (1984). Long-lasting potentiation in hippocampus is not due to an increase in glutamate receptors. *Life Sci.* **34**, 1497–501.

Sastry, B. R., Goh, J. W., and Auyeung, A. (1986). Associative induction of post-tetanic and long-term potentiation in CA1 neurons of rat hippocampus. *Science*, **232**, 988–90.

Sato, S. and Smith, P. D. (1985). Magnetoencephalography. *J. Clin. Neurophysiol.* **2**, 173–92.

Saunders, R. C. (1983). Impairment in recognition memory after mammillary body lesions in monkeys. *Soc. Neurosci. Abst.* **9**, 28.

Saxe, L., Dougherty, D., and Cross, T. (1985). Scientific validity of polygraph testing. *Am. Psychol.* **40**(3), 355–66.

Schacher, S. and Proshansky, E. (1983). Neurite regeneration by *Aplysia* neurons in dissociated cell culture: modulation by *Aplysia* hemolymph and the presence of the initial axonal segment. *J. Neurosci.* **3**, 2403–13.

Schacher, S., Castellucci, V. F., and Kandel, E. R. (1988). cAMP evokes long-term facilitation in *Aplysia* sensory neurons that requires new protein synthesis. *Science*, **240**, 1667–9.

Scharma, L. H., de Graan, P. N. E., Zwiers, H., and Gispen, W. H. (1986). Comparison of a 52-kDa phosphoprotein from synaptic plasma membranes related to long-term potentiation and the major coated vesicle phosphoprotein. *J. Neurochem.* **47**, 1843–8.

Scheich, H., Bonke, B. A., Bonke, D., and Langner, G. (1979). Functional organization of some auditory nuclei in the guinea fowl demonstrated by the 2-deoxyglucose technique. *Cell Tiss. Res.* **204**, 17–27.

Schmaltz, L. W. and Theios, J. (1972). Acquisition and extinction of a classically conditioned response in hippocampectomized rabbits (*Oryctolagus cuniculus*). *J. Comp. Physiol. Psychol.* **79**, 328–33.

Schneiderman, N., Fuentes, I., and Gormezano, I. (1962). Acquisition and extinction of the classically conditioned eyelid response in the albino rabbit. *Science*, **136**, 650–2.

Scholz, K. P. and Byrne, J. H. (1987). Long-term sensitization in *Aplysia*: biophysical correlates in tail sensory neurons. *Science*, **235**, 685–7.

Scholz, K. P. and Byrne, J. H. (1988). Intracellular injections of cAMP induces a long-term reduction of neuronal K⁺ currents. *Science*, **240**, 1664–6.

Schramm, M. and Selinger, Z. (1984). Message transmission: receptor controlled adenylate cyclase system. *Science*, **225**, 1350–6.

Schuman, E. M., Farley, J., and Richards, W. G. (1986). Neural substrates of locomotion in *Hermissenda. Soc. Neurosci. Abst.* **12**, 1205.

Schwartz, J. H. and Greenberg, S. M. (1987). Molecular mechanisms for memory: second-messenger induced modifications of protein kinases in nerve cells. *Ann. Rev. Neurosci.* **10**, 459–76.

Schwartz, J. H., Castellucci, V., and Kandel, E. R. (1971). Functioning of identified neurons and synapses in abdominal ganglion of *Aplysia* in absence of protein synthesis. *J. Neurophysiol.* **34**, 939–53.

Schwartz, J. H., Bernier, L., Castellucci, V. F., Palazzolo, M., Saito, T., Stapleton, A., and Kandel, E. R. (1983). What molecular steps determine the time course of the memory for short-term sensitization in *Aplysia? Cold Spring Harbor Symp. Quant. Biol.* **48**, 811–19.

Schwartzkroin, P. A. and Wester, K. (1975). Long-lasting facilitation of a synaptic potential following tetanization in the *in vitro* hippocampal slice. *Brain Res.* **89**, 107–19.

Schworer, C. M., Colbran, R. J., and Soderling, T. R. (1986). Reversible generation of a Ca²⁺-independent form of Ca²⁺ (calmodulin)-dependent protein kinase II by an autophosphorylation mechanism. *J. Biol. Chem.* **261**, 8581–4.

Scott, J. P. (1962). Critical periods in behavioral development. *Science*, **138**, 949–58.

Scoville, W. B. and Milner, B. (1957). Loss of recent memory after bilateral hippocampal lesions. *J. Neurol. Neurosurg. Psychiat.* **20**, 11–21.

Sechenov, I. (1862). *Reflexes of the brain.* (English translation (1965), MIT Press, Cambridge, Mass.)

Segal, M. (1973). Flow of conditioned responses in limbic telencephalic system of the rat. *J. Neurophysiol.* **35**, 840–54.

Segal, M. (1977*a*). Excitability changes in rat hippocampus during conditioning. *Exp. Neurol.* **55**, 67–73.

Segal, M. (1977*b*). Changes of interhemispheric hippocampal responses during conditioning in the awake rat. *Exp. Brain Res.* **29**, 553–65.

Segal, M. and Olds, J. (1972). The behavior of units in the hippocampal circuit of the rat during learning. *J. Neurophysiol.* **35**, 680–90.

Selverston, A. I. (ed.) (1985). *Model neural networks and behavior.* Plenum Press, New York.

Seubert, P., Larson, J., Oliver, M., Jung, M. W., Baudry, M., and Lynch, G. (1988). Activation of NMDA receptors induces proteolysis of spectrin in hippocampus. *Soc. Neurosci. Abst.* **14**, 476.

Sharp, P. E., McNaughton, B. I., and Barnes, C. A. (1985). Enhancement of hippocampal field potentials in rats exposed to a novel, complex environment. *Brain Res.* **339**, 361–5.

Sharpless, S. and Jasper, H. (1956). Habituation of the arousal reaction. *Brain*, **79**, 655–82.

Shashoua, V. E. (1985). The role of extracellular proteins in learning and memory. *Am. Sci.* **73**, 364–70.

Shepherd, G. M. (1988). *Neurobiology* (2nd edn). Oxford University Press, New York.

Sherman, S. M. and Spear, P. D. (1982). Organization of visual pathways in normal and visually deprived cats. *Physiol. Rev.* **62**, 738–855.

Sherrington, C. S. (1906). *The integrative action of the nervous system.* Reprinted by Yale University Press, New Haven, 1977.

Shettleworth, S. J. (1983). Memory in food-hoarding birds. *Sci. Amer.* **248** (3), 86–95.

Shimahara, T. and Peretz, B. (1978). Soma potential of an interneurone controls transmitter release in a monosynaptic pathway in *Aplysia. Nature*, **273**, 158–60.

Shimamura, A. P. and Squire, L. R. (1986). Korsakoff's syndrome: a study of the relation between anterograde amnesia and remote memory impairment. *Behav. Neurosci.* **100**, 165–70.

Shimamura, A. P., Jernigan, T. L., and Squire, L. R. (1988). Korsakoff's syndrome: radiological (CT) findings and neuropyschological correlates. *J. Neurosci.* **8**, 4400–10.

Shoemaker, H. H. (1939). Effect of testosterone propionate on the behavior of the female canary. *Proc. Soc. Exp. Biol. Med.* **41**, 299–302.

Shuster, M. J., Camardo, J. S., Siegelbaum, S. A., and Kandel, E. R. (1985). Cyclic AMP-dependent protein kinase closes the serotonin-sensitive K^+ channels of *Aplysia* sensory neurones in cell-free membrane patches. *Nature*, **313**, 392–5.

Siegel, G. J., Agranoff, B. W., Albers, R. W., and Molinoff, P. (ed.), (1989). *Basic neurochemistry* (4th edn). Raven Press, New York.

Siegel, R. W. and Hall, J. (1979). Conditioned responses in courtship behaviour of normal and mutant *Drosophila. Proc. Natl. Acad. Sci. USA*, **76**, 3430–4.

Siegelbaum, S. A. and Tsien, R. W. (1983). Modulation of gated ion channels as a mode of transmitter action. *Trends in Neurosci.* **6**, 307–13.

Siegelbaum, S. A., Camardo, J. S., and Kandel, E. R. (1982). Serotonin and cyclic AMP close single K^+ channels in *Aplysia* sensory neurons. *Nature*, **299**, 413–17.

Signoret, J. L. (1985). Memory and amnesias. In *Principles of behavioral neurology* (ed. M. M. Mesulam), pp. 169–92. Davis, Philadelphia.

Siman, R. and Noszek, J. C. (1988). Excitatory amino acids activate calpain I and induce structural protein breakdown *in vivo. Neuron*, **1**, 279–87.

Simons, D. J. and Land, P. W. (1987). Early experience of tactile stimulation influences organization of somatic sensory cortex. *Nature*, **326**, 694–7.

Singer, W. (1987). Activity-dependent self-organization of synaptic connections as a substrate of learning. In *The neuronal and molecular bases of learning* (ed. J.-P. Changeux and M. Konishi), pp. 301–36. Wiley, New York.

Skarda, C. A. and Freeman, W. J. (1987). How brains make chaos in order to make sense of the world. *Behav. Brain Sci.* **10**, 161–95.

Skelton, R. W., Scarth, A. S., Wilkie, D. M., Miller, J. J., and Philips, A. G. (1987). Long-term increases in dentate granule cell responsivity accompany operant conditioning. *J. Neurosci.* **7**, 3081–7.

Skinner, B. F. (1938). *The behavior of organisms.* Appleton-Century-Crofts, New York.

Skinner, B. F. (1948). Superstition in the pigeon. *J. Exp. Psychol.* **38**, 168–72.

Skinner, B. F. (1957). *Verbal behavior.* Appleton-Century-Crofts, New York.

Skinner, B. F. (1960). Pigeons in a pelican. *Am. Psychol.* **15**, 28–37.

Skinner, B. F. (1984). Cannonical papers. *Behav. Brain Res.* **7**, 473–724.

Skoyles, J. R. (1988). Training the brain using neural-networks models. *Nature*, **333**, 401.

Skrede, K. K. and Malthe-Sorenssen, D. (1981). Increased resting and evoked release of transmitter following repetitive electrical tetanization in hippocampus: a biochemical correlate to long-lasting synaptic potentiation. *Brain Res.* **208**, 436–41.

Small, W. S. (1901). Experimental study of the mental processes of the rat. II. *Am. J. Psychol.* **12**, 206–39.

Smith, A. M. (1970). The effects of rubral lesions and stimulation on conditioned forelimb flexion responses in the cat. *Physiol Behav.* **5**, 1121–6.

Smith, F. V. and Bird, M. W. (1963). The relative attraction for the domestic chick of combinations of stimuli in different sensory modalities. *Anim. Behav.* **11**, 300–5.

Smith, R. F., Choi, K.-W., Tully, T., and Quinn, W. G. (1986). Deficient protein kinase C activity in *turnip*, a *Drosophila* learning mutant. *Soc. Neurosci. Abst.* **12**, 399.

Smith, S. J. and Augustine, G. J. (1988). Calcium ions, active zones and synaptic transmitter release. *Trends in Neurosci.* **11**, 458–64.

Smith, T. G., Barker, J. L., Smith, B. M., and Colburn, T. R. (1980). Voltage clamping with microelectrodes. *J. Neurosci. Meth.* **3**, 105–28.

Snipes, G. J., Freeman, J. A., Costello, B., Chan, S., and Routtenberg, A. (1986). Evidence that the growth-associated protein, GAP-43, and plasticity-associated protein, protein F1, are identical. *Soc. Neurosci. Abst.* **12**, 500.

Sokoloff, L., Reivich, M., Kennedy, C., Des Rosiers, M. H., Patlak, C. S., Pettigrew, K. D., Sakurada, O., and Shinohara, M. (1977). The [^{14}C]deoxyglucose method for the measurement of local cerebral glucose utilization: theory, procedure, and normal values in the conscious and anesthetized albino rat. *J. Neurochem.* **28**, 897–916.

Sokolov, E. N. (1963). Higher nervous functions: the orienting reflex. *Ann. Rev. Physiol.* **25**, 545–80.

Solomon, P. R. (1977). Role of the hippocampus in blocking and conditioned inhibition of the rabbit's nictitating membrane response. *J. Comp. Physiol. Psychol.* **91**, 407–17.

Solomon, P. R. (1980). A time and a place for everything? Temporal processing views of hippocampal function with special reference to attention. *Physiol. Psychol.* **8**, 254–61.

Solomon, P. R. and Moore, J. W. (1975). Latent inhibition and stimulus generalization of the classically conditioned nictitating membrane response in rabbits (*Oryctolagus cuniculus*) following dorsal hippocampal ablation. *J. Comp. Physiol. Psychol.* **89**, 1192–203.

Solomon, P. R., Lewis, J. L., LoTurco, J., Steinmetz, J. E., and Thompson, R. F. (1986). The role of the middle cerebellar peduncle in acquisition and retention of the rabbit's classically conditioned nictitating membrane response. *Bull. Psychon. Soc.* **24**, 75–8.

Sorabji, R. (1972). *Aristotle on memory.* Duckworth, London.

Spalding, D. A. (1873). Instinct, with original observations on young animals. *Macmillan's mag.* **27**, 282–93. (Reprinted (1954) in *Br. J. Anim. Behav.* **2**, 2–11.)

Sparks, D. L. and Nelson, J. S. (1987). Sensory and motor maps in the mammalian superior colliculus. *Trends in Neurosci.* **10**, 312–17.

Spiegler, B. J. and Mishkin, M. (1981). Evidence for the sequential participation of inferior temporal cortex and amygdala in the acquisition of stimulus-reward associations. *Behav. Brain Res.* **3**, 303–17.

Spitzer, H., Desimone, R., and Moran, J. (1988). Increased attention enhances both behavioral and neuronal performance. *Science*, **240**, 338–40.

Squire, L. R. (1981). Two forms of human amnesia: analysis of forgetting. *J. Neurosci.* **1**, 635–40.

Squire, L. R. (1982). The neuropsychology of human memory. *Ann. Rev. Neurosci.* **5**, 241–73.

Squire, L. R. (1987). *Memory and brain.* Oxford University Press, New York.

Squire, L. R. and Barondes, S. H. (1976). Amnesic effect of cycloheximide not due to depletion of a constitutive brain protein with short half-life. *Brain Res.* **103**, 183–9.

Squire, L. R. and Moore, R. Y. (1979). Dorsal thalamic lesion in a noted case of human memory disfunction. *Ann. Neurol.* **6**, 503–6.

Squire, L. R., Cohen, N. J., and Zouzounis, J. A. (1984). Preserved memory in retrograde amnesia: sparing of a recently acquired skill. *Neuropsychologia*, **22**, 145–52.

Stanton, P. K., Sarvey, J. M., and Moskal, J. R. (1987). Inhibition of the production and maintenance of long-term potentiation in rat hippocampal slices by a monoclonal antibody. *Proc. Natl. Acad. Sci. USA*, **84**, 1684–8.

Stanton, P. K., Jester, J., Chattarji, S., and Sejnowski, T. J. (1988). Associative long-term potentiation (LTP) or depression (LTD) is produced in hippocampus dependent upon the phase of rhythmically active inputs. *Soc. Neurosci. Abst.* **14**, 19.

Steinmetz, J. E., Lavond, D. G., and Thompson, R. F. (1985). Classical conditioning of the rabbit eyelid response with mossy fiber stimulation as the conditioned stimulus. *Bull. Psychon. Soc.* **23**, 245–8.

Steinmetz, J. E., Rosen, D. J., Chapman, P. F., Lavond, D. G., and Thompson, R. F. (1986). Classical conditioning of the rabbit eyelid response with a mossy fiber stimulation CS. I. Pontine nuclei and middle cerebellar peduncle stimulation. *Behav. Neurosci.* **100**, 878–87.

Stent, G. S. (1973). A physiological mechanism for Hebb's postulate of learning. *Proc. Natl. Acad. Sci. USA*, **70**, 997–1001.

Steward, O. and Levy, W. B. (1982). Preferential localization of polyribosomes under the base of dendritic spines in granule cells of the dentate gyrus. *J. Neurosci.* **2**, 284–91.

Stillings, N. A., Feinstein, M. H., Garfield, J. L., Rissland, E. L., Rosenbaum, D. A., Weisler, S. E., and Baker-Ward, L. (1987). *Cognitive science. An introduction.* MIT Press, Cambridge, Mass.

Stuss, D. T. and Benson, D. F. (1986). *The frontal lobes.* Raven Press, New York.

Sugishita, M., Ettlinger, G., and Ridley, R. M. (1978). Disturbance of cage finding in the monkey. *Cortex*, **14**, 431–8.

Sullivan, W. E. and Konishi, M. (1986). Neural map of interaural phase differences in the owl's brainstem. *Proc. Natl. Acad. Sci. USA*, **83**, 8400–4.

Suzuki, H. (1985). Distribution and organization of visual and auditory neurons in the monkey prefrontal cortex. *Vision Res.* **25**, 465–9.

Suzuki, H. and Azuma, M. (1977). Prefrontal neuronal activity during gazing at a light spot in the monkey. *Brain Res.* **126**, 497–508.

Swanson, L. W. (1987). Limbic system. In *Encyclopedia of Neuroscience* (ed. G. Adelman), pp. 589–91. Birkhauser, Boston.

Sweatt, D., Volterra, A., Siegelbaum, S., and Kandel, E. (1988). Molecular consequences of presynaptic facilitation and presynaptic inhibition on common substrate proteins of individual sensory neurons in *Aplysia*. In *The molecular biology of signal transduction*, p. 75. Cold Spring Harbor, New York.

Swindale, N. V. (1981). Dendritic spines only connect. *Trends in Neurosci.* **4**, 240–1.

Tank, D. W. and Hopfield, J. J. (1987). Collective computation in neuronlike circuits. *Sci. Amer.* **257** (12), 62–70.

Tarpy, R. M. (1975). *Basic principles of learning.* Scott, Foresman, Glenview, Ill.

Tauc, L. (1965). Presynaptic inhibition in the abdominal ganglion of *Aplysia*. *J. Physiol.* **181**, 282–307.

Tauc, L. (1966). Physiology of the nervous system. In *Physiology of mollusca*, Vol. 2 (ed. K. M. Wilbur and C. M. Yonge), pp. 387–454. Academic Press, New York.

Technau, G. (1984). Fiber number in the mushroom bodies of adult *Drosophila melanogaster* depends on age, sex and experience. *J. Neurogenet.* **1**, 113–26.

Tempel, B. L., Bonini, N., Dawson, D. R., and Quinn, W. G. (1983). Reward learning in normal and mutant *Drosophila. Proc. Natl. Acad. Sci. USA*, **80**, 1482–6.

Tempel, B. L., Livingstone, M. S., and Quinn, W. G. (1984). Mutations in the dopa decarboxylase gene affect learning in *Drosophila. Proc. Natl. Acad. Sci. USA*, **81**, 3577–81.

Terrace, H. S. (1984). Animal learning, ethology, and biological constraints. In *The biology of learning* (ed. P. Marler and H. S. Terrace), pp. 15–45. Springer-Verlag, Berlin.

Teuber, H. L. (1965). Postscript: some needed revisions of the classical views of agnosias. *Neuropsychologia*, **3**, 371–8.

Teuber, H. L., Milner, B., and Vaughan, H. G., Jr (1968). Persistent anterograde amnesia after stab wound of the basal brain. *Neuropsychologia*, **6**, 267–82.

Teyler, T. J. and DiScenna, P. (1986). The hippocampal memory indexing theory. *Behav. Neurosci.* **100**, 147–54.

Teyler, T. J. and DiScenna, P. (1987). Long-term potentiation. *Ann. Rev. Neurosci.* **10**, 131–61.

Thach, W. T. (1978). Correlation of neural discharge with pattern and force of muscular activity, joint position, and direction of intended next movement in motor cortex and cerebellum. *J. Neurophysiol.* **41**, 654–76.

Thompson, R. F. (1976). The search for the engram. *Am. Psychol.* **31**, 209–27.

Thompson, R. F. (1986). The neurobiology of learning and memory. *Science*, **233**, 941–7.

Thompson, R. F. and Spencer, W. A. (1966). Habituation: a model phenomenon for the study of neuronal substrates of behavior. *Psychol. Rev.* **73**, 16–43.

Thompson, R. F., Barchas, J. D., Clark, G. A., Donegan, N., Kettner, R. E., Lavond, D. G., Madden, J., Mauk, M. D., and McCormick, D. A. (1984). Neuronal substrates of associative learning in the mammalian brain. In *Primary neural substrates of learning and behavioural change* (ed. D. L. Alkon and J. Farley), pp. 71–99. Cambridge University Press, Cambridge.

Thomson, A. (1986). A magnesium-sensitive post-synaptic potential in rat cerebral cortex resembles responses to *N*-methylaspartate. *J. Physiol.* **370**, 531–49.

Thorndike, E. L. (1911). *Animal intelligence.* Macmillan, New York.

Thorndike, L. (1923). *A history of magic and experimental science during the first thirteen centuries of our era*, Vols I and II. Columbia University Press, New York.

Thorpe, W. H. (1954). The process of song-learning in the chaffinch as studied by means of the sound spectrograph. *Nature*, **173**, 465.

Thorpe, W. H. (1956). *Learning and instinct in animals.* Harvard University Press, Cambridge, Mass.

Thorpe, W. (1958). The learning of song patterns by birds, with special reference to the song of the chaffinch, *Fringilla coelebs. Ibis*, **100**, 535–70.

Tilloston, D. (1979). Inactivation of Ca^{2+} conductance dependent on entry of Ca^{2+} ions in molluscan neurones. *Proc. Natl. Acad. Sci. USA*, **76**, 1497–500.

Tinbergen, N. (1969). *The study of instinct.* Oxford University Press, Oxford.

Tischler, M. D. and Davis, M. (1983). A visual pathway that mediates fear-conditioned enhancement of acoustic startle. *Brain Res.* **276**, 55–71.

Tolman, E. C. (1932). *Purposive behavior in animals and man.* Appleton-Century-Crofts, New York. (Reprinted (1967) in Century psychology series.)

Tolman, E. C. (1938). The determiners of behavior at a choice point. *Psychol. Rev.* **45**, 1–41.

Tramontano, D., Chin, W. W., Moses, A. C., and Ingbar, S. H. (1986). Thyrotropin and dibutyryl cyclic AMP increase levels of *c-myc* and *c-fos* mRNAs in cultured rat thyroid cells. *J. Biol. Chem.* **261**, 3919–22.

Tsukahara, N. (1981). Synaptic plasticity in the mammalian central nervous system. *Ann. Rev. Neurosci.* **4**, 351–79.

Tsukahara, N. (1984). Classical conditioning mediated by the red nucleus: an approach beginning at the cellular level. In *Neurobiology of learning and memory* (ed. G. Lynch, J. L. McGaugh, and N. M. Weinberger), pp. 165–80. Guilford, New York.

Tsukahara, N., Oda, Y., and Notsu, T. (1981). Classical conditioning mediated by the red nucleus in the cat. *J. Neurosci.* **1**, 72–9.

Tsumoto, T., Hagihara, K., Sato, H., and Hata, Y. (1987). NMDA receptors in the visual cortex of young kittens are more effective than those of adult cats. *Nature*, **327**, 513–14.

Tully, T. (1987). *Drosophila* learning and memory revisited. *Trends in Neurosci.* **10**, 330–5.

Tully, T. and Quinn, W. G. (1985). Classical conditioning and retention in normal and mutant *Drosophila melanogaster. J. Comp. Physiol.* **157**, 263–77.

Tulving, E. (1972). Episodic and semantic memory. In *Organization of memory* (ed. E. Tulving and W. Donaldson), pp. 381–403. Academic Press, New York.

Tulving, E. (1983). *Elements of episodic memory.* Oxford University Press, Oxford.

Tulving, E. (1984). Précis of *Elements of episodic memory* (with open peer commentary). *Behav. Brain Sci.* **7**, 223–68.

Turner, R. W., Baimbridge, K. G., and Miller, J. J. (1982). Calcium-induced long-term potentiation in the hippocampus. *Neurosci.* **7**, 1411–16.

Tusa, R. J., Palmar, L. A., and Rosenquist, A. C. (1978). The retinotopic organization of area 17 (striate cortex) in the cat. *J. Comp. Neurol.* **177**, 213–35.

Ullman, S. (1980). Against direct perception. *Behav. Brain Sci.* **3**, 373–415.

Ullman, S. (1986). Artificial intelligence and the brain: computational studies of the visual system. *Ann. Rev. Neurosci.* **9**, 1–26.

Ungar, G. (1970). Molecular mechanisms in information processing. *Int. Rev. Neurobiol.* **13**, 223–50.

Valenstein, E. S. (ed.) (1973). *Brain stimulation and motivation.* Scott, Foresman, Glenview, Ill.

Van Hoesen, G. W. (1985). Neural systems of the non-human primate forebrain implicated in memory. *Ann. NY Acad. Sci.* **444**, 97–112.

Van Tienhoven, A. and Juhasz, L. P. (1962). The chicken telencephalon, diencephalon and mesencephalon in stereotaxic coordinates. *J. Comp. Neurol.* **118**, 185–98.

Verma, I. M. (1986). Proto-oncogene *fos*: a multifaceted gene. *Trends in Genet.* **2**, 93–6.

Victor, M. (1976). The Wernicke-Korsakoff syndrome. In *Handbook of clinical neurology* (ed. P. J. Vinken and G. W. Bruyn), Vol. 28: Metabolic and deficiency diseases of the nervous system, Part II, pp. 243–70. North-Holland, Amsterdam.

Vidal, J. M. (1980). The relations between filial and sexual imprinting in the domestic fowl: effects of age and social experience. *Anim. Behav.* **28**, 880–91.

von der Malsburg, C. (1987). Synaptic plasticity as basis of brain organization. In *The neural and molecular bases of learning* (ed. J.-P. Changeux and M. Konishi), pp. 411–32. Wiley, Chichester.

Wagner, H., Takahashi, T., and Konishi, M. (1987). Representation of interaural time difference in the central nucleus of the barn owl's inferior colliculus. *J. Neurosci.* **7**, 3105–16.

Walk, R. D. (1978). Perceptual learning. In *Handbook of perception*, Vol. IX: Perceptual processing (ed. C. Carterette and M. P. Friedman), pp. 257–97. Academic Press, New York.

Walker, J. A. and Olton, D. S. (1984). Fimbria-fornix lesions impair spatial working memory but not cognitive mapping. *Behav. Neurosci.* **98**, 226–42.

Wall, J. T., Gibbs, C. M., Broyles, J. L., and Cohen, D. H. (1985). Modification of neuronal discharge along the ascending tectofugal pathway during visual conditioning. *Brain Res.* **342**, 67–76.

Wall, J. T., Kaas, J. H., Sur, M., Nelson, R. J., Felleman, D. J., and Merzenich, M. M. (1986). Functional reorganization in somatosensory cortical areas 3b and 1 of adult monkeys after median nerve repair: possible relationships to sensory recovery in humans. *J. Neurosci.* **6**, 218–33.

Walsh, J. P. and Byrne, J. H. (1984). Forskolin mimics and blocks a serotonin-sensitive decreased K$^+$ conductance in tail sensory neurons of *Aplysia*. *Neurosci. Lett.* **52**, 7–11.

Walters, E. T. and Byrne, J. H. (1983). Associative conditioning of single sensory neurons suggests a cellular mechanism for learning. *Science*, **219**, 405–8.

Walters, E. T. and Byrne, J. H. (1985). Long-term enhancement produced by activity-dependent modulation of *Aplysia* sensory neurons. *J. Neurosci.* **5**, 662–72.

Walters, E. T., Byrne, J. H., Carew, T. J., and Kandel, E. R. (1983a). Mechanoafferent neurons innervating the tail of *Aplysia*. I. Response properties and synaptic connections. *J. Neurophysiol.* **50**, 1522–42.

Walters, E. T., Byrne, J. H., Carew, T. J., and Kandel, E. R. (1983b). Mechanoafferent neurons innervating the tail of *Aplysia*. II. Modulation by sensitizing stimulation. *J. Neurophysiol.* **50**, 1543–59.

Warrington, E. K. (1982). Neuropsychological studies of object recognition. *Phil. Trans. R. Soc. Lond. B*, **298**, 15–33.

Watanabe, M. (1986). Prefrontal unit activity during delayed conditional go/no-go discrimination in the monkey. I. reaction to the stimulus. *Brain Res.* **382**, 1–14.

Watanabe, T. and Niki, H. (1985). Hippocampal unit activity and delayed response in the monkey. *Brain Res.* **325**, 241–54.

Watson, J. B. (1914). *Behavior: an introduction to comparative psychology*. Holt, Rinehart and Winston, New York.

Watson, J. D., Hopkins, N. H., Roberts, J. W., Steitz, J. A., and Weiner, A. M. (1987). *Molecular biology of the gene* (4th edn). Benjamin/Cummings, Menlo Park, California.

Wechsler, D. (1945). A standardized memory scale for clinical use. *J. Psychol.* **19**, 87–95.

Wechsler, D. (1958). *The measurement and appraisal of adult intelligence* (4th edn). Williams and Wilkins, Baltimore.

Weinberg, R. A. (1985). The action of oncogenes in the cytoplasm and nucleus. *Science*, **230**, 770–6.

Weiskrantz, L. (1956). Behavioural changes associated with ablation of the amygdaloid complex in monkeys. *J. Comp. Physiol. Psychol.* **49**, 381–91.

Weiskrantz, L. (1985). On issues and theories of the human amnesic syndrome. In *Memory systems of the brain* (ed. N. M. Weinberger, J. L. McGaugh, and G. Lynch), pp. 380–415. Guilford, New York.

Weiskrantz, L. and Warrington, E. K. (1979). Conditioning in amnesic patients. *Neuropsychologia*, **17**, 187–94.

Weisz, D. J., Clark, G. A., and Thompson, R. F. (1984). Increased responsivity of dentate granule cells during nictitating membrane response conditioning in rabbit. *Behav. Brain Res.* **12**, 145–54.

Welker, E. and Van der Loos, H. (1986). Quantitative correlation between barrel-field size and the sensory innervation of the whiskerpad: a comparative study in six strains of mice bred for different patterns of mystacial vibrissae. *J. Neurosci.* **6**, 3355–73.

Welsh, J. P. and Harvey, J. A. (1989). Cerebellar lesions and the nictitating membrane reflex: performance deficits of the conditioned and unconditioned response. *J. Neurosci.* **9**, 299–311.

West, M. J. and King, A. P. (1988). Female visual displays affect the development of male song in the cowbird. *Nature*, **334**, 244–6.

Westrum, L. E., Hugh Jones, D., Gray, E. G., and Barron, J. (1980). Microtubules, dendritic spines and soma apparatuses. *Cell. Tissue Res.* **208**, 171–81.

Wetzel, C. D. and Squire, L. R. (1982). Cued recall in anterograde amnesia. *Brain and Language*, **15**, 70–81.

Wetzel, W., Ott, T., and Matthies, H. (1976). Is actinomycin D suitable for the investigation of memory processes? *Pharmacol. Biochem. Behav.* **4**, 515–19.

White, G., Levy, W. B., and Steward, O. (1988). Evidence that associative interactions between synapses during induction of long-term potentiation occur within local dendritic domains. *Proc. Natl. Acad. Sci. USA*, **85**, 2368–72.

Wiesel, T. N. (1982). Postnatal development of the visual cortex and the influence of environment. *Nature*, **299**, 583–91.

Wigstrom, H. and Gustafsson, B. (1983). Facilitated induction of hippocampal long-lasting potentiation during blockade of inhibition. *Nature*, **301**, 603–4.

Wigstrom, H. and Gustafsson, B. (1984). A possible correlate of the postsynaptic condition for long-lasting potentiation in the guinea pig hippocampus *in vitro. Neurosci. Lett.* **44**, 327–32.

Wigstrom, H. and Gustafsson, B. (1985). On long-lasting potentiation in the hippocampus: a proposed mechanism for its dependence on coincident pre- and postsynaptic activity. *Acta Physiol. Scand.* **123**, 519–22.

Wigstrom, H., Swann, J. W., and Andersen, P. (1979). Calcium dependency of long-lasting potentiation in the hippocampal slice. *Acta Physiol. Scand.* **105**, 126–8.

Wigstrom, H., McNaughton, B. L., and Barnes, C. A. (1982). Long-term synaptic enhancement in hippocampus is not regulated by postsynaptic membrane potential. *Brain Res.* **233**, 195–9.

Wild, J. M. and Cohen, D. H. (1985). Invariance of retinal output during visual learning. *Brain Res.* **331**, 127–35.

Williams, H. and Nottebohm, F. (1985). Auditory responses in avian vocal motor neurons: a motor theory for song perception in birds. *Science*, **229**, 279–82.

Williams, J. M., Hamilton, L. W., and Carlton, P. L. (1975). Ontogenetic dissociation of two classes of habituation. *J. Comp. Physiol. Psychol.* **89**, 733–7.

Willmund, R., Mitschulat, H., and Schneider, K. (1986). Long-term modulation of Ca^{2+}-stimulated autophosphorylation and subcellular distribution of the Ca^{2+}/cadmodulin-dependent protein kinase in the brain of *Drosophila. Proc. Natl. Acad. Sci. USA*, **83**, 9789–93.

Wilson, C. J., Groves, P. M., Kitai, S. T., and Linder, J. C. (1983). Three-dimensional structure of dendritic spines in the rat neostriatum. *J. Neurosci.* **3**, 383–98.

Wilson, D. A., Sullivan, R. M., and Leon, M. (1987). Single-unit analysis of postnatal olfactory learning. Modified olfactory bulb output response patterns to learned attractive odors. *J. Neurosci.* **7**, 3154–62.

Winocur, G., Oxbury, S., Roberts, R., Agnetti, V., and Davis, C. (1984). Amnesia in a patient with bilateral lesions to the thalamus. *Neuropsychologia* **22**, 123–43.

Woodruff-Pak, D. S., Lavond, D. G., and Thompson, R. F. (1985). Trace conditioning: abolished by cerebellar nuclear lesions but not lateral cerebellar cortex aspiration. *Brain Res.* **348**, 249–60.

Woody, C. D. (1970). Conditioned eye blink: gross potential activity at coronal-precruciate cortex of the cat. *J. Neurophysiol.* **33**, 838–50.

Woody, C. D. (1982). *Memory, learning, and higher function.* Springer-Verlag, New York.

Woody, C. D. (1984). Studies of pavlovian eye-blink conditioning in awake cats. In *Neurobiology of learning and memory* (ed. G. Lynch, J. L. McGaugh, and N. M. Weinberger), pp. 181–96. Guilford, New York.

Woody, C. D. and Black-Cleworth, P. (1973). Differences in excitability of cortical neurons as a function of motor projection in conditioned cats. *J. Neurophysiol.* **36**, 1104–16.

Woody, C. D. and Brozek, G. (1969). Changes in evoked responses from facial nucleus of cat with conditioning and extinction of an eye blink. *J. Neurophysiol.* **32**, 717–26.

Woody, C. D. and Engel, J. (1972). Changes in unit activity and thresholds to electrical micro-stimulation at coronal-pericruciate cortex of cat with classical conditioning of different facial movements. *J. Neurophysiol.* **35**, 230–41.

Woody, C. D. and Yarowsky, P. J. (1972). Conditioned eye blink using electrical stimulation of coronal-precruciate cortex as conditional stimulus. *J. Neurophysiol.* **35**, 242–52.

Woody, C. D., Vassilevsky, N. N., and Engel, J. (1970). Conditioned eye blink: unit activity at coronal-precruciate cortex of the cat. *J. Neurophysiol.* **33**, 851–64.

Woody, C. D., Yarowsky, P., Owens, J., Black-Cleworth, P., and Crow, T. (1974). Effect of lesions of cortical motor areas on acquisition of eyeblink in the cat. *J. Neurophysiol.* **37**, 385–94.

Woody, C. D., Knispel, J. D., Crow, T. J., and Black-Cleworth, P. A. (1976). Activity and excitability to electrical current of cortical auditory receptive neurons of awake cats as affected by stimulus association. *J. Neurophysiol.* **39**, 1045–61.

Woody, C. D., Swartz, B. E., and Gruen, E. (1978). Effects of acetylcholine and cyclic GMP on input resistance of cortical neurons in awake cats. *Brain Res.* **158**, 373–95.

Woody, C. D., Kim, E. H.-J., and Berthier, N. E. (1983). Effects of hypothalamic stimulation on unit responses recorded from neurons of sensorimotor cortex of awake cats during conditioning. *J. Neurophysiol.* **49**, 780–91.

Woody, C. D., Alkon, D. L., and Hay, B. (1984). Depolarization-induced effects of Ca^{2+}-calmodulin-dependent protein kinase injection, in vivo, in single neurons of cat motor cortex. *Brain Res.* **321**, 192–7.

Woody, C. D., Aou, S., Gruen, E., Birt, D., Melamed, O., and Wangwongvivat, J. (1988). Short and long latency blink CRs are supported by activity of units in the motor cortex and thalamus of cats. *Soc. Neurosci. Abst.* **14**, 861.

Wu, J.-Y., Zecevic, D. P., London, J. A., Rioult, M., and Cohen, L. B. (1987). Optical measurement of neuron activity during the gill withdrawal reflex in *Aplysia*. *Soc. Neurosci. Abst.* **13**, 817.

Yakovlev, P. I. (1948). Motility, behavior and the brain. *J. Nerv. Ment. Dis.* **107**, 313–35.

Yamamoto, K. K., Gonzalez, G. A., Biggs, W. H. III., and Montiminy, M. R. (1988). Phosphorylation-induced binding and transcriptional efficacy of nuclear factor CREB. *Nature*, **334**, 494–8.

Yeo, C. H. and Hardiman, M. J. (1988). Loss of conditioned responses following cerebellar cortical lesions is not a performance deficit. *Soc. Neurosci. Abst.* **14**, 3.

Yeo, C. H., Hardiman, M. J., Moore, J. W., and Steele Russell, I. (1983). Retention of conditioned inhibition of the nictitating membrane response in decorticate rabbits. *Behav. Brain Res.* **10**, 383–92.

Yeo, C. H., Hardiman, M. J., and Glickstein, M. (1984). Discrete lesions of the cerebellar cortex abolish the classically conditioned nictitating membrane response of the rabbit. *Behav. Brain Res.* **13**, 261–6.

Yeo, C. H., Hardiman, M. J., and Glickstein, M. (1985*a*). Classical conditioning of the nictitating membrane response of the rabbit. I. Lesions of the cerebellar nuclei. *Exp. Brain Res.* **60**, 87–98.

Yeo, C. H., Hardiman, M. J., and Glickstein, M. (1985*b*). Classical conditioning of the nictitating membrane response of the rabbit. II. Lesions of the cerebellar cortex. *Exp. Brain Res.* **60**, 99–113.

Yerkes, R. M. and Yerkes, D. N. (1928). Concerning memory in the chimpanzee. *J. Comp. Psychol.* **8**, 237–71.

Yoshimasa, T., Sibley, D. R., Bouvier, M., Lefkowitz, R. J., and Caron, M. C. (1987). Cross-talk between cellular signalling pathways suggested by phorbol-ester induced adenylate cyclase phosphorylation. *Nature*, **327**, 67–70.

Young, J. Z. (1965). The organization of a memory system. *Proc. R. Soc. Lond. B*, **163**, 285–320.

Young, J. Z. (1978). *Programs of the brain.* Oxford University Press, Oxford.

Young, J. Z. (1979). Learning as a process of selection and amplification. *J. R. Soc. Med.* **72**, 801–14.

Young, W. S., Bonner, T. I., and Brann, M. R. (1986). Mesencephalic dopamine neurons regulate the expression of neuropeptide mRNAs in the rat forebrain. *Proc. Natl. Acad. Sci. USA*, **83**, 9827–31.

Yovell, Y. and Abrams, T. W. (1988). Order dependence in the activation of adenylate cyclase by serotonin and Ca^{2+}/calmodulin: a possible molecular mechanism for associative learning in *Aplysia. Soc. Neurosci. Abst.* **14**, 910.

Yovell, Y. and Dudai, Y. (1987). The possible involvement of adenylate cyclase in learning and short-term memory: experimental data and some theoretical considerations. *Isr. J. Med. Sci.* **23**, 49–60.

Yovell, Y., Kandel, E. R., Dudai, Y., and Abrams, T. W. (1987). Biochemical correlates of short-term sensitization in *Aplysia*: temporal analysis of adenylate cyclase stimulation in a perfused-membrane preparation. *Proc. Natl. Acad. Sci. USA*, **84**, 9285–9.

Yovell, Y., Dudai, Y., Kandel, E., and Abrams, T. W. (1988). The Ca^{2+}-calmodulin responsiveness of adenylate cyclase: quantitative analysis of the enzyme from *Aplysia*, *Drosophila*, and rat brain. (Submitted.)

Zeki, S. and Shipp, S. (1988). The functional logic of cortical connections. *Nature*, **335**, 311–17.

Zigmund, P. E., Detrick, R. A., and Pfaff, D. W. (1980). An autoradiographic study of the localization of androgen concentrating cells in the chaffinch. *Brain Res.* **182**, 369–81.

Zipser, D. and Andersen, R. A. (1988). A back-propagation programmed network that simulates response properties of a subset of posterior parietal neurons. *Nature*, **331**, 679–84.

Zola-Morgan, S. and Squire, L. R. (1984). Preserved learning in monkeys with medial temporal lesions: sparing of motor and cognitive skills. *J. Neurosci.* **4**, 1072–85.

Zola-Morgan, S. and Squire, L. R. (1985*a*). Medial temporal lesions in monkeys impair memory on a variety of tasks sensitive to human amnesia. *Behav. Neurosci.* **99**, 22–34.

Zola-Morgan, S. and Squire, L. R. (1985*b*). Amnesia in monkeys after lesions of the mediodorsal nucleus of the thalamus. *Ann. Neurol.* **17**, 558–64.

Zola-Morgan, S. and Squire, L. R. (1986). Memory impairment in monkeys following lesions limited to the hippocampus. *Behav. Neurosci.* **100**, 155–60.

Zola-Morgan, S., Squire, L. R., and Mishkin, M. (1982). The neuroanatomy of amnesia: amygdala-hippocampus versus temporal stem. *Science*, **218**, 1337–9.

Zola-Morgan, S., Squire, L. R., and Amaral, D. G. (1986). Human amnesia and the medial temporal region: enduring memory impairment following a bilateral lesion limited to field CA1 of the hippocampus. *J. Neurosci.* **6**, 2960–7.

Zwiers, H., Jolles, J., Aloyo, V. J., Oestreicher, A. B., and Gispen, W. H. (1982). ACTH and synaptic membrane phosphorylation in rat brain. *Prog. Brain Res.* **56**, 405–17.

Index